Fundamentals of Periodic Table and Elements

FUNDAMENTALS OF PERIODIC TABLE AND ELEMENTS

Serhii Solodin

ARCLER PRESS

www.arclerpress.com

Fundamentals of Periodic Table and Elements

Serhii Solodin

Arcler Press
224 Shoreacres Road
Burlington, ON L7L 2H2
Canada
www.arclerpress.com
Email: orders@arclereducation.com

ISBN: 978-1-77469-941-6 (Hardcover)

ABOUT THE AUTHOR

Serhii Solodin is a PhD in Chemistry (2021) and the author of 14 scientific publications rated in Scopus database. His thesis and scientific interest concerned the study of defects in semiconductor crystals, particularly Cadmium Telluride, which was carried out in Poland, the Czech Republic and Austria. Additionally, he spent the past years researching and developing products to solve consumer issues in the personal care and cosmetic industry. Among the products he has developed (340+) there are products for skin care (creams, serums, serums, gels), hair care (shampoos, conditioners, masks, sprays), oral care (toothpaste and tablets), laundry care (gels, tablets, sheets). He likes solving extraordinary, new and challenging issues and always strives for self-improvement and constantly monitor the market to keep abreast of current trends and approaches to product creation. In particular, recently he has been developing products aimed at environmental friendliness and biodegradability, in particular, products without plastic: effervescent tablets and powders for oral cavity, laundry, skin- and hair-care as well as detergent sheets.

TABLE OF CONTENTS

LIST OF FIGURES

LIST OF TABLES

LIST OF ABBREVIATIONS

Ac	Actinium
AGB	Asymptotic Giant Branch
AgCl	Silver Chloride
Al	Aluminum
$AlBr_3$	Aluminum Bromide
$AmCl_2$	Americium Chloride
aq	Aqueous Solution
Ar	Argon
B	Boron
Ba^{2+}	Barium
$BaCrO_4$	Barium Chromate
$BaSO_4$	Barium Sulfate
BCC	Body-Centered Cubic
$BeBr_2$	Beryllium Bromide
$BeCl_2$	Beryllium Chloride
BeO	Beryllium Oxide
BrCl	Bromine Chloride
Ca	Calcium
Ca^{2+}	Calcium Ion
$CaCO_3$	Calcium Carbonate
CaF_2	Calcium Fluoride
CaH_2	Calcium Hydride
$CaSO_4$	Calcium Sulfate
Ce	Cerium
CFCs	Chlorofluorocarbons

ClO^-	Hypochlorite
CO_3^{2-}	Carbonate Ions
Co_3O_4	Tricobalt Tetroxide
$CoCO_3$	Cobaltous Carbonate
CoF_3	Cobaltic Fluoride
CoO	Cobaltous Oxide
CoS	Cobaltous Sulfide
$CoSO_4$	Cobaltous Sulfate
$CoZnO_2$	Cobalt Zincate
CrO	Chromous Oxide
DDT	Dichlorodiphenyltrichloroethane
DOTA	1,4,7,10-Tetraazacyclododecane-1,4,7,10-Tetraacetic Acid
EPA	Environmental Protection Agency
EPR	Electron Paramagnetic Resonance
EU	European Union
Eu	Europium
$EuBr_3$	Europium Bromide
$EuCl_2$	Europium Chloride
EuF_2	Europium Fluoride
Fe_2O_3	Ferric Oxide
$FeCr_2O_4$	Ferrous Chromite
$FeSO_4$	Ferrous Sulfate
Fl	Flerovium
g	Gas
Ga	Gallium
Gd	Gadolinium
Ge	Germanium
GI	Gastrointestinal
H	Hydrogen
H_2O	Water
H_2O_2	Hydrogen Peroxide
H_2S	Hydrogen Sulfide
HBr	Hydrobromic Acid
HCl	Hydrochloric Acid

He	Helium
HF	Hydrofluoric Acid
HFCs	Hydrofluorocarbons
HI	Hydroiodic Acid
HNO	Nitric Acid
In	Indium
Ir	Iridium
ISM	Interstellar Medium
IUPAC	International Union of Pure and Applied Chemistry
JINR	Joint Institute for Nuclear Research
K_2CO_3	Potassium Carbonate
K_2SO_4	Potassium Sulfate
KBr	Potassium Bromide
KCl	Potassium Chloride
KI	Potassium Iodide
KNO_3	Potassium Nitrate
KOH	Potassium Hydroxide
Kr	Krypton
l	Liquid
La	Lanthanum
LEDs	Light-Emitting Diodes
Li	Lithium
Li_2CO	Lithium Carbonate
Li_2O	Lithium Oxide
LiCl	Lithium Chloride
LiH	Lithium Hydride
Lns	Lanthanides
Lr	Lawrencium
Lv	Livermorium
Mg	Magnesium
Mg^{2+}	Magnesium Ion
MgC_2O	Magnesium Oxalate
MgO	Magnesium Oxide
MMT	Million Metric Tons

MnO	Manganous Oxide
MnO_2	Manganese Dioxide
MO_2O_5	Molybdenum Pentoxide
$MoCl_5$	Molybdenum Pentachloride
MRI	Magnetic Resonance Imaging
Na	Sodium
Na_2CO_3	Sodium Carbonate
$Na_2H_2V_6O_{17}$	Sodium Hexavanadate
Na_3VO_4	Sodium Orthovanadate
$Na_4V_2O_7$	Sodium Pyrovanadate
NaCl	Sodium Chloride
NaClO	Sodium Hypochlorite
NaCN	Sodium Cyanide
NaF	Sodium Fluoride
$NaHCO_3$	Sodium Bicarbonate
NaN_3	Sodium Azide
NaOH	Sodium Hydroxide
Nb_2O_5	Niobium Pentoxide
Nb_3N_5	Niobium Nitride
$NbCl_5$	Niobium Pentachloride
Nd	Neodymium
$NdBr_3$	Neodymium Bromide
$NdCl_3$	Neodymium Chloride
NdF_3	Neodymium Fluoride
Ne	Neon
Nh	Nihonium
NH_3	Ammonia
NH_4Br	Ammonium Bromide
NiAs	Nickel Arsenide
NMR	Nuclear Magnetic Resonance
Np_2O_5	Neptunium Pentoxide
$NpOF_3$	Neptunium Oxyfluoride
O	Oxygen
Os	Osmium

PBr_3	Phosphorus Tribromide
PCBs	Polychlorinated Biphenyls
Pd	Palladium
Pm	Promethium
Po	Polonium
PP	Proton-Proton
ppb	Parts Per Billion
ppq	Parts Per Quadrillion
Pr	Praseodymium
PrN	Praseodymium Nitride
Pt	Platinum
PuN	Plutonium Nitride
PVC	Polyvinyl Chloride
REEs	Rare Earth Elements
REMs	Rare Earth Metals
REs	Rare Earths
Rh	Rhodium
Rn	Radon
Ru	Ruthenium
s	Solid
S	Sulfur
Sc	Scandium
Se	Selenium
SF_6	Sulfur Hexafluoride
Si	Silicon
SmO	Samarium Monoxide
SmS	Samarium Sulfide
SmSe	Samarium Selenide
SmTe	Samarium Telluride
SmX_3	Samarium Trihalides
$SnCl_2$	Stannous Chloride
SnF	Stannous Fluoride
Sr^{2+}	Strontium
$SrCO_3$	Strontianite

$SrCrO_4$	Strontium Chromate
$SrSO_4$	Minerals Celestite
Te	Tellurium
Th_3N_4	Thorium Nitride
$ThCl_4$	Thorium Chloride
ThF_4	Thorium Fluoride
ThO_2	Thorium Dioxide
$TiCl_2$	Titanium Dichloride
$TiCl_3$	Titanium Trichloride
$TiCl_4$	Titanium Tetrachloride
TiO	Titanium Oxide
TiO_2	Titanium Dioxide
Tl	Thallium
UCl_4	Uranium Tetrachloride
UCl_6	Uranium Hexachloride
UF_6	Uranium Hexafluoride
UO_2	Uranium Dioxide
UO_2S	Uranyl Sulfide
UO_3	Uranium Trioxide
Uuh	Ununhexium
Uuo	Unuoctium
Uuq	Ununquadium
VO	Vanadous Oxide
VSO_4	Vanadous Sulfate
Xe	Xenon
XeO_4	Xenon Tetroxide
Y	Yttrium
YF_3	Yttrium Fluoride
YPO_4	Yttrium Phosphate

PREFACE

The periodic table is one of the most important tools in chemistry. It is a tabular arrangement of chemical elements organized by their atomic number, electron configuration, and chemical properties. The periodic table allows chemists to quickly and easily see the patterns and trends in the elements, which can be used to predict their properties and reactivity.

The periodic table was first created in 1869 by Dmitri Mendeleev. Mendeleev arranged the elements by their atomic weight, and he noticed that certain elements with similar properties fell into groups. He called these groups "families" of elements. Over the years, the periodic table has been modified to reflect our growing understanding of the elements. Today, the periodic table is organized by atomic number, which is the number of protons in the nucleus of an atom. The elements are arranged in rows called periods and in columns called groups.

The elements in each group have similar chemical properties. For example, the alkali metals (group 1) are all very reactive and have a single valence electron. The alkaline earth metals (group 2) are also reactive but have two valence electrons. The elements in each period also have similar properties. For example, the elements in the first period all have one electron shell. The elements in the second period all have two electron shells, and so on.

The periodic table is a valuable tool for understanding the elements. It can be used to predict an element's properties, identify elements that are likely to react with each other, and understand the chemical behavior of matter.

This book, Fundamentals of Periodic Table and Elements, provides a comprehensive introduction to the periodic table and its elements. The book begins with an overview of the periodic table, including its history, organization, and elements. The following chapters then discuss the elements in detail, grouped by their chemical properties.

Chapter 1 introduces the periodic table and its basic concepts. Chapter 2 discusses hydrogen and alkali metals, which are the most reactive elements in the periodic table. Chapter 3 covers the alkaline earth metals, which are less reactive than the alkali metals but still quite reactive. Chapter 4 discusses transition metals, which are a diverse group of elements with a wide range of properties. Chapter 5 covers lanthanides, which are a group of rare-earth elements with similar chemical properties. Chapter 6 covers actinides, which are a group of radioactive elements that are even rarer than lanthanides. Chapter 7 discusses the boron, carbon, nitrogen, and oxygen groups, which are a group of elements that share some common properties. Chapter 8 covers the halogens and noble gases, which are the two groups of elements on the right side of the periodic table.

This book is intended for students and professionals who want to learn more about the periodic table and its elements. It is also a valuable resource for anyone who is interested in chemistry or science in general.

CHAPTER 1

Introduction to Periodic Table and Elements

Contents

1.1. INTRODUCTION

Elements are ubiquitously present in various forms, with certain elements being perceptible to the naked eye, such as gold, while others exist in a nearly imperceptible state, as exemplified by oxygen gas. An element comprises a fundamental substance that is indivisible and cannot be decomposed into simpler constituents (Schwerdtfeger et al., 2020). Every individual entity is composed of minuscule constituents known as atoms, which possess distinct characteristics specific to each element. The majority of chemical elements undergo a combination with additional elements to form compounds, which are produced through the amalgamation of two or more elements. Water is a chemical compound composed of hydrogen as well as oxygen (Weinhold & Bent, 2007).

The periodic table consists of a total of 118 elements, with 92 occurring naturally and the remaining elements being artificially synthesized by human intervention. Each element possesses distinct characteristics that set it apart from others. The majority of the elements in the periodic table exist in a solid state, with characteristics similar to those of metals. At standard ambient temperature and pressure, there are 11 elements that exist in the gaseous state, whereas bromine alongside mercury are the sole elements that exhibit liquid properties (Xie et al., 2012). The concept of elements has a long historical lineage, tracing its origins approximately 2,600 years ago to ancient Greece. Nevertheless, Greek philosophers held the belief that the cosmos consisted solely of four fundamental elements, namely earth, water, fire, as well as air. Empedocles, a prominent scholar, was the initial proponent of the notion that these fundamental elements constitute the fundamental constituents of all structures. It was not until a later period that scientists acquired the knowledge that not any of these entities can be classified as elements. Throughout millennia, individuals including ancient Egyptian religious leaders to medieval European practitioners of alchemy engaged in contemplation regarding the precise delineation and categorization of an elemental entity (Watson et al., 2020).

Approximately 99% of the body of a person is comprised of six fundamental elements, which are subsequently amalgamated to generate a myriad of diverse compounds. In contrast, the composition of Earth's atmosphere primarily consists of a combination of gases, the majority of which are elemental in nature. Approximately 99% of the Earth's atmosphere consists of nitrogen along with oxygen (Bernal & Railsback, 2008).

Did You Know?

Robert Boyle, an Irish researcher, and inventor, is credited as the pioneer who employed scientific methods to comprehend the fundamental constituents of matter. The individual in question engaged in the pursuit of scientific knowledge by employing rationality. During the 1660s, he conducted pioneering experiments in the field of chemistry, thereby demonstrating the fallacy of numerous beliefs held by practitioners of alchemy.

The periodic table serves as a valuable tool for the systematic arrangement of chemical elements. The arrangement of elements is based on their atomic number, a distinct value representing the total number of protons inside an atom's nucleus. The table further categorizes the elements into horizontal rows, referred to as "periods," and vertical columns, referred to as "groups." The table of elements, formulated by chemist Dmitri Mendeleev, was organized according to the resemblance of specific physical and chemical characteristics exhibited by the elements (Bierenstiel & Snow, 2019).

1.2. THE ELEMENTS

During the ancient Greek time period, philosophers acknowledged a limited set of four fundamental elements: earth, water, air, as well as fire. Interestingly, these elements persist in the astrological categorization of the 12 signs within the zodiac. Certain philosophers held the belief that the diverse elements were comprised of minuscule constituents possessing distinct shapes, which purportedly accounted for the manifold properties exhibited by these elements. The aforementioned designs or structures were hypothesized to exhibit the characteristics of Platonic solids (Figure 1.1), which are composed entirely of a single two-dimensional shape (Gordy & Thomas, 1956). According to Greek philosophy, the composition of the earth was postulated to consist of minuscule cubic particles, a notion that was put forth to account for the observed resistance encountered when attempting to displace the earth. In the interim, the concept of water's liquidity was elucidated through an argument centered around the more streamlined structure inherent in the icosahedron. Conversely, the tactile discomfort associated with fire was attributed to its composition comprising of acute particles arranged in the configuration of tetrahedra (Kelly & Street, 2006). The prevailing belief was that air was composed of octahedra, as it was considered to be the sole remaining Platonic solid. Subsequently, the identification of the dodecahedron referred to as a 5th Platonic solid prompted the conjecture of the existence of a fifth element, commonly referred to as "quintessence" or ether. While the concept that elements consist of Platonic solids seems inaccurate in contemporary understanding, it serves as the foundation for the highly productive idea that the macroscopic characteristics of substances are determined by the arrangement of their microscopic constituents. These "elements" persisted throughout the Middle Ages and subsequent periods, supplemented by additional elements that were uncovered by the alchemists, who served as precursors to contemporary chemists. The transformation of elements appears to have been among the numerous objectives pursued by the alchemists. It is not unexpected that the endeavor that particularly captivated individuals was the pursuit of transmuting the elemental substance lead, which possesses common attributes, into the esteemed precious metal gold. The distinctive hue, scarcity, and chemical stability of gold have rendered it a highly esteemed material since the earliest stages of human civilization (Almquist et al., 2016).

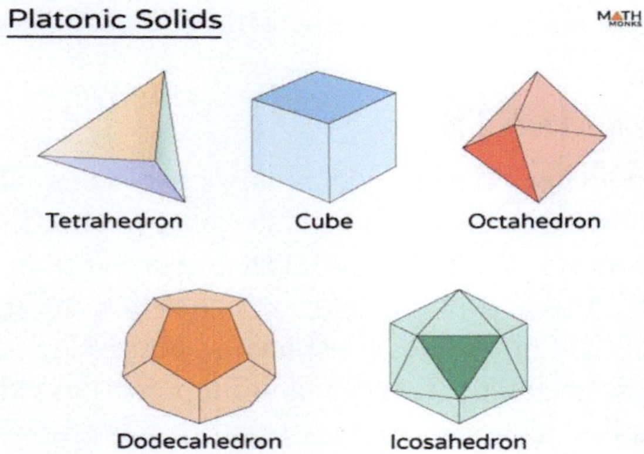

Figure 1.1. The five platonic solids.

Source: Eric R. Scerri, creative commons license.

The initial conceptualization of the term "element" by Greek philosophers was centered around the notion of a "tendency" or "potentiality" that served as the foundation for the discernible characteristics exhibited by said element. The nuanced differentiation between the abstract manifestation of an element and its perceptible manifestation has largely faded from prominence in contemporary society. However, it has nevertheless functioned as an essential guiding principle for notable contributors in developing the periodic system, including Dimitri Mendeleev, who is recognized as its primary discoverer. According to the majority of textbook narratives, the field of chemistry made significant progress when it distanced itself from alchemy and its seemingly mystical comprehension of elemental properties. The success of contemporary science is commonly attributed to its reliance on direct experimentation along with the embrace of an empiricist perspective, which asserts that only phenomena that may be observed ought to be considered valid. As expected, the more nuanced and potentially more fundamental interpretation of the notion of elements was met with rejection. Robert Boyle as well as Antoine Lavoisier shared the perspective that the definition of an element should rely on empirical observations, thereby rejecting the involvement of abstract elements. The recommendation posits that an element can be characterized as a substance of material nature that remains unbroken into any further fundamental constituents through chemical processes (Lavoisier, 1806).

A number of such compounds would meet the criteria for classification as elements according to contemporary standards. However, certain substances, like lumière (light) as well as caloric (heat), are no longer considered to be elements. The future holds great potential for chemists to enhance and broaden their repertoire of separation as well as characterization methods for chemical compounds, thus facilitating further advancements in this field. Spectroscopy, a significant scientific method, involves the measurement of absorption as well as emission spectrum of diverse forms of radiation. This technique has the potential to provide a highly precise approach to identifying individual elements based on their distinct spectral characteristics. In contemporary society, there is an acknowledgment of the existence of 91 elements that occur naturally. Furthermore, advancements have enabled the expansion of the elemental range above those that are naturally found (Justino et al., 2015).

1.2.1. The Discovery of the Elements

The narrative surrounding the exploration and identification of the elements seems an intriguing topic that has garnered significant attention, with numerous notable accounts dedicated to its exploration. The timeline of the discoveries is presented in Table 1.1. The narrative pertaining to this particular story is not comprehensively examined within the current publication, albeit allusions to prognostications and the unearthing of elements are intermittently mentioned (Stark et al., 2007). Throughout the annals of chemistry, there are several significant occurrences wherein a cluster of approximately six elements was unearthed in rapid succession, either concurrently or within a brief timeframe spanning a few years. Naturally, certain elements, including iron, copper, gold, along with various other metallic substances, have been recognized since ancient times. Historians and archaeologists commonly utilize specific designations such as the Age of Iron or the Age of Copper to delineate distinct periods in the history of mankind. The alchemists expanded the roster of elements by incorporating additional substances, such as sulfur, mercury, as well as phosphorus (Weeks, 1932). During a more contemporary era, the advent of electricity facilitated the isolation of numerous highly reactive elements that were previously unattainable through the conventional method of heating their ores with carbon, unlike copper and iron.

Humphry Davy, a British chemist, employed the utilization of electricity, particularly electrolysis, to successfully isolate a notable number of elements, encompassing calcium, barium, magnesium, sodium, as well as chlorine. After the identification of radioactivity as well as nuclear fission, coupled with advancements in radiochemistry methodologies, researchers were able to address the remaining discrepancies within the periodic table (Karol et al., 2016). The final void that required addressing pertained to the element with atomic number 43, subsequently designated as technetium, derived from the Greek term "techne" denoting artificial or synthetic. The substance was "synthesized" through a series of radiochemical reactions that were not previously achievable prior to the emergence of nuclear physics. Previously, there was a prevailing belief that this particular element lacked a natural occurrence.

However, a recent reassessment of historical evidence has indicated that it does indeed exist naturally (Zhang & Gerstein, 2003). Furthermore, it has been proposed that the initial reports regarding its discovery in the year 1925 might have been unfairly discredited. The recent surge in elemental discoveries can be attributed to advancements in technology, specifically the ability to generate and manipulate beams consisting solely of atoms or fundamental particles like neutrons. The aforementioned particles may be directed toward one another with exceptional accuracy in order to induce reactions that cause nuclear fusion, resulting in the formation of novel elements characterized by exceedingly high atomic numbers (Munzenberg, 1988). The individual who pioneered this discipline was Glenn Seaborg, a US-based chemist renowned for his achievement in synthesizing plutonium in 1943. Furthermore, Seaborg led research teams that successfully synthesized numerous additional transuranium elements (Blanchette & Tompa, 2002).

1.2.2. Names and Symbols of the Modern Elements

The allure of the chemical periodic table stems from the distinctive characteristics of the elements along with their nomenclature. The individual in question is a chemist who also endured the harrowing experience of being incarcerated in a concentration camp. Primo Levi initiated the commencement of every chapter in his highly acclaimed literary work, The Periodic Table (Vandevelde, 1980).

Table 1.1. Discovery Timeline for the Elements and Approximate Dates of Contributions

Era of Contribution	**Elements**
Antiquity	Au, Ag, Cu, Fe, Sn, Pb, Sb, Hg, S, C
Middle Ages	As, Bi, Zn, P, Pt
1700	_
1710	_
1720	_
1730	Co
1740	_
1750	Ni, Mg
1760	H
1770	N, O, Cl, Mn, Ba
1780	Mo, W, Te, Zr, U
1790	Ti Y, Be
1800	V, Nb, Ta, Rh, Pa, Os, Ir, Ce K, Na, B, Ca, Sr, Ru, Ba
1810	I, Th, Li, Se, Cd
1820	Si, Al, Br
1830	La
1840	Er
1850	_
1860	Cs, Rb, Tl, In, He
1870	Ga, Ho, Yb, Sc, Tm
1880	Gd, Pr, Nd, Ge, F, Dy
1890	Ar, He, Kr, Ne, Xe, Po, Ra, Ac

1900	Rn, Eu, Lu
1910	Pa
1920	Hf, Re, Tc
1930	Fr
1940	Np, At, Pu, Cm, Am, Pm, Bk
1950	Cf, Es, Fm, Md, No
1960	Lr, Rf, Db
1970	Sg
1980	Bh, Mt, Hs
1990	Ds, Rg, 112, 114, 116
2000	–

By providing a detailed portrayal of an elemental substance, like gold, lead, or oxygen, one can enhance the clarity and comprehension of its characteristics. The primary focus of the book revolves around the author's interpersonal connections and acquaintances. However, each anecdote presented in the book is driven by Levi's profound affection for a specific chemical element. In a more contemporary context, the esteemed neurologist and accomplished writer Oliver Sacks has authored a literary work entitled "Uncle Tungsten," wherein he recounts his childhood captivation with the field of chemistry, specifically focusing on his profound interest in the periodic table (Zhi-Jun & Cheng-Ai, 2006). Throughout the extensive timeline of elemental discovery spanning numerous centuries, a diverse array of thematic frameworks has been employed to designate and assign names to these chemical entities. Merely perusing a compilation of the nomenclature of chemical elements has the potential to evoke recollections of narratives derived from ancient Greek mythology. The element with atomic number 61, known as promethium, derives its nomenclature from the mythological figure Prometheus. In Greek mythology, Prometheus defied the gods by clandestinely obtaining fire from the heavens and bestowing it upon humanity. As a consequence of his audacious act, Prometheus faced retribution from Zeus, the supreme deity. The correlation between this narrative and element 61 appears to lie in the arduous endeavor required for its isolation, mirroring the challenging and perilous task undertaken by Prometheus (Aksholakova, 2014). Promethium is among the select group of chemical elements that are not naturally present in terrestrial environments. The substance in question was acquired as a byproduct of the process of nuclear fission involving uranium, another chemical element. Certain elements have been named after planets as well as other celestial bodies. As an illustration, the element palladium, first identified in 1803, derives its name from Pallas, the 2nd asteroid to be found in 1802, a mere year prior. The element helium derives its name from Helios, the Greek deity associated with the sun. The presence of this phenomenon was initially detected within the solar spectrum in 1868, and its identification in terrestrial specimens was not achieved until 1895. Numerous chemical elements are named after various colors (Manuel, 1995).

The element cesium derives its name from the Latin term "cesium," denoting the color gray-blue, owing to the notable presence of gray-blue spectral lines in its emission spectrum. The etymology of the term "chlorine" can be traced back to its Greek origin, specifically the word "kleros," which is associated with the hue yellow-green. The salts derived from rhodium, an element within the periodic table, frequently exhibit a pink hue. This characteristic serves as the basis for the selection of the element's name, which originates from the Greek term "Rhodon," meaning rose. In instances involving elements that have been synthesized more recently, the nomenclature is derived via the discoverer's name or from someone to whom the discoverers intend to pay tribute to. The elements bohrium, curium, einsteinium, fermium, gadolinium, lawrencium, meitnerium, mendelevium, nobelium, roentgenium, rutherfordium, as well as seaborgium, are included in the periodic table for specific reasons. Fifteen elements, namely americium, berkelium, californium, darmstadtium, europium, francium, germanium, hassium, polonium, gallium, hafnium, lutetium, rhenium, ruthenium, and scandium, have been derived from the geographical location where their discoverer resided or sought to pay tribute (Hofmann & Münzenberg, 2000).

Furthermore, certain element names originate from geographic regions that are associated with the minerals where they were discovered. This category encompasses the notable instance of four elements that have been designated with names derived from the Swedish town of Ytterby, located in proximity to Stockholm. Erbium, terbium, ytterbium, as well as yttrium were discovered in ores that were geographically situated in the vicinity of this village. Additionally, a fifth element, holmium, derived its name from the Latin term for Stockholm. The process of naming the subsequent trans-uranium elements constitutes an independent narrative, encompassing nationalistic disagreements and, in certain instances, contentious disagreements regarding the initial synthesis of the element and the subsequent attribution of the privilege to designate its name. In an effort to address such conflicts, the International Union of Pure and Applied

Chemistry (IUPAC) established a mandate stipulating that the nomenclature of elements ought to be conducted in an unbiased and methodical manner, utilizing Latin numerals corresponding to atomic numbers assigned to each respective element (Meija, 2014).

For instance, the element with atomic number 105 would be designated as un-nil-Pentium, whereas the element with atomic number 106 would be referred to as un-nil-Nexium. In recent times, subsequent to extensive deliberation regarding the rightful discoverers of certain superheavy elements, the IUPAC has reinstated the prerogative of naming these elements to the individuals or research groups who have been deemed to have developed priority in their discovery or synthesis. Elements 105 and 106 have been assigned names such as bohrium as well as seaborgium, accordingly, replacing their previous designations based on impersonal Latin nomenclature. Seaborgium presents an intriguing scenario as the committee, for an extended period, withheld approval for the designation of this element after the American scientist Glenn Seaborg, despite his significant contributions in synthesizing approximately ten novel elements, which includes element 106 (Constable, 2019). The official rationale appears to have been based on an antiquated regulation stipulating that no chemical element might be designated after an individual who remains alive. After extensive advocacy efforts by scientists in the US and various regions globally, Seaborg was ultimately awarded the recognition of his element during his lifetime. Another intriguing instance pertains to Otto Hahn, a German chemist, whose appellation was informally assigned to the element hahnium, only to subsequently be rescinded and replaced with the name dubnium, derived from the location where numerous trans-uranium elements had been synthesized. In the interim, an element is being designated with the appellation of Lise Meitner, a former associate of Hahn (Elding, 1994). According to numerous observers, this action can be considered justifiable, as Otto Hahn was bestowed with the prestigious award for his significant contribution to the discoveries of nuclear fission, while Lise Meitner, who played a pivotal role in various key aspects of the research, was not granted the same recognition. To various individuals, it signifies an overabundance of adherence to politically correct principles. The symbols employed to represent individual elements within the periodic table possess a compelling and intricate narrative. During the era of alchemy, it was common for the symbols representing the elements to align alongside the symbols representing the planets from where they derived their names as well as those to which they were connected (Figure 1.2).

Metals of antiquity				
Metal	**Celestial body**	**Characteristics**	**Symbol**	**Day of the week**
gold	Sun	perfection bright yellow	☉	Sunday
silver	Moon	near perfection silvery	☽	Monday
copper	Venus	resembles gold looking-glass of Venus	♀	Friday
iron	Mars	warriors' metal shield and spear	♂	Tuesday
mercury (quicksilver)	Mercury	moves quickly messenger of gods	☿	Wednesday
tin	Jupiter	lower form of silver	♃	Thursday
lead	Saturn	least desired tarnished/dark	♄	Saturday

Figure 1.2. Names and symbols of the ancient metals compared to names of celestial bodies and days of the week.

Source: Rignes, V., (1989).

For instance, the chemical element mercury is denoted by the identical symbol as the planet Mercury, which is the innermost celestial body in the solar system. Copper was historically linked to the celestial body Venus, as both the chemical element as well as the planet were represented by the identical symbol. In the year 1805, Dalton presented his atomic theory, wherein he chose to preserve a number of alchemical symbols representing the elements. Yet, these objects were found to be unwieldy and not conducive to being easily replicated in scholarly articles and books (Sargeson et al., 1994). The utilization of basic letter symbols in contemporary contexts was introduced subsequent to 1813 by Jacob Berzelius, a scientist hailing from Sweden. Within the contemporary arrangement of the periodic table, a limited subset of elements is denoted by a solitary alphabetical character. The elements hydrogen, carbon, oxygen, nitrogen, sulfur, as well as fluorine are denoted by the symbols H, C, O, N, S, as well as F, respectively. The majority of chemical elements are represented by a two-letter symbol, wherein the initial letter is capitalized and the subsequent letter is in lowercase. This phenomenon leads to the emergence of elemental symbols like Li, Be, Ne, Ca, and Sc, denoting lithium, beryllium, neon, calcium, as well as scandium, correspondingly. Several of the two-letter symbols mentioned, namely Cu, Na, Fe, Pb, Hg, Ag, and Au, lack intuitive clarity as they originate from the Latin nomenclature of the corresponding elements, specifically copper, sodium, iron, lead, mercury, silver, as well as gold (Franco Mariscal et al., 2012). The chemical element tungsten is denoted by the symbol W, derived from its German name "wolfram." During a limited timeframe, certain heavy elements were represented by three-letter symbols, as previously indicated. However, a significant number of these elements have since been assigned specific names, like seaborgium over element 106. However, the elements that have been recently discovered, including Element 112, Element 114, as well as Element 116, remain subject to ongoing debate regarding the official discoverers. These elements are currently represented by three-letter symbols, specifically Uub, Uuq, and Uuh (Corish & Rosenblatt, 2004).

1.3. THE MODERN PERIODIC TABLE

The arrangement of elements in rows as well as columns within the contemporary periodic table, commonly referred to as the medium-long shape (Figure 1.3), unveils numerous interrelationships among them. Certain relationships are widely recognized, whereas others remain to be uncovered. As an illustrative instance, during the 1990s, researchers made a significant finding pertaining to elevated temperatures superconductivity, which entails the unimpeded flow of electric current. This phenomenon was observed to manifest at comparably elevated temperatures of approximately 100 Kelvin. This discovery was partially fortuitous. The compound exhibited elevated temperatures of superconductivity while lanthanum, copper, oxygen, as well as barium were incorporated in a specific manner. Subsequently, a global surge of efforts ensued to elevate the threshold temperature where the phenomenon could be sustained. The objective was to attain superconductivity at ambient temperature, a phenomenon that would facilitate significant advancements in technology, including the seamless movement of levitating trains upon superconducting tracks (Seaborg, 1996). The periodic table of elements served as a primary guiding principle within this endeavor. The utilization of the table facilitated the substitution of certain constituents within the compound with alternative elements that exhibit comparable behaviors, thereby enabling the investigation of their impact on the phenomenon of superconductivity. The present study elucidates the method by which the element yttrium was integrated into a novel series of superconducting compounds, resulting in the synthesis of the compound $YBa_2Cu_3O_7$ exhibiting a superconducting transition temperature of 93 Kelvin. This knowledge, along with potentially additional information, remains latent within the periodic system, awaiting exploration and practical application. The conventional representation of the periodic table comprises horizontal rows, known as periods, and vertical columns, referred to as groups (Scerri, 2012). Observable patterns can be discerned when examining the elements both horizontally and vertically within the table. Each horizontal row within the table corresponds to a distinct period. When traversing a period, individuals transition from leftward metals like potassium and calcium, proceed through transitional metals that include iron, cobalt, as well as nickel, subsequently encounter certain semi-metallic elements like germanium, and ultimately reach nonmetals like arsenic, selenium, as well as bromine situated on the right-hand side of the periodic table. Typically, there exists a gradual progression in physical as well as chemical features as one moves across a period. However, numerous exceptions to this prevailing pattern are prevalent, rendering the exploration of chemistry an intriguing and inherently intricate discipline. Metals exhibit a broad range of physical properties, encompassing both soft, lusterless solids like sodium and potassium, as well as hard, reflective substances like chromium, platinum, as well as gold. In contrast, nonmetals typically exist in the form of solids or gases, exemplified by oxygen as well as carbon, respectively (Supanchaiyamat & Hunt, 2019). Differentiating between solidified metals as well as solid nonmetals can be challenging based solely on their physical characteristics. From the perspective of individuals without specialized knowledge in metallurgy, it may be perceived that a metal possessing a hard and lustrous appearance exhibits a greater degree of metallic characteristics compared to a softer metal like sodium. In a chemical context, elements with lower ionization energy levels are considered to possess a higher degree of metallic character. Chemists consider sodium to possess a higher degree of metallic character compared to elements like iron or copper. The periodic trend observed in the transition from metallic substances to nonmetallic substances is recurrent in each period, resulting in the formation of columns or groups consisting of elements with similar properties when the rows are arranged in a stacked manner.

Elements belonging to the same group typically exhibit numerous significant physical and chemical characteristics in common, albeit with certain notable exceptions (Scerri, 2019). The nomenclature of groups in the contemporary periodic table is intricate and subject to debate. The main-group elements, also known to be representative elements, are situated on the far left and far right of the contemporary periodic table. In the US, it is customary to designate these groups using Roman numerals ranging from I to VIII, occasionally appending the letter A to distinguish them from transitional metals or groups IB to VIIIB, which occupy the central region of the periodic table. In contrast, the standard in Europe deviates from this pattern by repeatedly marking all groups from left to right as IA to VIIIA, until reaching the group led by copper. At this point, the labeling shifts to IB up toward the noble gases, which are designated as group VIIIB (Figure 1.4) (Joag, 2014). Both of these systems

H																	He
Li	Be											B	C	N	O	F	Ne
Na	Mg											Al	Si	P	S	Cl	Ar
K	Ca	Sc	Ti	V	Cr	Mn	Fe	Co	Ni	Cu	Zn	Ga	Ge	As	Se	Br	Kr
Rb	Sr	Y	Zr	Nb	Mo	Tc	Ru	Rh	Pd	Ag	Cd	In	Sn	Sb	Te	I	Xe
Cs	Ba	Lu	Hf	Ta	W	Re	Os	Ir	Pt	Au	Hg	Tl	Pb	Bi	Po	At	Rn
Fr	Ra	Lr	Rf	Db	Sg	Bh	Hs	Mt	Ds	Rg	Cn						

La	Ce	Pr	Nd	Pm	Sm	Eu	Gd	Tb	Dy	Ho	Er	Tm	Yb
Ac	Th	Pa	U	Np	Pu	Am	Cm	Bk	Cf	Es	Fm	Md	No

Figure 1.3. The modern or medium-long form table.

Source: Eric R. Scerri, creative commons license.

H																	He
Li	Be											B	C	N	O	F	Ne
Na	Mg											Al	Si	P	S	Cl	Ar
K	Ca	Sc	Ti	V	Cr	Mn	Fe	Co	Ni	Cu	Zn	Ga	Ge	As	Se	Br	Kr
Rb	Sr	Y	Zr	Nb	Mo	Tc	Ru	Rh	Pd	Ag	Cd	In	Sn	Sb	Te	I	Xe
Cs	Ba	Lu	Hf	Ta	W	Re	Os	Ir	Pt	Au	Hg	Tl	Pb	Bi	Po	At	Rn
Fr	Ra	Lr	Rf	Db	Sg	Bh	Hs	Mt	Ds	Rg	Cn	Nh					

La	Ce	Pr	Nd	Pm	Sm	Eu	Gd	Tb	Dy	Ho	Er	Tm	Yb
Ac	Th	Pa	U	Np	Pu	Am	Cm	Bk	Cf	Es	Fm	Md	No

Figure 1.4. Diagram of conventional periodic table format with alternative numbering system for groups.

Source: Eric R. Scerri, creative commons license.

Both of these systems employ the identical Roman numeral for every column, which, in the context of main-group elements, additionally represents the quantity of outer-shell electrons. Due to the perplexity engendered by these conventions, considerable focus has been devoted to the acquisition of a cohesive system. In a recent development, the IUPAC has put forth a recommendation advocating for the sequential numbering of groups using Arabic numerals, specifically from left to right. This numbering system would encompass groups 1 to 18, eliminating the utilization of the characters

A or B. One regrettable consequence of this idea is the loss of the direct relationship across the number of electrons from the outer shell within main-group element atoms as well as the group labels within the previous U.S. as well as European systems. As an illustration, the oxygen atom possesses a total of six electrons in its outer shell, thereby classifying it as a member of group VI (followed through an A or B) in previous classification systems. However, according to the IUPAC structure, the oxygen atom is categorized as belonging to group 16^{-} (Bryce, 2019).

Group I, located on the far left side of the periodic table, consists of metallic elements like sodium, potassium, as well as rubidium. Such substances exhibit atypical softness and reactivity, deviating from the conventional characteristics associated with metals, including iron, chromium, gold, as well as silver. The elements belonging to group I in the periodic table exhibit such high reactivity that the mere introduction of a small fragment of any of these metals to pure water initiates an energetic reaction, resulting in the generation of hydrogen gas as well as the formation of a transparent alkaline solution. The group II elements, namely magnesium, calcium, as well as barium, exhibit a lower level of reactivity compared to the group I elements across various aspects. (Suryelita et al., 2019). As one progresses toward the right side of the periodic table, a prominent feature is the presence of a central rectangular grouping of elements referred to as transition metals. This category encompasses various elements, including but not limited to iron, copper, as well as zinc. The elements mentioned were originally positioned within the groups that comprised the main-group elements found in early short forms periodic tables (Figure 1.6). The modern arrangement of the periodic table has resulted in the separation of certain elements from the primary structure, causing the loss of several significant features in their chemistry. However, despite these losses, the benefits of this revised organization outweighed the aforementioned drawbacks. Adjacent to the transition metals, there exists an additional set of representative elements commencing from group III and concluding with group VIII, which ultimately culminates in the noble gases positioned at the far right of the periodic table (No et al., 2022).

Occasionally, the characteristics that a collective entity possesses may not be readily apparent. The aforementioned group, known as Group IV, comprises carbon, silicon, germanium, tin, as well as lead. A significant variation is observed when descending the group, indicating a notable diversity. Carbon, positioned at the forefront of the chemical group, is a solid nonmetallic that manifests itself in three distinct structural configurations, namely diamond, graphite, as well as fullerenes (McMinn et al., 1994). It constitute the fundamental framework of all biological organisms. The subsequent element, silicon, is classified as a semimetal and holds significance in the realm of synthetic life or "artificial intelligence" due to its integral role in computer systems. The subsequent element in the periodic table, germanium, constitutes a semimetal that has been identified relatively recently. Its existence was initially predicted by Mendeleev, and subsequent research confirmed that germanium possesses several properties that were anticipated by Mendeleev. When considering the transition to tin as well as lead, one encounters two metals that have been recognized since ancient times. Despite the considerable diversity observed across the elements within group IV with regard to their metal-nonmetal characteristics, they do share a significant chemical similarity (Howe, 1900). Specifically, all elements within this group exhibit an optimal valence of 4, indicating their highest potential for combining with other elements. The observed variation among the elements within group VII is further accentuated. The two elements, fluorine, as well as chlorine, which occupy the top position in the group, are characterized by their toxic gaseous nature. Bromine, the subsequent element within the periodic table, is among the limited set of two recognized elements that exhibit a liquid state under standard room temperature conditions. The other element sharing this characteristic is the metallic element mercury (Bazhin et al., 2019).

Continuing along the periodic table, the subsequent element encountered is iodine, which is characterized as a solid with a violet-black appearance. If an inexperienced chemist were tasked with categorizing these elements based on their visual characteristics, it would be highly unlikely for them to consider grouping fluorine, chlorine, bromine, as well as iodine together. This particular case exemplifies the utility of discerning between the tangible and conceptual interpretations of the notion of an element. The primary basis for their similarity resides in the abstract nature of their elements, rather than the elements themselves being tangible and observable entities. Upon reaching the far right side of the periodic table, one encounters a notable collection of elements known as noble gases (Sweedler, 2019). These elements were initially isolated in the late 19th and early 20th centuries, either just before or around the time of the turning point of the 20th century. The primary characteristic of these entities, which initially appeared contradictory, was their absence of chemical properties upon isolation. The elements helium, neon, argon, as well as krypton were absent from early periodic tables due to their lack of discovery and unforeseen nature. Upon their discovery, the presence of these elements presented a significant obstacle to the periodic system. However, this challenge was ultimately resolved through the expansion of the table of elements to incorporate a novel group, denoted as Group VIII or Group 18 according to the IUPAC system. Situated at the lower portion of the contemporary periodic table, there exists an additional group of elements known as the rare earths, which are often visually represented as being physically detached from the rest of the table. However, it should be noted that the aforementioned characteristic is merely a superficial aspect of the commonly employed representation of the periodic table (Kragh, 2019). Similar to the manner in which transition metals are typically incorporated as a distinct block within the central region of the periodic table, it is indeed feasible to apply a comparable arrangement for rare earth elements (REEs). Undoubtedly, numerous instances of extensive presentations have been published. Although the inclusion of long-form tables, as depicted in Figure 1.5, provides a more cohesive representation of the REEs within the periodic table, their practicality is limited due to their cumbersome nature and lack of suitability for creating effortlessly shaped wall representations for the periodic system. The periodic table exists in various forms, yet the fundamental principle that underlies its structure, regardless of how it is presented, is the periodic law (Maret, 2022).

1.4. THE PERIODIC LAW

The periodic law posits that chemical elements exhibit a recurring pattern in their properties, occurring at regular yet variable intervals. As an illustration, fluorine, chlorine, as well as bromine, which are classified within group VII of the periodic table, exhibit the common characteristic of producing white crystalline salts using the general formula NaX when combined with the element sodium (Jensen, 2009). The fundamental principle that underlies every aspect associated with the periodic system involves the recurring pattern of properties (Erduran, 2007).

To begin with, it is important to note that periodicity within the elements is not characterized by a constant or precise pattern. In the conventional representation of the periodic table, the initial row comprises two elements, while the subsequent second and third rows consist of eight elements each. The 4th and 5th rows, in turn, encompass 18 elements, as well as this pattern continues throughout the table (Kaji, 2003).

H																														He
Li	Be																								B	C	N	O	F	Ne
Na	Mg																								Al	Si	P	S	Cl	Ar
K	Ca														Sc	Ti	V	Cr	Mn	Fe	Co	Ni	Cu	Zn	Ga	Ge	As	Se	Br	Kr
Rb	Sr														Y	Zr	Nb	Mo	Tc	Ru	Rh	Pd	Ag	Cd	In	Sn	Sb	Te	I	Xe
Cs	Ba	La	Ce	Pr	Nd	Sm	Eu	Gd	Tb	Dy	Ho	Er	Tm	Yb	Lu	Hf	Ta	W	Re	Os	Ir	Pt	Au	Hg	Tl	Pb	Bi	Po	At	Rn
Fr	Ra	Ac	Th	Pa	U	Pu	Am	Cm	Bk	Cf	Es	Fm	Md	No	Lr	Rf	Db	Sg	Bh	Hs	Mt	Ds	Rg							

Figure 1.5. Long-form periodic table.

Source: Eric R. Scerri, creative commons license.

This suggests the presence of a fluctuating pattern characterized by a sequence of numbers: 3, 9, 9, 19, and so forth, which significantly deviates from the regular periodicity observed in phenomena such as the days for the week or just the notes within a musical scale. In the aforementioned instances, the duration of the period remains consistent, exemplified by the fixed length of eight of the days during the week and the total number of notes upon a Western melodious scale (Ostrovsky, 2001). Within the realm of elements, it is worth noting that the duration of their respective periods not only exhibits variability but also lacks precise regularity. The elements inside a given column in a periodic table do not exhibit identical properties. In this regard, the periodicity of these phenomena bears a resemblance to the structure of the musical scale. Similar to how the musical scale repeats with a note represented by the exact same letter, albeit at a higher octave, these phenomena also exhibit a recurring pattern that echoes the original but is not identical to it (Brush, 1996). The diverse durations of the periods of elements along with the approximate pattern of their repetition has led certain chemists to refrain from using the term "law" in reference to chemical periodicity. The degree of lawfulness exhibited by chemical periodicity, in comparison to the laws governing physics, is a subject of contention regarding its significance. One could posit that chemical periodicity serves as an illustrative instance of a conventional chemical law, characterized by its approximate and intricate nature, yet fundamentally exhibiting behavior consistent with the principles of law (Kauffman, 1969). This location may serve as an appropriate forum for the examination and discussion of additional aspects pertaining to terminology. What are the distinctions between a periodic table and a periodic system? The concept of "periodic system" can be considered as the broader of the two terms. The periodic arrangement is a conceptual framework that posits the existence of an essential connection among the elements. When representing the periodic structure, one has the option to utilize a three-dimensional setup, a circular configuration, or various two-dimensional tables. Naturally, the term "table" inherently denotes a representation that is limited to two dimensions. While the expression "periodic table" is widely recognized among the 3 terms – law, system, as well as a table – it is, in fact, the most limited in scope (Restrepo & Pachón, 2007).

1.5. REACTING ELEMENTS AND ORDERING THE ELEMENTS

A significant portion of the knowledge pertaining to the elements is being acquired through the examination of their interactions with other elements as well as the exploration of their bonding characteristics. The metallic elements located on the left side of the traditional periodic table can be considered complementary counterparts to the nonmetallic elements, which are predominantly situated towards the right side (Wells, 2011). This phenomenon occurs due to the fact that, in contemporary terminology, metals undergo electron loss to generate positive ions, whereas nonmetals acquire electrons to produce negative ions. Oppositely charged ions undergo a process of combination to yield salts that possess a neutral charge, like sodium chloride or just calcium bromide (Bause & Schwegler, 2013). There exist additional complementary characteristics between metals as well as nonmetals. Metal oxides and hydroxides have the ability to undergo dissolution in water, resulting in the formation of bases. Conversely, nonmetal oxides and hydroxides exhibit the capacity to dissolve in water, leading to the formation of acids. In a neutralization process, an acid, as well as a base, undergo a chemical reaction resulting in the formation of water as well as salt. Bases as well as acids, akin to the metallic and nonmetallic substances from which they originate, exhibit contrasting yet interdependent characteristics (Bu et al., 2020).

The relationship between acids and bases is closely tied to the beginnings of the periodic system, as they played a significant role in the development of the idea of equivalent weights, that were initially employed to arrange the elements. The determination of the corresponding weight of a specific metal was initially derived by measuring the quantity of metal that undergoes a chemical reaction with a predetermined quantity of a selected standard acid. The concept of "equivalent weight" was later expanded to refer to the quantity of an element that undergoes a reaction with a standardized quantity of oxygen. In the course of history, the arrangement of elements within periods was initially established based on equivalent weight, subsequently transitioning to atomic weight, and ultimately being established by atomic number (Fermi, 1946). Chemists initially initiated the practice of conducting quantitative assessments to compare the quantities of acids as well as bases involved in chemical reactions. Subsequently, this methodology was expanded to encompass reactions involving acids and metals. This enabled chemists to arrange the metals in a numerical sequence based on their comparable weight, which, as previously stated, represents the quantity of metal that reacts with a standardized quantity of acid. The notion of comparable weights can be considered empirical in nature, as it does not necessarily rely on the theoretical postulation that elements are fundamentally comprised of atoms (Schneider & Reklaitis, 1975).

The determination of atomic weights, in contrast to equivalent weights, was initially achieved during the early 19th century by John Dalton. Dalton derived these values indirectly by measuring the combined masses of the respective elements involved. However, Dalton encountered complexities within this seemingly straightforward approach, which compelled him to make certain assumptions regarding the chemical formulas underlying the compounds under investigation. The fundamental aspect to consider in relation to this inquiry pertains to valence, which refers to the capacity of an element to combine with other elements. As an illustration, a monovalent atom forms chemical bonds with hydrogen atoms within a 1:1 ratio. In contrast, divalent atoms, like oxygen, blend with hydrogen atoms within a 2:1 ratio. This pattern continues for other valence states as well (Okamoto, 1988).

The idea of equivalent weight, as previously stated, is often considered to be a purely empirical notion as it appears to be independent of one's belief in the presence of atoms. After the introduction of atomic weights, a considerable number of chemists who harbored reservations regarding the theory of atoms sought to return to the previous notion of comparable weights. It was posited by the researchers that the determination of equivalent weights might be solely based on empirical evidence, thus rendering them more dependable. However, numerous scholars, including the recent work of Alan Rocke, have contended that such aspirations were merely a mirage. This is due to the fact that equivalent weights were also predicated on the supposition of specific compound formulas, which are fundamentally theoretical concepts (Onyejekwe, 1996).

Over an extended period, a significant amount of confusion arose due to the interchangeable utilization of equivalent weight as well as atomic weight. Dalton postulated that water was composed of a single hydrogen atom mixed with a single oxygen atom, thereby suggesting that its atomic weight, as well as its equivalent weight, were identical. However, his hypothesis regarding the valence of oxygen was ultimately proven to be inaccurate. The interchangeable use of the terms "equivalent weight" and "atomic weight" by numerous authors has contributed to the existing confusion (Yadav & Jiwari, 2019). The comprehensive understanding of the interconnections among equivalent weight, atomic weight, as well as valency was definitively established during the inaugural significant scientific symposium in 1860, hosted in Karlsruhe, Germany. The establishment of clear atomic weights and the subsequent widespread acceptance of their consistency paved the way for the autonomous identification of the periodic system. This breakthrough was achieved by multiple individuals across different nations, each proposing their own versions of the periodic table, which achieved varying levels of success. The elements were typically arranged in ascending order of atomic weight by each individual (Corcella et al., 2001).

The atomic number is the third ordering concept discussed, and it is considered the most contemporary. Following the comprehension of the atomic number, it superseded atomic weight to serve as an organizing concept for the elements. The reliance on weight combinations is no longer necessary for assigning atomic numbers, as they can be straightforwardly interpreted at a microscopic level based on the atomic arrangement of any particular element. The atomic number for an element is determined by the number of protons, which are positively charged particles, residing within the nucleus of its atoms. Hence, it can be observed that every element in the periodic table possesses an additional proton in comparison to the element that precedes it. An upsurge in the total number of neutrons within the nucleus as one progresses along the periodic table results in an associated rise in both atomic numbers as well as atomic weight. However, it is important to note that it seems the atomic number serves as the primary identifier for each element. It can be stated that atoms belonging to a specific element consistently possess an identical quantity of protons while exhibiting variability in the number of neutrons they encompass (Huang et al., 2018).

1.6. DIFFERENT REPRESENTATIONS OF THE PERIODIC SYSTEM

The contemporary periodic system effectively organizes the elements based on their atomic number, resulting in the formation of coherent groups. However, it is worth noting that there exist multiple representations of this system (Bittanti & Colaneri, 2000). Hence, a variety of periodic table formats exist, each tailored to specific applications. While a chemist may prioritize a format that emphasizes the reactive nature of elements, a professional in electrical engineering may have a preference for highlighting similarities as well as patterns in electrical conductivities (Casassa et al., 2015).

The manner in which the method of periodicity is presented is a captivating topic that particularly captivates the general public's imagination. Numerous endeavors have been made to acquire the definitive periodic table, starting from the early iterations developed by John Newlands, Julius Lothar Meyer, as well as Dimitri Mendeleev. It has been estimated that around 700 distinct iterations of the periodic table had been released within a century following the introduction of Mendeleev's renowned table in 1869 (Aleixo et al., 2007). The aforementioned alternatives encompass a wide range of options, such as 3-D tables, helices, concentric rings, spirals, zigzags, step tables, as well as mirror image tables. Presently, the Journal of Chemical Education continues to publish papers on a regular basis that claim to present novel and enhanced iterations of the periodic system (Faber et al., 2015).

The crux of all these endeavors lies in the singular form of the periodic law. This aspect associated with the periodic system remains unchanged across all the various displays. Numerous chemists emphasize that the physical representation of this law holds no significance, as long as specific fundamental criteria are fulfilled (Stewart, 2018). However, taking a philosophical perspective into account, it remains important to explore the basic representations of the elements or just the final shape of the periodic system. This is particularly relevant in the context of determining whether the law of periodicity should be understood as a realistic concept or just an extension of the convention. The commonly held belief that representation is solely the domain of convention appears to contradict the realist perspective, which posits the existence of objective truths regarding the occurrences of property repetitions (Kruchinin et al., 2018).

1.7. CHANGES IN THE PERIODIC TABLE

In the year 1945, Glenn Seaborg proposed the notion that the elements commencing with actinium, specifically element number 89, ought to be classified as a rare earth series. This proposition challenged the prevailing belief that the latest generation of rare earths was to start after element 92, namely uranium (as depicted in Figure 1.6) (Rahm et al., 2019). The novel periodic table proposed by Seaborg unveiled a correlation between europium (atomic number 63) and gadolinium (atomic number 64), as well as their hypothetical counterparts, elements 95 as well as 96, which had not yet been discovered at the time. Seaborg accomplished the synthesis and identification of two novel elements, subsequently denoted as americium as well as curium, drawing upon the aforementioned analogies. Several additional trans-uranium elements were synthesized subsequently (Johnson, 2019).

Minor modifications have been made to the standard format of the periodic table with respect to the elements that initiate both the 3rd and 4th rows of transitional elements. In contrast to earlier versions of periodic tables, which designate lanthanum (57) as well as actinium (89) as the elements in question, contemporary scientific investigations and analytical findings have established that lutetium (71), as well as lawrencium (103), now occupy their respective positions. It is noteworthy to observe that certain earlier versions of periodic tables, predicated on macroscopic attributes, had foreseen these alterations (Niaz et al., 2014).

The following instances illustrate ambiguities within the domain of secondary categorization, which lacks the same level of clarity as primary categorization, specifically in terms of the sequential arrangement of elements. In the realm of classical chemistry, the concept of secondary categorization pertains to the chemical resemblances exhibited by different elements within a given group. In contemporary parlance, the explanation of secondary categorization is elucidated through the utilization of electronic arrangements. Irrespective of whether an individual adopts a traditional qualitative chemical methodology or an improved physical approach centered on electronic arrangements, it can be argued that secondary categorization of this nature is less definitive than primary categorization and lacks a categorical establishment (Steffen, 1996). The establishment of secondary categorization, as delineated in this context, exemplifies the contemporary conflict between employing chemical properties versus physical properties for the purpose of classification. The specific positioning of an element throughout the groups of the periodic table may exhibit variation based on the degree of emphasis placed on its electronic arrangement, a physical attribute, or its chemical characteristics. Indeed, numerous contemporary discussions regarding the positioning of helium within the periodic table primarily center around the relative significance that ought to be attributed to these two methodologies (Cao et al., 2021). In recent years, there has been a notable expansion in the overall number of elements, surpassing the 100 marks, due to the successful synthesis of synthetic elements. As of the present moment, there exists definitive evidence regarding the existence of element 111. These elements are commonly characterized by a high degree of instability, resulting in the production of only a limited number of atoms at any given moment. Nevertheless, researchers have developed innovative chemical methods to investigate the chemical characteristics of such superheavy elements, enabling them to verify the consistency of chemical properties when applied to extremely heavy atoms (Schwarz et al., 2022).

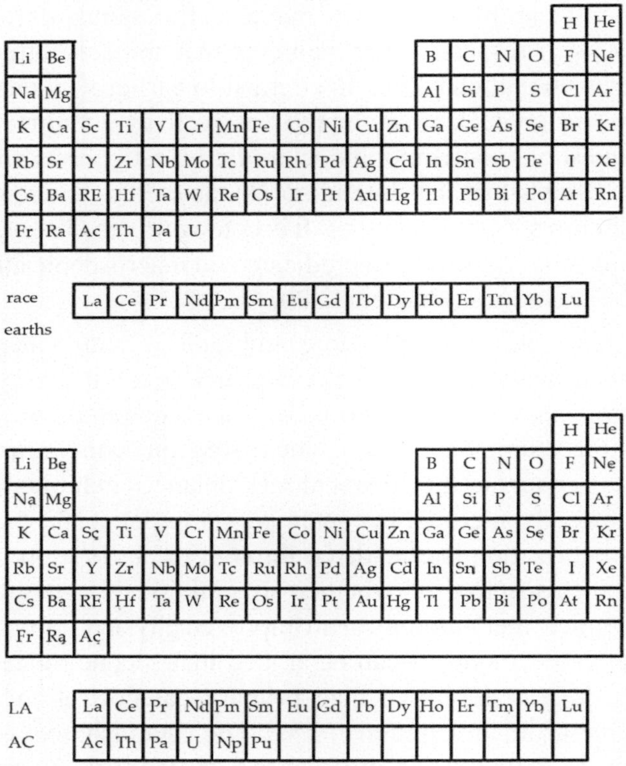

Figure 1.6. Pre-Seaborg (a) and post-Seaborg (b) periodic tables.

Source: Robin Garrell, creative commons license.

From a philosophical standpoint, the synthesis of these elements provides an opportunity to investigate whether the periodic law adheres strictly as a universal law, akin to Newton's law of gravitation, or if deviations from anticipated patterns in chemical properties may occur once the atomic number surpasses a certain threshold. To date, no unexpected findings have been reported. However, the inquiry into whether certain superheavy elements exhibit the anticipated chemical properties remains unresolved to a significant extent. A notable complication that emerges within this specific domain of the periodic table pertains to the escalating (Allen & Knight, 2002). Relevance of relativistic effects attributable to the high velocity of electrons.52 The aforementioned effects give rise to the adoption of unforeseen electronic configurations in certain atoms, potentially leading to equally unforeseen chemical properties.

1.8. SUMMARY

The periodic table is a tabular arrangement of chemical elements, organized based on their atomic number, electron configuration, and recurring chemical properties This is a systematic arrangement of the chemical elements, presented in an ascending order of their atomic weights. The periodic table exhibits a recurring pattern of specific properties, a phenomenon initially observed by Dmitry I. Mendeleyev in 1869. Entities that are positioned within the exact same column (group) of a table are typically organized in a manner that reflects their shared characteristics. During the 20th century, as scientists gained a comprehensive understanding of atomic structure, it became evident that the periodic table accurately represents the ascending order of atomic numbers. Individuals belonging to a given group within the periodic table exhibit identical quantities of electrons within the outermost energy levels of their respective atoms. Consequently, these individuals tend to form bonds of a consistent nature, typically characterized by the same valence. It is worth noting that noble gases, which possess complete outer energy levels, generally do not engage in bond formation. The periodic table has significantly enhanced our comprehension of bonding as well as chemical behavior. Additionally, this theory facilitated the anticipation of novel elements, a considerable number of which were subsequently observed or artificially produced.

REFERENCES

1. Aksholakova, A., (2014). Proper name as a clue symbol of identity. *Procedia-Social and Behavioral Sciences, 112*, 465–471.
2. Aleixo, J. C., Polderman, J. W., & Rocha, P., (2007). Representations and structural properties of periodic systems. *Automatica, 43*(11), 1921–1931.
3. Allen, L. C., & Knight, E. T., (2002). The Löwdin challenge: Origin of the n+ ℓ, n (Madelung) rule for filling the orbital configurations of the periodic table. *International Journal of Quantum Chemistry, 90*(1), 80–88.
4. Almquist, E., Senior, J., & Bloch, N., (2016). The elements of value. *Harvard Business Review, 94*(9), 47–53.
5. Analytical Methods Committee, AMCTB, No. (2022). A periodic table for liquid chromatography separation modes. *Analytical Methods, 14*(31), 3033–3037.
6. Bause, M., & Schwegler, K., (2013). Higher order finite element approximation of systems of convection–diffusion–reaction equations with small diffusion. *Journal of Computational and Applied Mathematics, 246*, 52–64.
7. Bazhin, V. Y., Aleksandrova, T. A., Kotova, E. L., & Suslov, A. P., (2019). A modern view of anomalies in the metal groups of the periodic system of DI Mendeleev. Записки Горного Института, *239*, 520–527.
8. Bernal, J. P., & Railsback, L. B., (2008). An introduction to the periodic table of the elements and their ions for the earth sciences. *Revista Mexicana de Ciencias Geológicas, 25*(2), 236–246.
9. Bierenstiel, M., & Snow, K., (2019). Periodic universe: A teaching model for understanding the periodic table of the elements. *Journal of Chemical Education, 96*(7), 1367–1376.
10. Bittanti, S., & Colaneri, P., (2000). Invariant representations of discrete-time periodic systems. *Automatica, 36*(12), 1777–1793.
11. Blanchette, M., & Tompa, M., (2002). Discovery of regulatory elements by a computational method for phylogenetic foot printing. *Genome Research, 12*(5), 739–748.
12. Brush, S. G., (1996). The reception of Mendeleev's periodic law in America and Britain. *ISIS, 87*(4), 595–628.
13. Bryce, D. L., (2019). New frontiers for solid-state NMR across the

periodic table: A snapshot of modern techniques and instrumentation. *Dalton Transactions, 48*(23), 8014–8020.

14. Bu, W., Ji, L., Tang, Y., & Zhou, J., (2020). Space-time finite element method for the distributed-order time fractional reaction diffusion equations. *Applied Numerical Mathematics, 152*, 446–465.
15. Cao, C., Vernon, R. E., Schwarz, W. H., & Li, J., (2021). Understanding periodic and non-periodic chemistry in periodic tables. *Frontiers in Chemistry, 8*, 549296.
16. Casassa, S., Erba, A., Baima, J., & Orlando, R., (2015). Electron density analysis of large (molecular and periodic) systems: A parallel implementation. *Journal of Computational Chemistry, 36*(26), 1940–1946.
17. Constable, E. C., (2019). Evolution and understanding of the d-block elements in the periodic table. *Dalton Transactions, 48*(26), 9408–9421.
18. Corcella, G., Knowles, I. G., Marchesini, G., Moretti, S., Odagiri, K., Richardson, P., & Webber, B. R., (2001). HERWIG 6: An event generator for hadron emission reactions with interfering gluons (including supersymmetric processes). *Journal of High Energy Physics, 2001*(01), 010.
19. Corish, J., & Rosenblatt, G. M., (2004). Name and symbol of the element with atomic number 111 (IUPAC recommendations 2004). *Pure and Applied Chemistry, 76*(12), 2101–2103.
20. Elding, L. I., (1994). Names and symbols of transfermium elements (IUPAC recommendations 1994). *Pure and Applied Chemistry, 66*(12), 2419–2421.
21. Erduran, S., (2007). Breaking the law: Promoting domain-specificity in chemical education in the context of arguing about the periodic law. *Foundations of Chemistry, 9*, 247–263.
22. Faber, F., Lindmaa, A., Von, L. O. A., & Armiento, R., (2015). Crystal structure representations for machine learning models of formation energies. *International Journal of Quantum Chemistry, 115*(16), 1094–1101.
23. Fermi, E., (1946). The development of the first chain reacting pile. *Proceedings of the American Philosophical Society, 90*(1), 20–24.
24. Franco, M. A. J., Oliva, M. J. M., & Bernal, M. S., (2012). An educational card game for learning families of chemical elements.

Journal of Chemical Education, 89(8), 1044–1046.

25. Gordy, W., & Thomas, W. O., (1956). Electronegativities of the elements. *The Journal of Chemical Physics, 24*(2), 439–444.
26. Hofmann, S., & Münzenberg, G., (2000). The discovery of the heaviest elements. *Reviews of Modern Physics, 72*(3), 733.
27. Howe, J. L., (1900). The eighth group of the periodic system and some of its problems. II. *Science, 12*(288), 20–34.
28. Huang, C., Duraisamy, K., & Merkle, C., (2018). Challenges in reduced order modeling of reacting flows. In: *2018 Joint Propulsion Conference* (Vol. 2, p. 4675).
29. Jensen, W. B., (2009). Misapplying the periodic law. *Journal of Chemical Education, 86*(10), 1186.
30. Joag, S. D., (2014). An effective method of introducing the periodic table as a crossword puzzle at the high school level. *Journal of Chemical Education, 91*(6), 864–867.
31. Johnson, J. A., (2019). Populating the periodic table: Nucleosynthesis of the elements. *Science, 363*(6426), 474–478.
32. Justino, C. I., Freitas, A. C., Pereira, R., Duarte, A. C., & Santos, T. A. R., (2015). Recent developments in recognition elements for chemical sensors and biosensors. *TrAC Trends in Analytical Chemistry, 68*, 2–17.
33. Kaji, M., (2003). Mendeleev's discovery of the periodic law: The origin and the reception. *Foundations of Chemistry, 5*(3), 189–214.
34. Karol, P. J., Barber, R. C., Sherrill, B. M., Vardaci, E., & Yamazaki, T., (2016). Discovery of the elements with atomic numbers Z= 113, 115 and 117 (IUPAC technical report). *Pure and Applied Chemistry, 88*(1, 2), 139–153.
35. Kauffman, G. B., (1969). American forerunners of the periodic law. *Journal of Chemical Education, 46*(3), 128.
36. Kelly, G. M., & Street, R., (2006). Review of the elements of 2-categories. In: *Category Seminar: Proceedings Sydney Category Theory Seminar 1972/1973* (Vol. 2, pp. 75–103). Berlin, Heidelberg: Springer Berlin Heidelberg.
37. Kragh, H., (2019). The periodic system and the idea of a chemical element: From Mendeleev to superheavy elements. *Centaurus, 61*(4), 329–344.
38. Kruchinin, S. Y., Krausz, F., & Yakovlev, V. S., (2018). Colloquium:

Strong-field phenomena in periodic systems. *Reviews of Modern Physics, 90*(2), 021002.

39. Lavoisier, A. L., (1806). *Elements of Chemistry in a New Systematic Order Containing all the Modern Discoveries* (Vol. 1, pp. 987–990). Evert Duyckinck, no. 110 Pearl-Street, and James and Thomas Ronalds, no. 188 Pearl-Street.
40. Manuel, P., (1995). Music as symbol, music as simulacrum: Postmodern, pre-modern, and modern aesthetics in subcultural popular musics. *Popular Music, 14*(2), 227–239.
41. Maret, W., (2022). The quintessence of metallomics: A harbinger of a different life science based on the periodic table of the bioelements. *Metallomics, 14*(8), 051.
42. McMinn, D. G., Nakamaye, K. L., & Smieja, J. A., (1994). Enhancing undergraduate education: Curriculum modification and instrumentation. *Journal of Chemical Education, 71*(9), 755.
43. Meija, J., (2014). Symbols of the elements. *Chemistry International, 36*(1), 20–21.
44. Munzenberg, G., (1988). Recent advances in the discovery of transuranium elements. *Reports on Progress in Physics, 51*(1), 57.
45. Niaz, M., Luiggi, M., Niaz, M., & Luiggi, M., (2014). *Facilitating Conceptual Change in Students' Understanding of the Periodic Table* (Vol. 2, pp. 1–49). Springer International Publishing.
46. Okamoto, N., (1988). Analysis of convective diffusion problem with first-order chemical reaction by boundary element method. *International Journal for Numerical Methods in Fluids, 8*(1), 55–64.
47. Onyejekwe, O. O., (1996). Green element description of mass transfer in reacting systems. *Numerical Heat Transfer, 30*(4), 483–498.
48. Ostrovsky, V. N., (2001). What and how physics contributes to understanding the periodic law. *Foundations of Chemistry, 3*(2), 145–181.
49. Rahm, M., Cammi, R., Ashcroft, N. W., & Hoffmann, R., (2019). Squeezing all elements in the periodic table: Electron configuration and electronegativity of the atoms under compression. *Journal of the American Chemical Society, 141*(26), 10253–10271.
50. Restrepo, G., & Pachón, L., (2007). Mathematical aspects of the periodic law. *Foundations of Chemistry, 9*, 189–214.

51. Rignes, V., (1989). *Journal of Chemical Education, 66,* 731–738.
52. Ross, L. N., (2021). Causal explanation and the periodic table. *Synthese, 198*(1), 79–103.
53. Sargeson, A. M., McKleverty, J. A., Koppenol, W. H., Arduengo, A. J., Coucouvanis, D. N., De Bolster, M. W. G., & De Oliviera, C. J. D., (1994). Names and symbols of transfermium elements (IUPAC recommendations 1994). *Pure and Applied Chemistry, 66*(12), 2419–2421.
54. Scerri, E. R., (2012). The periodic table. In: *Philosophy of Chemistry* (Vol. 2, pp. 329–338). North-Holland.
55. Scerri, E. R., (2019). Happy 150th birthday to the periodic table. *Chemistry–A European Journal, 25*(31), 7410–7415.
56. Schneider, D. R., & Reklaitis, G. V., (1975). On material balances for chemically reacting systems. *Chemical Engineering Science, 30*(2), 243–247.
57. Schwarz, W. E., Müller, U., & Kraus, F., (2022). The good reasons for a standard periodic table of the chemical elements. *Zeitschrift für Anorganische und Allgemeine Chemie, 648*(13), 321–325.
58. Schwerdtfeger, P., Smits, O. R., & Pyykkö, P., (2020). The periodic table and the physics that drives it. *Nature Reviews Chemistry, 4*(7), 359–380.
59. Seaborg, G. T., (1996). Evolution of the modern periodic table. *Journal of the Chemical Society, Dalton Transactions*, (20), 3899–3907.
60. Stark, A., Lin, M. F., Kheradpour, P., Pedersen, J. S., Parts, L., Carlson, J. W., & Kellis, M., (2007). Discovery of functional elements in 12 Drosophila genomes using evolutionary signatures. *Nature, 450*(7167), 219–232.
61. Steffen, W. L., (1996). A periodic table for ecology? A chemist's view of plant functional types. *Journal of Vegetation Science, 7*(3), 425–430.
62. Stewart, P. J., (2018). Tetrahedral and spherical representations of the periodic system. *Foundations of Chemistry, 20*, 111–120.
63. Supanchaiyamat, N., & Hunt, A. J., (2019). Conservation of critical elements of the periodic table. *ChemSusChem, 12*(2), 397–403.
64. Suryelita, S., Guspatni, G., & Defriati, P., (2019). Description of learning difficulties on atomic structure and periodic table topics of tenth grade students in SMAN 7 Padang. In: *Journal of Physics:*

Conference Series (Vol. 1317, pp. 12147–12149). IOP Publishing.

65. Sweedler, J. V., (2019). The periodic table of elements at 150. *Analytical Chemistry, 91*(9), 5469–5469.
66. Vandevelde, K. J., (1980). The new property of the nineteenth century: The development of the modern concept of property. *Buff. L. Rev., 29*, 325.
67. Watson, G. S., Green, D. W., & Watson, J. A., (2020). Introducing students to the periodic table using a descriptive approach of superheroes, meats, and fruits and nuts. *Journal of Chemical Education, 98*(2), 669–672.
68. Weeks, M. E., (1932). The discovery of the elements. XVI. The rare earth elements. *Journal of Chemical Education, 9*(10), 1751.
69. Weinhold, F., & Bent, H. A., (2007). News from the periodic table: An introduction to "periodicity symbols, tables, and models for higher-order valency and donor–acceptor kinships." *Journal of Chemical Education, 84*(7), 1145.
70. Wells, G. N., (2011). Analysis of an interface stabilized finite element method: The advection-diffusion-reaction equation. *SIAM Journal on Numerical Analysis, 49*(1), 87–109.
71. Xie, L. H., Yin, C. R., Lai, W. Y., Fan, Q. L., & Huang, W., (2012). Polyfluorene-based semiconductors combined with various periodic table elements for organic electronics. *Progress in Polymer Science, 37*(9), 1192–1264.
72. Yadav, O. P., & Jiwari, R., (2019). A finite element approach for analysis and computational modelling of coupled reaction diffusion models. *Numerical Methods for Partial Differential Equations, 35*(2), 830–850.
73. Zhang, Z., & Gerstein, M., (2003). Of mice and men: Phylogenetic foot printing aids the discovery of regulatory elements. *Journal of Biology, 2*(2), 1–4.
74. Zhi-jun, W., & Cheng-Ai, N., (2006). Application of traditional culture symbol in modern product design. In: *2006 7th International Conference on Computer-Aided Industrial Design and Conceptual Design* (Vol. 2, pp. 1–5). IEEE.

CHAPTER 2

Hydrogen and Alkali Metals

Contents

2.1. INTRODUCTION

Hydrogen (H) is positioned as the first element, situated above the alkali metals within the primary column that makes up the periodic table. Nevertheless, due to its distinct characteristics in comparison to the elements positioned beneath it, hydrogen is excluded from its designated group. This gaseous substance possesses the most basic atomic structure of all elements, consisting of a single electron and a single proton. The element in question exhibits a high degree of reactivity and readily combines with several other elements to produce compounds (Cronyn, 2003; Chand & Biswal, 2020).

As the atomic mass of the alkali metals grows, certain trends in their characteristics become evident. Lithium exhibits a significantly higher specific heat in comparison to the other alkali elements. Furthermore, it is observed that specific heats generally drop as one descends down the column. Lithium, being the least dense among the alkali metals, has a trend of increasing density as one moves down the column. The decline in electron affinities with increasing atomic weight can be attributed to the diminishing impact of the nuclear charge on the valence electron, primarily due to the shielding effect exerted by the core electrons. The alkali metals exhibit a high degree of reactivity when exposed to water. Nevertheless, the reaction with water exhibits an escalated level of violence as the atomic weight of the alkali element rises. The chemical properties of alkali metals are predominantly influenced by the anions at which they form chemical bonds. The primary factors contributing to variances in behavior are primarily attributed to disparities in ion sizes and divergent temperatures of hydration. A diverse array of inorganic compounds has been synthesized utilizing alkali metals, such as hydrides, hydroxides, nitrates, nitrides, oxides, peroxides, permanganates, phosphates, and silicates (Alkorta et al., 2020). Furthermore, a significant quantity of organo-alkali compounds has been synthesized, wherein sodium or potassium mostly react with diverse hydrocarbons. The study of alkalis in analytical chemistry poses significant challenges. Several intricate reagents have the capability to form precipitates with ions such as Na^+ and K^+. However, it is advisable to conduct these reactions utilizing ether or alcohol to serve as solvent instead of water. The identification of alkali metals within solutions or compounds is typically accomplished through the observation of the distinct hues shown by alkali metal ions when subjected to flames. When arranged in a descending manner, lithium salts exhibit a carmine hue, sodium salts display a yellow hue, potassium salts manifest a violet hue, rubidium salts present a bluish-red hue, and cesium salts demonstrate a blue hue (Sacks, 2006; Laing, 2007) (Table 2.1).

Table 2.1. Different Transition Elements (Hydrogen Element is in Italics, and Alkali Metals are in Bold Type)

H				
Li	Be			
Na	Mg			
K	Ca	Sc	T	V
Rb	Sr	Y	Zr	Nb
Cs	Ba	La	Hf	Ta
Fr	Ra	Ac	Th	Pa

2.2. HYDROGEN

Hydrogen, a chemical element with an atomic number of 1, occupies a unique position on the periodic chart. This element does not pertain to any specific family of elements. Despite being a nonmetal, hydrogen is positioned on the leftmost side of the periodic table with the alkali metals in Group 1. However, it is important to note that hydrogen does not fall into the alkali metals category. Hydrogen, being the most prevalent element in the cosmos, is commonly observed in the form of interstellar gas and serves as the primary component of main-sequence stars. Helium, akin to hydrogen in terms of its simplicity and abundance, occupies a position in the same row towards the right-hand side of the periodic table (Petruševski, & Cvetković, 2018). Despite helium being classified as a noble gas and so exhibiting chemical inertness, hydrogen displays reactivity with a wide range of elements, resulting in the formation of diverse and practical compounds. An example of a significant chemical interaction involves the formation of covalent bonds between hydrogen and carbon atoms, resulting in the creation of a fundamental structure found in a wide range of organic compounds referred to as hydrocarbons. In a similar vein, the chemical interaction between hydrogen and oxygen results in the formation of water, which is widely recognized as the most crucial component on our planet. Despite being the most prevalent element, hydrogen constitutes a mere 0.14% of the Earth's crust in terms of weight. However, it is found abundantly as a constituent of the water present in oceans, ice packs, lakes, rivers, as well as the atmosphere. Hydrogen is ubiquitously found in numerous carbon compounds, permeating animal and vegetable tissue, as well as petroleum (Jaksic, 2000; Midilli et al., 2005). Despite the common assertion that carbon possesses the highest number of known compounds among all elements, it is important to acknowledge that the inclusion of hydrogen in nearly all carbon compounds, as well as its ability to form multiple compounds with every element except certain noble gases, raises the possibility that hydrogen compounds may actually outnumber carbon compounds. The primary industrial utilization of elemental hydrogen is observed in the production of ammonia, a substance formed by the combination of hydrogen and nitrogen. Additionally, hydrogen is employed in the process of hydrogenation, which involves the conversion of carbon monoxide as well as organic compounds into a diverse range of chemicals and fuels (Kaesz & Atkins, 2003). Hydrogen has been widely regarded as a promising candidate for power generation and transportation, and it holds the potential to serve as a plentiful and environmentally friendly energy resource in the future (Figure 2.1).

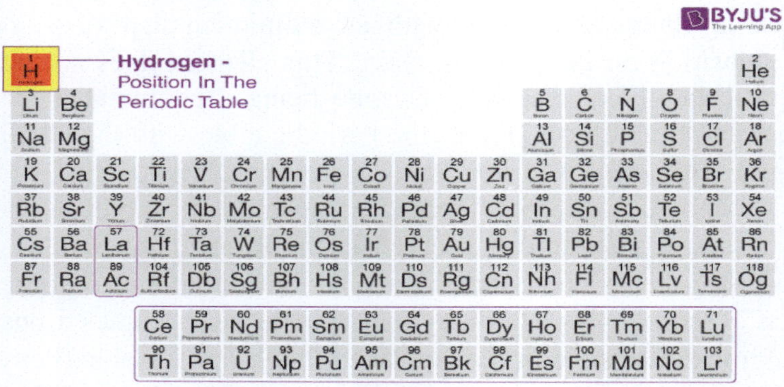

Figure 2.1. Hydrogen position in the periodic table.

Source: M Laing, creative commons license.

2.2.1. Physical and Chemical Properties of Hydrogen

Hydrogen is inherently present in its molecular state. The gas in question is characterized by its lack of color, flavor, and taste. The transition from a gaseous state to a liquid state occurs at a temperature of -252.77 ^{0}C (-422.99 ^{0}F), while the transition from a liquid state to a solid state occurs at a temperature of -259.2 ^{0}C (-434.6 ^{0}F). The substance exhibits limited solubility in both water and alcohol. With a density of 0.08999 g/L, this element possesses the lowest density among all known elements. Furthermore, it exhibits the unique characteristic of dispersing rapidly and ascending swiftly to the high atmosphere. Hydrogen gas has a high degree of flammability (Sacks, 2006; Petruševski & Cvetković, 2018). The substance undergoes combustion in the presence of air and oxygen, resulting in the formation of water. Under some conditions, the combination of this substance with air and chlorine has the potential to undergo a spontaneous explosion triggered by a spark, heat, or exposure to sunlight. The process of molecular hydrogen breakdown, when it separates into atomic hydrogen, is characterized by a dissociation energy of around 435 kJ/mol. Atomic hydrogen has a high degree of reactivity.

It has the ability to form chemical bonds with other atoms, resulting in the formation of either covalent or ionic bonds (Petruševski & Cvetković, 2017). Hydrogen exhibits an atypical behavior in the formation of covalent bonds. While most atoms adhere to the octet rule, which entails acquiring eight valence electrons, hydrogen deviates from this pattern and adheres to the duet rule. Consequently, hydrogen possesses only two electrons available for bonding. Several examples of covalent bonds involving hydrogen may be found in various substances. These involve water (H_2O), hydrogen sulfide (H_2S), as well as ammonia (NH_3), as well as other organic molecules that are generated on a hydrogen-carbon backbone. In order to form an ionic bond, the atom acquires an additional electron, resulting in the formation of a negatively charged ion known as the hydride ion (H^-). Subsequently, it has the capability to form a chemical bond with a positively charged metallic ion, also known as a cation. Ionic hydrides serve as practical reservoirs of hydrogen gas (Cvetković & Petruševski, 2017). For example, commercially available calcium hydride (CaH_2) offers a highly convenient method for the production of hydrogen. The hydrogen gas generated from the chemical reaction between calcium hydride and water possesses the potential for utilization in the inflation of life rafts. Metals that lack the ability to form stable hydrides, such as platinum, exhibit catalytic properties on their surfaces, facilitating the recombination process of hydrogen atoms into hydrogen molecules. As a result of this exothermic reaction, the metals become heated to the point of incandescence (Richards, 1956). There are two distinct forms of molecular hydrogen, namely ortho as well as para hydrogen. The observed variations arise from the distinct magnetic interactions shown by the protons, which can be attributed to their rotational movements. In the case of ortho-hydrogen, the spins from the two protons exhibit parallel alignment, meaning they are oriented in the same direction. In contrast, para-hydrogen exhibits spins that are oriented in opposing directions, resulting in an antiparallel configuration. The magnetic characteristics of atoms are determined by the connection of spin alignments

(Bańkowski, 1966; Cronyn, 2003). Typically, interconversions between ortho as well as para molecules, specifically in relation to hydrogen, are not observed, hence establishing orthohydrogen and para-hydrogen as different variations of hydrogen. The two forms, however, have the potential to undergo interconversion under specific conditions. There are multiple methods by which equilibrium may be maintained between the two types. One approach involves the utilization of catalysts, like charcoal that has been activated or other paramagnetic compounds. Another technique is subjecting the gas to an electrical discharge or elevating its temperature to a high degree. The equilibrium level for para-hydrogen within a mixture is temperature-dependent. At absolute zero temperature (0 K or -273 ^{0}C), hydrogen predominantly exists in the form of para-hydrogen, which exhibits greater thermodynamic stability. At the temperature corresponding to the liquefaction of air, approximately 80 K or -193 ^{0}C, the proportion for ortho to para-hydrogen appears to be 1:1. However, as the temperature rises to room temperature, this ratio increases to 3:1. In order to obtain para-hydrogen in its pure form, it is possible to achieve this by subjecting the mixture to charcoal at a specific temperature for liquid hydrogen (Hernández & Novaro, 2014). This process effectively turns all ortho-hydrogen present into para-hydrogen. In contrast, the direct preparation of ortho-hydrogen from the mixture is unfeasible due to the consistent presence of para-hydrogen at a minimum concentration of 25% (Figure 2.2).

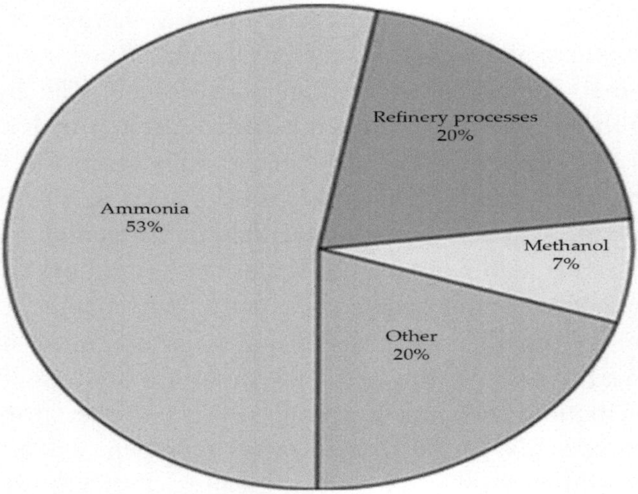

Figure 2.2. Uses of hydrogen in the chemical industry.

Source: H. Idriss, creative commons license.

There exist modest variations in physical characteristics between the two distinct types of hydrogen. The melting temperature of para-hydrogen exhibits a decrease of 0.10 ^{0}C in comparison to a 3:1 blend of ortho-hydrogen as well as para-hydrogen. At a temperature of -252.77 ^{0}C, the vapor pressure of liquid para-hydrogen is measured to be 1.035 atm, while the vapor concentration of the 3:1 ortho-para combination is found to be 1.000 atm. The distinct vapor pressures shown by para-hydrogen and ortho-hydrogen allow for their separation through the utilization of low temperatures gas chromatography. This analytical technique effectively segregates several atomic as well as molecular species by exploiting their discrepant volatilities.

2.3. ALKALI METALS

The ancient civilizations possessed knowledge of sodium as well as potassium salts. The symbol Na, denoting the chemical element sodium, originates from its Latin appellation, natrium. The Latin term "kalium" is the origin of the symbol K used to represent the chemical element potassium. During the first decade for the 19th century, Sir Humphrey Davy emerged as a meticulous researcher within the realm of electrochemistry. In 1807, the researcher conducted initial experiments involving the electrolysis of aqueous solutions comprising soda ash (NaOH) as well as potash (KOH) in an effort to break down these compounds. Nevertheless, his efforts resulted solely in the separation of water into hydrogen as well as oxygen gases (Molitor & Gessner, 2016). Subsequently, Davy proceeded to apply an electric current to a piece of potash that had been slightly wet. The researcher made observations of gas bubbles emanating from the positive electrode, also known as the anode. The researcher observed the production of globules at the negatively charged electrode, specifically the cathode, which exhibited a visual resemblance to mercury. Certain globules underwent combustion and detonated, exhibiting the same flammable reaction when submerged in water. Subsequent globules underwent rapid tarnishing. Davy made the observation that when the metal came into contact with water, it underwent a reaction that resulted in the release of hydrogen gas. Subsequently, it was the combustion of this hydrogen gas that generated the flame (Ohse, 1985; Sigel et al., 2016). The new metal was dubbed "potassium" by Davy, as it was derived from potash. It was observed that a more potent electrical current was necessitated, and in a short span of time subsequent to the isolation of potassium, the discovery of sodium was made (termed such due to its derivation from soda ash). Initially, there was skepticism among fellow scientists over the veracity of Davy's claim that sodium, as well as potassium, were genuine elements. It was hypothesized that Davy's "elements" could potentially be compounds consisting of potassium or sodium as well as hydrogen. Davy, however, was capable of to successfully establishing the elemental basis of their findings. Jöns Jakob Berzelius (1779–1848) of Sweden emerged as a prominent figure in the field of chemistry during the 19th century. In the year 1817, Berzelius assigned the task of investigating the mineral petalite to one of his assistants, namely the Swedish chemist Johan August Arfwedson (1792–1841). Arfwedson was found to constitute 96% of the composition of petalite, leaving the remainder of four percent unexplained. By the year 1818, Berzelius and Arfwedson had reached the conclusion that petalite, a mineral, must possess an alkali metal of an unidentified kind.

The composition of petalite was shown to be lithium aluminum silicate, as well as Arfwedson is credited with the discovery of lithium. The etymology of the term "lithium" can be traced back to its Greek origin, specifically the word "lithos," which translates to "stone." In the same year, the identification of the red coloration caused by the presence of lithium salts in flames was made. Arfwedson's attempts to achieve the isolation of pure lithium metal through electrolysis proved to be fruitless. However, it was W. T. Brande along with Davy who ultimately achieved the successful isolation of white lithium metal through the electrolysis of lithia, independently from each other. For an extended period, researchers were only able to acquire minuscule amounts of lithium. Significant quantities of lithium were not obtained through the electrolysis for molten lithium chloride (LiCl) until the year 1855.

The discovery of cesium, the subsequent alkali metal following lithium, occurred 42 years later. In the years 1850, a spectroscope known as the Kirchhoff-Bunsen spectroscope was constructed by German chemist Robert Wilhelm Bunsen (1811–99) along with German physicist Gustav Robert Kirchhoff (1824–87). This instrument was utilized by scientists to ascertain the chemical makeup of minerals. Minerals have the capacity to emit light when subjected to elevated temperatures. Bunsen, indeed, developed a burner (often referred to as the Bunsen burner) during the years 1854–1855, characterized by a flame that lacks coloration. The spectroscope employs a prism to disperse the emitted light, resulting in the manifestation of distinct colors as luminous lines contrasted against a backdrop of darkness (Schilling, 2006; Rich, 2007). The identification of elements within a mineral is possible due to the presence of discrete spectral patterns associated with each element, analogous to the individualized fingerprints found in individuals. In the year 1860, Bunsen and Kirchhoff made the observation of previously unidentified blue spectral lines, which were not found to match with any known chemical element. The researchers ascertained the novel chemical element to possess alkali metal properties and designated it as "cesium," deriving its name from the Greek term "caesius," denoting the azure hue observed in the higher regions of the atmosphere. The process of isolating cesium metal spanned a period of two decades, culminating in its successful isolation within Bunsen's laboratory using the application of electrolysis to cesium cyanide. In the year 1861, Bunsen and Kirchhoff made a subsequent announcement following their initial finding of cesium, whereby they reported the identification of rubidium within the mineral lepidolite (Varma & Rempe, 2006; Dye, 2015).

The existence of rubidium was discerned through the identification of two previously unidentified dark red lines within the spectrum of the material. The term "rubidium" derives its etymology from the Latin term "rubidus," denoting a shade of red that is very intense. Bunsen was successful in the isolation of metallic rubidium. Bunsen gained significant renown during his era and was held in high regard by his pupils due to his contributions to the advancement of spectrum analysis techniques. Subsequently, chemists made further discoveries of numerous additional elements through the utilization of Bunsen as well as Kirchoff's spectroscope. The aforementioned elements encompass gallium, helium, holmium, indium, lutetium, neodymium, praseodymium, thallium, thulium, as well as ytterbium. The subsequent identification of a metal that is alkali took place about eight decades subsequent to the aforementioned discovery. In the year 1939, Marguerite Perey, a scientist from Paris, made an observation regarding an atypical rate of radioactive breakdown in a sample of actinium salt, which is composed of element 89. The researcher successfully achieved the isolation of the novel element, afterward demonstrating its classification as an alkali metal (Vosko & Wilk, 1980; Lindemer et al., 1981). In a gesture of homage to her place of origin, France, she bestowed upon it the appellation of "francium." Due to the remarkably short half-life of its most stable isotope, which spans a mere 21 minutes, francium stands as an exceedingly scarce element within the periodic table, particularly in comparison to elements below atomic number 98. Consequently, the delayed discovery of francium relative to other radioactive elements occupying the same region of the periodic table can be attributed to this inherent rarity (Robertson et al., 2019).

2.3.1. Lithium

Lithium, classified as element number 3 in the periodic table, is a solid substance characterized by its softness, silvery-white appearance, and a density of approximately 0.535 g/cm³. This particular metal possesses the lowest density among all metals and may be combined with other metallic elements to produce structural materials that are characterized by their lightweight properties.

Did You Know?

Prawns along with other shellfish have the capacity to accumulate lithium from their surrounding marine environment.

Lithium is found in silicate ores, which are minerals mostly composed of silicon and oxygen, along with aluminum as well as other alkalis. The primary mineral of interest is spodumene, with the chemical formula $LiAlSi_2O_6$. Lithium mining predominantly takes place in America, Russia, China, as well as Australia, with an additional source being the extraction from salt deposits located in Chile (Koenig et al., 1984; Ohse, 1985; Dacorogna & Cohen, 1986). Lithium is an element with a high electropositivity and possesses the greatest oxidation potential among all known elements. As a result, metal exhibits a high degree of efficacy as a reducing agent. Similar to other alkali metals, compounds for lithium exclusively consist of lithium in the form of the +1 ion (Figure 2.3).

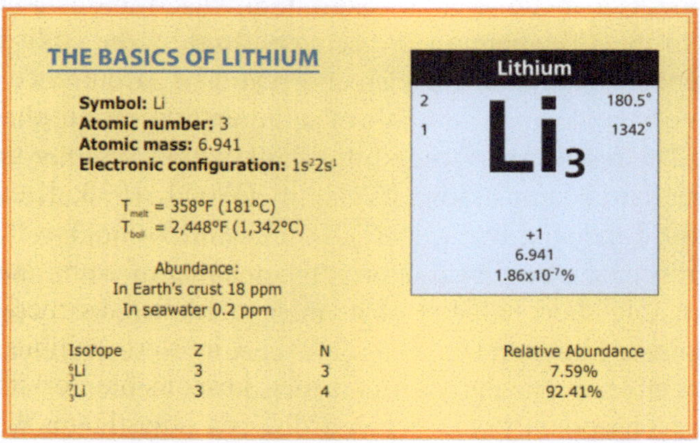

Figure 2.3. Basics of lithium.

Source: Monica Halka, creative commons license.

2.3.1.1. The Astrophysics of Lithium

Lithium, a relatively heavy nucleus, is synthesized during the initial stages of big-bang nucleosynthesis. It is created through the combination of deuterium as well as tritium, or through the fusion of both alpha particles, as described below:

$$2\,{}^2_1\mathrm{H} + {}^3_1\mathrm{H} \rightarrow {}^7_3\mathrm{Li}$$

$${}^4_2\mathrm{He} \aleph {}^4_2\mathrm{He} \quad {}^7_3\mathrm{Li} \quad {}^1_1\mathrm{p}.$$

Lithium nuclei in star cores undergo continuous bombardment with high-energy protons, resulting in similar processes. The production of lithium is effectively eradicated within star interiors due to the exceedingly high temperatures, typically reaching a few million kelvins (Varma & Rempe, 2006; Jin et al., 2010). Hence, the scrutiny of the element's presence within the spectra of specific star classifications or within the interstellar medium (ISM) holds significant importance for astronomers. Lithium exhibits two stable isotopes, namely ^{6}Li as well as ^{7}Li, which possess the capability to function as markers for significant astronomical phenomena. The isotopic ratio, specifically, can offer valuable insights into the creation of light elements in interstellar environments and can provide information regarding convection in star atmospheres (Dye, 1979).

The synthesis of lithium within the ISM, which is the region between stars, is attributed to a distinct process. In the present scenario, spallation, commonly characterized as the fragmentation of a substance, pertains to the disintegration of bigger interstellar nuclei (C, N, O) into smaller nuclei of lithium 6 or lithium 7. This process is facilitated by the collision between cosmic ray protons with adequate energy. The occurrence of similar collisional reactions can be observed in the exterior atmospheres of stars characterized by intense magnetic fields or within regions that undergo thermonuclear or just gravitational shocks (Wood, 1933; Knudsen et al., 1983). This phenomenon may provide an explanation for the observed variations in lithium spectra across various kinds of stars, such as those that exhibit chemical peculiarities and are classified as metal-poor halo celestial bodies. The abundance of ^{6}Li/ ^{7}Li in stars exhibits significant variation, and the underlying mechanisms responsible for its formation remain poorly comprehended. The present study investigates the respective contributions of convection, diffusion, as well as spallation in order to enhance comprehension of the process of star evolution (Nitta & Yushin, 2014; Wu & Yushin, 2017).

2.3.1.2. The Chemistry of Lithium

The utmost significant chemical characteristic exhibited by alkali metals, like lithium, pertains to their exceptional reactivity in chemical reactions. As an illustration, all of these substances exhibit a robust reaction when exposed to water, resulting in the production of hydrogen gas. This phenomenon is exemplified by the reaction between lithium and water:

$$2\,Li\,(s) + 2\,H_2O\,(l) \rightarrow 2LiOH\,(aq) + H_2\,(g).$$

The level of exothermicity exhibited by this reaction is insufficient to initiate the combustion of hydrogen. Due to lithium's inherent reactivity with water, it is imperative to ensure that any specimens of lithium metal are appropriately wrapped to prevent their exposure to water and atmospheric moisture (Ashcroft, 2002 Tarascon, 2010; Guillaume et al., 2011). All compounds formed by lithium will exhibit ionic characteristics, as lithium will exist in the form of lithium-ion (Li+). Furthermore, lithium exhibits a pronounced reactivity not only with water but also with oxygen gas present in the atmosphere, resulting in the formation of lithium oxide (Li_2O). Additionally, lithium quickly reacts with various compounds such as chlorides, sulfates, as well as carbonates, leading to the formation of lithium chloride (Li Cl), lithium sulfate ($Li_2\,SO_4$), as well as lithium carbonate (Li_2CO_3), respectively. Lithium has the ability to undergo a chemical reaction with hydrogen gas, resulting in the formation of lithium hydride (LiH). This product is characterized by its ionic nature, where hydrogen exists as a negatively charged ion with a -1 charge, deviating from its typical +1 charge (Platman & Fieve, 1968; Birch, 1970; Leroy et al., 1991). Lithium hydride, being a solid material characterized by a substantial hydrogen density, exhibits potential as a viable option for hydrogen storage in fuel cell applications. In practical implementation, the addition of a small amount of water will result in the liberation of hydrogen gas, as evidenced by the subsequent chemical equation:

$$LiH\,(s) + H_2\,O\,(l) \rightarrow LiOH\,(aq) + H_2\,(g).$$

Did You Know?

One notable advantage of utilizing lithium in thermonuclear weapons stems from its inherent stability. The storage of pure deuterium presents challenges due to the requirement of maintaining extremely low temperatures. Whenever lithium is coupled with deuterium, it forms lithium deuteride (LiD), which is a solid characterized by low reactivity. This compound enables the easy transportation of deuterium at room temperature. When utilized for this specific purpose, lithium is typically enriched by the lithium-6 isotope.

2.3.2. Sodium

Sodium, denoted by its atomic number 11, is a metallic element characterized by its soft and silvery gray appearance. It possesses a density of 0.97 g/cm^3. Given that the density for sodium metal is lower than that of water, it exhibits buoyancy (despite its highly reactive nature upon coming into contact with water). Sodium exhibits a remarkable degree of softness, to the extent that it may be readily manipulated and sculpted using a common household utensil such as a butter knife. Potassium is the most plentiful of the metals that are alkali, and it may be easily acquired from seawater (Pretzel et al., 1960; Dovesi, et al., 1991). This element finds extensive utilization in numerous applications. Among the several alkali metals, sodium metal stands out as the most extensively manufactured and utilized in industrial applications. Sodium is commonly recognized by individuals as the primary contributor to the taste sensation of "saltiness" in table salt, which has been a highly prized commodity in worldwide trade and business for countless centuries. Halite, also known as "rock salt," is the mineral that predominantly consists of sodium chloride. Rock deposits of salt are formed through the process of evaporation in inland saline bodies of water, where the salt is left behind as the water dissipates. The United States is home to several notable bodies of water, including the Salton Sea located in southern California, Mono Lake situated in eastern California, as well as the Great Salt Lake found in northern Utah. An illustrious illustration from the Old World has become the Dead Sea located in Israel (Knudsen et al., 1983; Tapia-Ruiz et al., 2013) (Figure 2.4).

Did You Know?

The ancient Egyptians held a belief in the existence of a hereafter, which led them to engage in the practice of preserving the bodies of their deceased individuals. The deceased individual underwent a process of cleansing and organ extraction, followed by the application of sodium compounds in crystal form to facilitate desiccation. Ultimately, the body was enveloped, therefore concluding the mummification procedure.

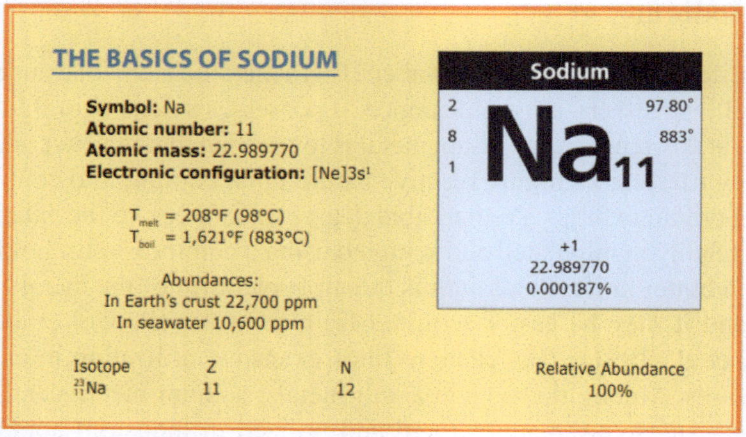

Figure 2.4. Basics of sodium.

Source: creative commons license, Howard Bryant.

2.3.2.1. The Astrophysics of Sodium

The production mechanisms of astrophysical sodium are contingent upon the mass and evolutionary stage of a particular star. Sodium 23 is generated in the course of carbon-burning within highly massive stars, thereafter being disseminated throughout the ISM during the supernova event that transpires upon the star's explosion.

$$^{12}_{6}C + ^{12}_{6}C \rightarrow ^{23}_{11}Na + ^{1}_{1}P$$

There seem to be other intriguing mechanisms involved in the generation of sodium within the atmospheres of red giant stars. The observed abundance of sodium suggests the occurrence of certain atmospheric mixing processes that are not yet fully comprehended (Krajewski et al., 2017; Raturi et al., 2021). Moreover, yellow supergiant stars exhibit an excessive concentration of sodium that is depending on temperature. The sodium abundance observed in the ISM exceeds the anticipated levels solely attributed to supernova activity, particularly when considering interplanetary dust particles. It is plausible that the contribution of star winds as well as planetary nebulae to the dispersion of sodium throughout the universe is significant. Further investigation into the production and distribution mechanisms of sodium inside interstellar space will contribute to the refinement of astrophysicists' comprehension of star phenomena (Tarascon, 2010; Qin et al., 2022).

2.3.2.2. The Chemistry of Sodium

The reaction between sodium metal and water is known to have a strong behavior, as depicted in the subsequent chemical equation:

$2\ Na\ (s) + 2\ H_2\ O\ (l) \rightarrow 2\ NaOH\ (aq) + H_2(g)$.

The exothermic nature of the reaction is such that the heat generated has the potential to initiate the ignition of hydrogen gas, resulting in an explosive event. The exhibition of such reaction is frequently observed in educational settings within the field of chemistry (Bauer et al., 2016). It is imperative to exercise caution and conduct the demonstration under proper supervision to prevent the occurrence of an explosion within the container where the reaction is transpiring, hence avoiding the potential release of scalding NaOH, a highly corrosive alkali, onto individuals in close proximity. To ensure safety, the demonstration is conducted exclusively using minuscule sodium pellets. Several significant molecules incorporate sodium, including the Na+ cation. Similar to other alkali metal ions, the +1 ion represents the sole often-seen ion. Rock salt, also known as table salt in its crystalline structure, is composed of sodium chloride (NaCl). Sodium hydroxide (NaOH) serves as a prevalent potent base that is frequently utilized as a constituent in commercial drain openers such as Drano. Baking powder, as well as baking soda, are two common chemical compounds known as sodium bicarbonate ($NaHCO_3$), which is often referred to as bicarbonate of soda. Washing soda, chemically known as sodium carbonate (Na_2CO_3), is commonly utilized in laundry applications as an effective cleansing agent. Sodium carbonate is commonly referred to as soda ash in academic literature. Sodium peroxide, denoted as Na_2O_2, exhibits similarities to hydrogen peroxide (H_2O_2) often available in pharmacies (Sharma & Seefurth, 1976). However, it is important to note that sodium peroxide exists in a solid state, while hydrogen peroxide is typically marketed in solution form, often at a concentration level of 3 percent. Sodium has the ability to generate several phosphate compounds (Jäckle & Groß, 2014). The chemical reaction between phosphoric acid ($H_3\ PO_4$) and sodium hydroxide (NaOH) results in the formation of sodium dihydrogen phosphate ($NaH_2\ PO_4$) as well as water. The reaction between sodium dihydrogen phosphate and sodium hydroxide results in the formation of sodium mono hydrogen phosphate ($Na_2\ HPO_4$) as well as water (Jain et al., 2016). In conclusion, the reaction between sodium mono hydrogen phosphate and sodium hydroxide results in the formation of sodium phosphate ($Na_3\ PO_4$). The reactions are illustrated by the subsequent equations:

H_3PO_4 (aq) + NaOH (aq) → NaH_2PO_4 (aq) + H_2O (l)

NaH_2PO_4 (aq) + NaOH (aq) → Na_2HPO_4 (aq) + H_2O (l)

Na_2HPO_4 (aq) + NaOH (aq) → Na_3PO_4 (aq) + H_2O (l)

Trisodium phosphate, commonly referred to as TSP or $Na_3 PO_4$, is commercially available in hardware as well as paint stores. It is utilized as an abrasive cleanser for many purposes, including the preparation of previously painted surfaces in anticipation of subsequent paint applications. Sodium cyanide (NaCN) is a toxic substance, as are all cyanides, and it finds application within the steel industry (Treptow, 2003). Sodium nitrate ($NaNO_3$) is a compound that occurs naturally in Chilean saltpeter deposits and is commonly employed in the production of fertilizers as well as explosives. Sodium sulfate has the potential to exist in two forms: an anhydrous state, denoted as $Na_2 SO_4$, or as a hydrate, represented as $Na_2 SO_4 \times 10H_2 O$. In the latter scenario, it is commonly referred to as Glauber's salt. Sodium thiosulfate ($Na_2 S_2O_3$), commonly referred to as sodium hyposulfite, is employed in the domain of film photography to halt the progression of the developing procedure (Guillaume et al., 2011; Bauer & Gitlin, 2016). The term "hypo" is commonly used by photo processors. Sodium azide (NaN_3) finds application in automotive airbags. Upon collision, an electrical discharge induces the fast decomposition of sodium azide into sodium metal as well as nitrogen gas, as represented by the following chemical equation:

$2NaN_3$ (s) → 2Na (s) + $3N_2$ (g)

The inflation process of the airbag involves the rapid introduction of nitrogen gas, which typically occurs within a time frame of around 40 milliseconds. The bag experiences a subsequent deflation prior to any contact between the passenger's head or body and the bag. In the event that the bag did not experience a minor deflation, the fully inflated bag would possess such rigidity as to potentially cause harm to the passenger (Raturi, et al., 2023). On the other hand, a bag that is slightly deflated offers the needed protective cushioning. The inclusion of potassium nitrate (KNO_3), as well as silicon dioxide (SiO_2) in the container, serves the purpose of expediting the conversion of sodium metal, which is generated during the reaction, into innocuous sodium compounds. Alternatively, there exists a potential hazard of the occurrence of an explosion when metallic sodium is exposed to water (Zhu et al., 2006; Syuy et al., 2018). Flame analysis is widely recognized as the predominant qualitative analytical method employed to identify the existence of sodium within aqueous solutions or soluble solids. A wire composed of a chemically unreactive substance such as platinum or nichrome becomes submerged into a solution and subsequently subjected to the heat of a flame. The emission of a strong yellow flame by sodium ions serves as a reliable means of confirming the existence of sodium inside the given material (Guillaume et al., 2011; Deemyad & Zhang, 2018).

2.3.3. Potassium

Potassium, denoted by its atomic number 19, exhibits a density of 0.86 g/cm^3 and possesses a remarkable softness that renders it susceptible to being cut using a butter knife. Potassium exhibits a higher reactivity towards water and oxygen in comparison to lithium or sodium, hence precluding its occurrence in a pure metallic state. Moreover, due to safety concerns, the utilization of pure metal is hardly practiced. Due to safety considerations, the need for potassium metal within the industry is limited, primarily for the purpose of producing sodium-potassium alloys (Knudsen et al., 1983). Potassium mostly originates from potash rock, which comprises a collection of sedimentary materials commonly occurring with salt from rocks, gypsum, as well as dolomite. The etymology of the phrase "potash" can be traced back to its original formation by the combination of seawater and wood ashes, resulting in the production of potashash upon boiling. Due to the significant role that potassium plays in facilitating plant growth, it is noteworthy that around 95% of the global potash supply undergoes conversion into fertilizer (Figure 2.5).

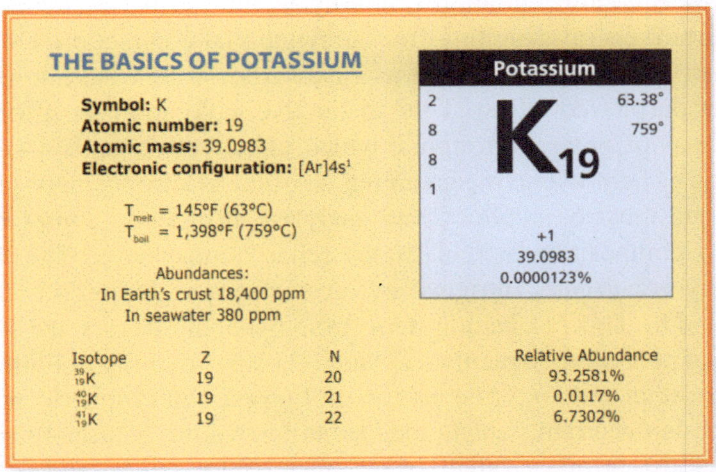

Figure 2.5. Basics of potassium.

Source: Greg Nobles, creative commons license.

2.3.3.1. The Astrophysics of Potassium

The majority of empirical evidence indicates that the potassium abundance within meteorites, stars within the galactic disk, as well as interstellar dust is approximately equivalent to the solar abundance. The presence of an element in such a widespread and abundant manner is a phenomenon that is not commonly observed and is understood by astrophysicists as an indication of the significant involvement of explosive oxygen burning in the synthesis of potassium (Scerri, 2019; Sigel, & Sigel, 2019). The phenomenon of explosive oxygen burning is exclusively observed in the highly energetic detonations occurring during Type II supernovae. This phenomenon entails that during the relatively short and intense period, the gaseous envelope around the star reaches a sufficiently high temperature for oxygen atoms to undergo fusion, subsequently interacting with surrounding protons as well as neutrons to generate potassium nuclei. These nuclei are subsequently dispersed into the universe due to the energy released by the explosion. One particular category of celestial objects referred to as "brown dwarfs," however, exhibits an abundance of potassium within their outer atmospheric layers. Brown dwarfs are celestial bodies composed primarily of gas and possess a diameter comparable to that of Jupiter, albeit far more massive (Kwatra, 2017; Hart, 2020).

These objects, due to their insufficient gravitational mass, are unable to sustain the necessary temperature for hydrogen fusion. In general, these celestial objects emit radiation primarily in the red wavelengths because to the thermal energy resulting from gravitational contraction. However, a subclass of these objects known as "T-dwarfs," which have lower mass, exhibit a purple-red glow. This distinctive coloration is attributed to the presence of potassium atoms, which effectively absorb a significant percentage of light within the green region of the electromagnetic spectrum. Further data must be obtained and analyzed in order to comprehend the potassium composition within these stellar atmospheres. The scientific community is currently intrigued by the surfaces of Jupiter's two moons, Europa, and Io. This fascination stems from two main factors: the surprising presence of potassium in plenty and their relative proximity to Earth (Scerri, 1998; Liu et al., 2016). The origin of Europa's atmospheric potassium abundance is uncertain, which adds to its interesting nature. Astronomers have previously proposed that the presence of potassium (and sodium) in the atmosphere of this celestial body can be attributed to its gravitational capture of these trace elements through the atmosphere of Io, one of Jupiter's moons. Recent findings, however, suggest that the atmospheric Na/K ratio of Europa is at least two times as elevated compared to that of Io. According to Michael Brown (2012), a researcher from the Division of Geological as well as Planetary Sciences at the California Institute of Technology, it can be inferred that the trace elements present in Europa's extended atmosphere originate from the surface of Europa itself. The use of potassium-argon dating methodologies has the potential to enhance our comprehension of the historical evolution of this celestial satellite orbiting Jupiter. Fortunately, astronomers have been fortunate in that Europa's surface has not undergone recent regeneration. With an age of a minimum of 10 million years, the surface of the object in question can provide valuable insights into its historical development. There is growing scientific interest from NASA and other research institutions in the study of Europa's potassium composition and its potential implications for the investigation of life processes. There is a possibility that Europa harbors a subsurface ocean. The consideration of lander missions holds significant potential for enhancing our understanding of among the most active celestial bodies inside our solar system (Edwards & Sienko, 1983).

2.3.3.2. The Chemistry of Potassium

Potassium exhibits a nearly identical capacity as sodium to produce chemical compounds. Potassium chloride (KCl), potassium bromide (KBr), potassium iodide (KI), potassium carbonate (K_2CO_3), potassium nitrate (KNO_3), potassium sulfate (K_2SO_4), potassium carbonate (K_2CO_3), as well as potassium hydroxide (KOH) are among the commonly encountered potassium compounds. It is worth noting that potassium hydroxide, similar to sodium hydroxide, is a potent caustic base. All of the aforementioned chemicals exhibit solubility in water. In aqueous solutions, it is generally observed that the potassium ion exhibits a lack of reactivity. Potassium nitrate, also known as niter or saltpeter, exists in its natural state and serves as a fundamental constituent of gunpowder, generally referred to simply as "black powder." This explosive substance comprises 75% KNO_3, 15% charcoal, as well as 10% sulfur. The three constituents are pulverized together into an exceedingly tiny particulate form (Fischer et al., 2006; Joag, 2014).

Potassium, denoted as K^+, is a vital element for humans as well as other animals, serving a crucial function in the process of metabolism. Its primary significance lies in its capacity to expedite chemical reactions within muscles and supply them with energy. The combined presence of sodium as well as potassium plays a crucial role in the regulation of water homeostasis within the human body. A diet comprising fruits, vegetables, and meat is expected to sufficiently fulfill the potassium requirements of a healthy individual (Fischer et al., 2006). The flame test is widely recognized as the predominant qualitative analytical method employed to detect the existence of potassium ions within an unfamiliar specimen. The presence of potassium in a solution can be detected by observing a wire which is immersed in the solution, which will exhibit a violet flame. Due to the probable coexistence of sodium, the more pronounced yellow flame of sodium will overshadow the violet flame of potassium, as sodium's hue exhibits greater intensity. Consequently, the flame is observed using a specialized cobalt-blue glass, which selectively absorbs the yellow color emitted by sodium while allowing the transmission of the violet color emitted by potassium.

2.3.4. Rubidium, Cesium, and Francium

Rubidium, cesium (spelled as "cesium" in England), as well as francium, are metallic elements characterized by their softness and silvery gray appearance. The relative abundances of all three of them are significantly lower compared to the relative abundances of lithium, sodium, as well as potassium. The below enumeration comprises three salient attributes of these items:

- Rubidium, classified as element number 37, possesses a density of 1.53 g/cm³, rendering it denser than water. This characteristic distinguishes rubidium from lithium, sodium, and potassium, as the latter elements exhibit a density lower than that of water (Parker et al., 1996).
- Cesium, denoted by its atomic number 55, exhibits a significantly lower abundance compared to rubidium, estimated to be approximately 30 times less. Its density is measured to be 1.88 g/cm³ (Knudsen et al., 1983).
- Francium is classified as the 87th element in the periodic table. Due to the radioactive nature and limited half-lives among all francium isotopes, the element is characterized by its significant scarcity, resulting in a dearth of comprehensive investigations of its properties. To yet, the quantity of francium available for study has been exceedingly limited, rendering the measurement of its density an unexplored endeavor (Westgaard et al., 1975).

The reactivity for alkali metals exhibits an upward trend as one moves down the group. When rubidium and cesium come into contact with atmospheric moisture, the resulting reaction can be quite explosive (Figures 2.6, 2.7, 2.8).

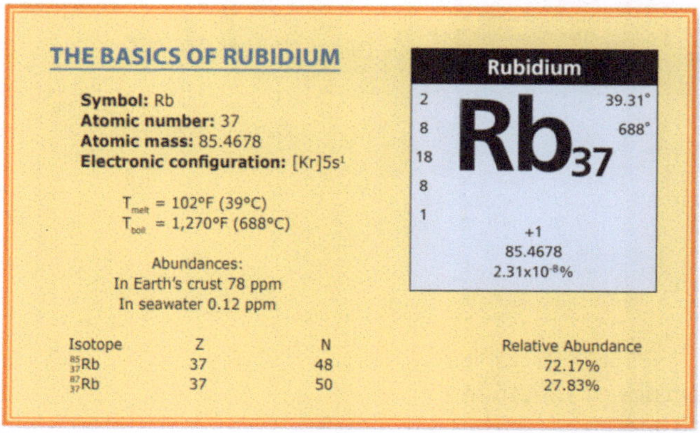

Figure 2.6. Basics of rubidium.

Source Nicole Leonard, creative common license.

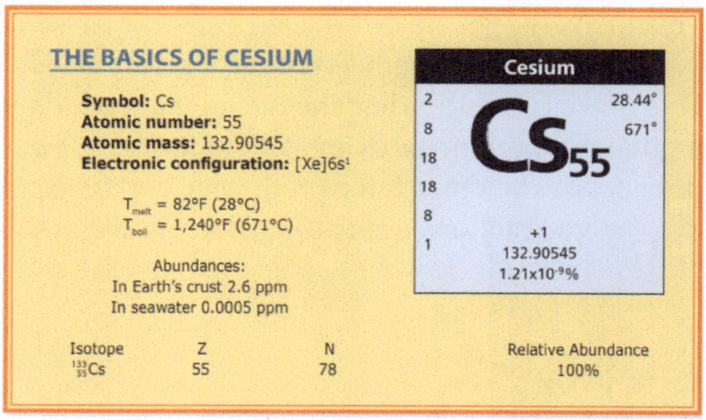

Figure 2.7. Basics of cesium.

Source: Nicole Leonard, creative common license.

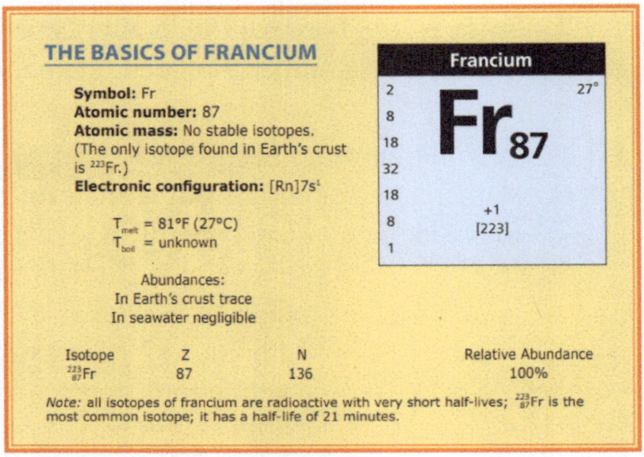

Figure 2.8. Basics of francium.

Source: Nicole Leonard, creative common license.

2.3.4.1. The Astrophysics of Heavy Alkali Metals

Francium is exclusively generated during supernova events, whereas rubidium as well as cesium are synthesized through a process known as the "s-process." This process occurs when the nuclei that contain iron and other elements with greater atomic masses in stars record free neutrons that are moving at a slow pace. The emission of free neutrons transpires as a result of alpha-particle record reactions transpiring within certain stellar interiors.

$$^{13}_{6}\text{C} + ^{4}_{2}\text{á} \rightarrow ^{16}_{8}\text{O} + ^{1}_{0}\text{n}$$

$$^{22}_{10}\text{Ne} + ^{4}_{2}\text{á} \rightarrow ^{25}_{12}\text{Mg} + ^{1}_{0}\text{n}$$

The dominant reaction is contingent upon the ambient temperature inside the contact zone. The carbon process is the dominant nuclear reaction below temperatures of around 100 million K, while at temperatures over this threshold, the neon process prefers to generate slow neutrons. The likelihood of the neon reaction occurring is higher in stars of greater mass due to the elevated temperatures resulting from increased gravitational force, which accelerates and warms the stellar material (Xu et al., 2021). Consequently, stars that are significantly more massive than the Sun are more prone to experiencing the neon reaction. However, it is not possible to accurately determine the temperature of a far star. The production of rubidium is influenced by neutron density, which is determined by the method of neutron production. Consequently, rubidium production can serve as a reliable indicator of the dominant reaction occurring in specific stars, particularly when compared to other s-process elements that include yttrium or strontium.

A notable surplus of rubidium has been detected in asymptotic giant branch (AGB) stars having masses ranging from around four to eight solar masses. The stars under consideration display atypical levels of activity and demonstrate thermal pulses that result in a heightened neutron density. The likelihood of the creation of a radioactive isotope such as 87Rb is higher in hotter stars compared to cooler ones. Rubidium-87 possesses a remarkably protracted half-life of around forty-seven billion years, so ensuring its enduring presence within a star until the culmination of its lifespan. Regrettably, the quantification of the 87Rb composition in celestial bodies has proven unattainable thus far. The determination of the 85Rb/87Rb proportion in the ISM has been successfully conducted; yet the obtained outcomes present perplexing observations. The determined interstellar value is approximately 50% lower than the value seen in meteoritic materials. The significance of this mismatch lies in the fact that the radioactive dating techniques employed for meteorites are predicated on the assumption that rubidium exchange did not occur during the period of formation of these meteorites inside the nascent solar system. The potential influence of neighboring AGB stars upon the rubidium isotope fraction has recently been contemplated by scientists (Westgaard et al., 1975).

In contrast, inside the disk and halo regions of the Milky Way, certain stars with low metal content exhibit an intriguing pattern of being enriched in rubidium compared to iron, while simultaneously displaying a modest deficiency in slightly heavier elements like yttrium and zirconium. This observation suggests that the rubidium synthesis in these stars predominantly takes place through the r-process, a nucleosynthesis mechanism that is limited in its ability to generate elements beyond a certain atomic mass. In general, there is still a significant amount of research and investigation required in order to achieve a comprehensive understanding of the astrophysical abundances pertaining to the heavy alkali elements (Hart et al., 2013; Yan et al., 2018).

2.3.4.2. The Chemistry of Rubidium and Cesium

Due to their high reactivity, rubidium, as well as cesium, are not naturally occurring in their elemental form. Rubidium exhibits an exceptionally high reactivity towards oxygen, to the extent that specimens of this metal have the propensity to spontaneously burn when exposed to an environment consisting solely of oxygen (Lavrukhina et al., 1960). Upon reaching its melting point, the molten metal exhibits a spontaneous ignition reaction when exposed to an oxygen concentration of 21% in the surrounding atmosphere. Rubidium exhibits a highly reactive nature, capable of undergoing explosive reactions even when in contact with ice at extremely low temperatures, reaching as low as -150°F (-101°C). The element rubidium has a high reactivity when exposed to bromine or chlorine, resulting in the generation of a visible flame. Due to its inherent dangers, rubidium is handled exclusively under stringent safety protocols enforced by both, Government as well as private laboratories (Jefferson, 2001; Sheng et al., 2010).

Cesium is characterized by its soft metallic properties and exhibits a relatively low melting point. It exhibits the highest level of chemical reactivity among metals and, surpassing rubidium, quickly forms compounds with oxygen, halogens, as well as water (Hart et al., 1973; Johnson et al., 1996). Cesium has the ability to undergo a chemical reaction with ice, even at extremely low temperatures reaching as low as -177 °F (-116 °C). Rubidium and cesium exclusively manifest as monovalent cations, denoted as Rb^+ and Cs^+, respectively, within compound formations. The compounds that are representative of the subject matter at hand encompass the following:

Chloride	Nitrate	Oxide	Carbonate	Sulfate
RbCl	$RbNO_3$	Rb_2O	Rb_2CO_3	Rb_2SO_4
CsCl	$CsNO_3$	$Cs_2\ O$	Cs_2CO_3	Cs_2SO_4

All of the aforementioned chemicals exhibit solubility in water.

2.4. SUMMARY

Hydrogen, denoted by the chemical symbol H and possessing an atomic number of 1, is recognized as the chemical element with the lowest mass. Hydrogen is a gaseous element that is characterized by its lack of color, flavor, taste, and flammability. It is commonly found in its diatomic form, H_2. The atomic structure of hydrogen comprises a single proton located in the nucleus and a solitary electron. However, the isotopes of hydrogen, namely deuterium as well as tritium, possess an additional 1 and 2 nuclear neutrons, respectively. Despite being the ninth most prevalent element on planet Earth, it constitutes roughly 75% of the total matter throughout the universe. In the past, hydrogen was utilized as a means to fill airships; however, it has since been replaced by the nonflammable gas helium. This compound finds application in the synthesis of ammonia, ethanol, aniline, as well as methanol. It is also utilized in the treatment of petroleum fuels, functioning as a reducing agent and providing a reducing environment. Additionally, it is employed in the production of hydrogen chloride and hydrogen bromide, as well as in hydrogenation processes, such as the hydrogenation of lipids. Liquid hydrogen, with a boiling point of -423 °F (-252.8 °C), finds utility in both scientific as well as commercial domains. It serves the purpose of generating highly reduced temperatures, as well as functioning as a propellant for rockets and a fuel source for fuel cells. The chemical reaction between hydrogen and oxygen results in the formation of water as the exclusive end product. The characteristics of the majority of acids, particularly when dissolved in water, are attributed to the presence of hydrogen ions (H^+), also known as hydronium ions (H_3O^+), which are the predominant type of hydrogen ions in an aqueous environment. Additionally, it is worth noting the presence of hydride and hydrocarbon compounds in this context.

The alkali metals include the chemical elements located within Group 1 within the periodic table. The group of elements known as alkali metals comprises lithium, sodium, potassium, rubidium, cesium, as well as francium. While hydrogen is commonly classified in Group 1 based on its electrical configuration, it does not strictly qualify as an alkali metal due to its infrequent manifestation of comparable characteristics. The term "alkali" was derived from the Arabic term "al qali," or "from ashes." This nomenclature is attributed to the fact that these substances, upon reacting with water, generate hydroxide ions, thereby resulting in the formation of alkaline solutions having a pH value greater than 7.

REFERENCES

1. Alkorta, I., Elguero, J., & Frontera, A., (2020). Not only hydrogen bonds: Other noncovalent interactions. *Crystals, 10*(3), 180.
2. Ashcroft, N. W., (2002). Putting the squeeze on lithium. *Nature, 419*(6907), 569–571.
3. Bańkowski, Z., (1966). Covalent radii of noble gas atoms and position of hydrogen in the periodic system of elements. *Nature, 209*(5018), 71–72.
4. Bauer, M., & Gitlin, M., (2016). *The Essential Guide to Lithium Treatment (No. 8775).* Basel: Springer International Publishing.
5. Bauer, M., Gitlin, M., Bauer, M., & Gitlin, M., (2016). What is lithium and how does it work? *The Essential Guide to Lithium Treatment,* 33–43.
6. Birch, N. J., (1970). Effects of lithium on plasma magnesium. *The British Journal of Psychiatry, 116*(533), 461–461.
7. Brammer, J. C., Blanke, G., Kellner, C., Hoffmann, A., Herres-Pawlis, S., & Schatzschneider, U., (2022). TUCAN: A molecular identifier and descriptor applicable to the whole periodic table from hydrogen to oganesson. *Journal of Cheminformatics, 14*(1), 66.
8. Brown, M. E., (2012). The compositions of Kuiper belt objects. *Annual Review of Earth and Planetary Sciences, 40*, 467–494.
9. Chand, A., & Biswal, H. S., (2020). Hydrogen bonds with chalcogens: Looking beyond the second row of the periodic table. *Journal of the Indian Institute of Science, 100*, 77–100.
10. Cronyn, M. W., (2003). The proper place for hydrogen in the periodic table. *Journal of Chemical Education, 80*(8), 947.
11. Cvetković, J., & Petruševski, V. M., (2017). On the placement of hydrogen and helium in the periodic system: A comment. *Chem. Bulg. J Sci. Educ., 26*, 167–170.
12. Dacorogna, M. M., & Cohen, M. L., (1986). First-principles study of the structural properties of alkali metals. *Physical Review B, 34*(8), 4996.
13. Dagotto, E., (2013). Colloquium: The unexpected properties of alkali metal iron selenide superconductors. *Reviews of Modern Physics, 85*(2), 849.
14. Deemyad, S., & Zhang, R., (2018). Probing quantum effects in lithium.

Physica C: Superconductivity and its Applications, 548, 68–71.

15. Dovesi, R., Roetti, C., Freyria-Fava, C., Prencipe, M., & Saunders, V. R., (1991). On the elastic properties of lithium, sodium, and potassium oxide. An ab initio study. *Chemical Physics, 156*(1), 11–19.
16. Dye, J. L., (1979). Compounds of alkali metal anions. *Angewandte Chemie International Edition in English, 18*(8), 587–598.
17. Dye, J. L., (2015). The alkali metals: 200 years of surprises. *Philosophical Transactions of the Royal Society A: Mathematical, Physical and Engineering Sciences, 373*(2037), 20140174.
18. Edwards, P. P., & Sienko, M. J., (1983). On the occurrence of metallic character in the periodic table of the elements. *Journal of Chemical Education, 60*(9), 691.
19. Fischer, C. F., Tachiev, G., & Irimia, A., (2006). Relativistic energy levels, lifetimes, and transition probabilities for the sodium-like to argon-like sequences. *Atomic Data and Nuclear Data Tables, 92*(5), 607–812.
20. Guillaume, C. L., Gregoryanz, E., Degtyareva, O., McMahon, M. I., Hanfland, M., Evans, S., & Mao, H. K., (2011). Cold melting and solid structures of dense lithium. *Nature Physics, 7*(3), 211–214.
21. Hart, M. M., (2020). *The Nuclei of Atoms in the Third Period of the Periodic Table ofElements (Monograph No. 3) (No. LLNL-TR-807338).* Lawrence Livermore National Lab. (LLNL), Livermore, CA (United States).
22. Hart, W. A., Beumel, O. F., & Whaley, T. P., (1973). The Chemistry of Lithium, Sodium, Potassium, Rubidium, Cesium, and Francium. *Pergamon Press, 1*, 1-19.
23. Hart, W. A., Beumel, O. F., & Whaley, T. P., (2013). *The Chemistry of Lithium, Sodium, Potassium, Rubidium, Cesium and Francium: Pergamon Texts in Inorganic Chemistry* (Vol. 7). Elsevier.
24. Hernández, R., & Novaro, O., (2014). The first metals in Mendeleiev's Table: Part II. A new argument against the placement of hydrogen atop the alkali metal column. *Foundations of Chemistry, 16*, 177–180.
25. Jäckle, M., & Groß, A., (2014). Microscopic properties of lithium, sodium, and magnesium battery anode materials related to possible dendrite growth. *The Journal of Chemical Physics, 141*(17).
26. Jain, A., Miyaoka, H., & Ichikawa, T., (2016). Destabilization of lithium hydride by the substitution of group 14 elements: A review.

International Journal of Hydrogen Energy, 41(14), 5969–5978.

27. Jaksic, M. M., (2000). Volcano plots along the periodic table, their causes, and consequences on electrocatalysis for hydrogen electrode reactions. *Journal of New Materials for Electrochemical Systems, 3*(2), 153–168.
28. Jefferson, R., (2001). Alkaline materials: Sodium, potassium, cesium, rubidium, francium, and lithium. *Patty's Toxicology*, 935–948.
29. Jin, K. H., Choi, S. M., & Jhi, S. H., (2010). Crossover in the adsorption properties of alkali metals on graphene. *Physical Review B, 82*(3), 033414.
30. Joag, S. D., (2014). An effective method of introducing the periodic table as a crossword puzzle at the high school level. *Journal of Chemical Education, 91*(6), 864–867.
31. Johnson, W. R., Liu, Z. W., & Sapirstein, J., (1996). Transition rates for lithium-like ions, sodium-like ions, and neutral alkali-metal atoms. *Atomic Data and Nuclear Data Tables, 64*(2), 279–300.
32. Kaesz, H., & Atkins, P., (2003). A central position for hydrogen in the periodic table. *Chemistry International--Newsmagazine for IUPAC, 25*(6), 14–14.
33. Knudsen, D., Peterson, G. A., & Pratt, P. F., (1983). Lithium, sodium, and potassium. *Methods of Soil Analysis: Part 2 Chemical and Microbiological Properties, 9,* 225–246.
34. Koenig, C., Christensen, N. E., & Kollar, J., (1984). Electronic properties of alkali-metal—Gold compounds. *Physical Review B, 29*(12), 6481.
35. Krajewski, M., Hamankiewicz, B., Michalska, M., Andrzejczuk, M., Lipinska, L., & Czerwinski, A., (2017). Electrochemical properties of lithium–titanium oxide, modified with Ag–Cu particles, as a negative electrode for lithium-ion batteries. *RSC Advances, 7*(82), 52151–52164.
36. Kwatra, B., (2017). Locator theory for elements in periodic table 'LEPT.' *Glob. J. Pure Appl. Chem. Res., 5*, 9–10.
37. Laing, M., (2007). Where to put hydrogen in a periodic table? *Foundations of Chemistry, 9*(2), 127–137.
38. Lavrukhina, A. K., Posdnyakov, A. A., & Rodin, S. S., (1960). Investigations on chemical properties of francium by means of Fr212. *The International Journal of Applied Radiation and Isotopes, 9*(1–4), 34–48.

39. Leroy, G., Sana, M., Wilante, C., & Van, Z. M. J., (1991). Revaluation of the bond energy terms for bonds between atoms of the first rows of the periodic table, including lithium, beryllium, and boron. *Journal of Molecular Structure, 247*, 199–215.
40. Lindemer, T. B., Besmann, T. M., & Johnson, C. E., (1981). Thermodynamic review and calculations—Alkali-metal oxide systems with nuclear fuels, fission products, and structural materials. *Journal of Nuclear Materials, 100*(1–3), 178–226.
41. Liu, Y., Merinov, B. V., & Goddard, III. W. A., (2016). Origin of low sodium capacity in graphite and generally weak substrate binding of Na and Mg among alkali and alkaline earth metals. *Proceedings of the National Academy of Sciences, 113*(14), 3735–3739.
42. Midilli, A., Ay, M., Dincer, I., & Rosen, M. A., (2005). On hydrogen and hydrogen energy strategies: I: Current status and needs. *Renewable and Sustainable Energy Reviews, 9*(3), 255–271.
43. Molitor, S., & Gessner, V. H., (2016). Alkali metal carbenoids: A case of higher stability of the heavier congeners. *Angewandte Chemie International Edition, 55*(27), 7712–7716.
44. Myers, J. B., Morgan, T. O., Carney, S. L., & Ray, C., (1980). Effects of lithium on the kidney. *Kidney International, 18*(5), 601–608.
45. Nitta, N., & Yushin, G., (2014). High-capacity anode materials for lithium-ion batteries: Choice of elements and structures for active particles. *Particle & Particle Systems Characterization, 31*(3), 317–336.
46. Ohse, R. W., (1985). *Handbook of Thermodynamic and Transport Properties of Alkali Metals* (p. 432). Oxford: Blackwell Scientific.
47. Parker, L. J., Atou, T., & Badding, J. V., (1996). Transition element-like chemistry for potassium under pressure. *Science, 273*(5271), 95–97.
48. Petruševski, V. M., & Cvetković, J., (2017). On the 'true position'of hydrogen in the periodic table. *Contributions, Section of Natural, Mathematical and Biotechnical Sciences, 38*(1), 83–90.
49. Petruševski, V. M., & Cvetković, J., (2018). On the 'true position'of hydrogen in the periodic table. *Foundations of Chemistry, 20*, 251–260.
50. Platman, S. R., & Fieve, R. R., (1968). Biochemical aspects of lithium in affective disorders. *Archives of General Psychiatry, 19*(6), 659–663.
51. Pretzel, F. E., Rupert, G. N., Mader, C. L., Storms, E. K., Gritton, G. V., & Rushing, C. C., (1960). Properties of lithium hydride I. Single

crystals. *Journal of Physics and Chemistry of Solids, 16*(1–2), 10–20.

52. Qin, Z., Wang, R., Li, S., Wen, T., Yin, B., & Wu, Z., (2022). MEAM interatomic potential for thermodynamic and mechanical properties of lithium allotropes. *Computational Materials Science, 214*, 111706.
53. Raturi, A., Mittal, P., & Choudhary, S., (2021). Effect of metal doping on visible light absorption and optical properties of lithium niobate ($LiNbO_3$) crystal: A first-principles calculation. *Bulletin of Materials Science, 44*(3), 237.
54. Raturi, A., Mittal, P., & Choudhary, S., (2023). Strain engineering for tuning the electronic and optical properties of lithium niobate for optoelectronic applications. *Solid State Communications, 361*, 115074.
55. Rich, R., (2007). *Hydrogen and the Alkali Metals* (pp. 25–35). Springer Berlin Heidelberg.
56. Richards, R. E., (1956). The location of hydrogen atoms in crystals. *Quarterly Reviews, Chemical Society, 10*(4), 480–497.
57. Robertson, S. D., Uzelac, M., & Mulvey, R. E., (2019). Alkali-metal-mediated synergistic effects in polar main group organometallic chemistry. *Chemical Reviews, 119*(14), 8332–8405.
58. Sacks, L. J., (2006). Concerning the position of hydrogen in the periodic table. *Foundations of Chemistry, 8*(1), 31–35.
59. Scerri, E. R., (1998). The evolution of the periodic system. *Scientific American, 279*(3), 78–83.
60. Scerri, E. R., (2019). Happy 150th birthday to the periodic table. *Chemistry–A European Journal, 25*(31), 7410–7415.
61. Schilling, J. S., (2006). Superconductivity in the alkali metals. *High Pressure Research, 26*(3), 145–163.
62. Sharma, R. A., & Seefurth, R. N., (1976). Thermodynamic properties of the lithium-silicon system. *Journal of the Electrochemical Society, 123*(12), 1763.
63. Sheng, D., Orozco, L. A., & Gomez, E., (2010). Preliminary studies for anapole moment measurements in rubidium and francium. *Journal of Physics B: Atomic, Molecular and Optical Physics, 43*(7), 074004.
64. Sigel, A., Sigel, H., & Sigel, R. K., (2016). *The Alkali Metal Ions: Their Role for Life* (Vol. 16). Cham, Switzerland: Springer.
65. Sigel, H., & Sigel, A., (2019). The bio-relevant metals of the periodic table of the elements. *Zeitschrift für Naturforschung B, 74*(6), 461–471.

66. Syuy, A. V., Sidorov, N. V., Palatnikov, M. N., Teplyakova, N. A., Shtarev, D. S., & Prokopiv, N. N., (2018). Optical properties of lithium niobate crystals. *Optik, 156*, 239–246.

67. Tapia-Ruiz, N., Segalés, M., & Gregory, D. H., (2013). The chemistry of ternary and higher lithium nitrides. *Coordination Chemistry Reviews, 257*(13, 14), 1978–2014.

68. Tarascon, J. M., (2010). Is lithium the new gold? *Nature Chemistry, 2*(6), 510.

69. Treptow, R. S., (2003). Lithium batteries: A practical application of chemical principles. *Journal of Chemical Education, 80*(9), 1015.

70. Varma, S., & Rempe, S. B., (2006). Coordination numbers of alkali metal ions in aqueous solutions. *Biophysical Chemistry, 124*(3), 192–199.

71. Vosko, S. H., & Wilk, L., (1980). Influence of an improved local-spin-density correlation-energy functional on the cohesive energy of alkali metals. *Physical Review B, 22*(8), 3812.

72. Westgaard, L., Aleklett, K., Nyman, G., & Roeckl, E., (1975). Beta-decay energies and masses of short-lived isotopes of rubidium, caesium, francium, and radium. *Zeitschrift für Physik A Atoms and Nuclei, 275*, 127–144.

73. Wood, R. W., (1933). Remarkable optical properties of the alkali metals. *Physical Review, 44*(5), 353.

74. Wu, F., & Yushin, G., (2017). Conversion cathodes for rechargeable lithium and lithium-ion batteries. *Energy & Environmental Science, 10*(2), 435–459.

75. Xu, J., Dou, S., Cui, X., Liu, W., Zhang, Z., Deng, Y., & Chen, Y., (2021). Potassium-based electrochemical energy storage devices: Development status and future prospect. *Energy Storage Materials, 34,* 85–106.

76. Yan, J., Lu, Y., Chen, G., Yang, M., & Gu, Z., (2018). Advances in liquid metals for biomedical applications. *Chemical Society Reviews, 47*(8), 2518–2533.

77. Zhu, W., Xiao, J., & Xiao, H., (2006). Density functional theory study of the structural and optical properties of lithium azide. *Chemical Physics Letters, 422*(1–3), 117–121.

CHAPTER 3

Alkaline Earth Metals

Contents

3.1. INTRODUCTION

The alkaline earth group exhibits distinct differences when compared to transition and post-transition metals. As an illustration, it can be observed that a significant number of metallic elements listed in the periodic table tend to produce precipitates that are insoluble when combined with the sulfide ion (S^{2-}). Consequently, sulfide ores containing transition and post-transition metals are abundantly found within the Earth's crust (Arzamendi et al., 2008).

In contrast, alkaline earth elements are frequently encountered in their natural state as sulfates (SO_4^{2-}), in contrast to transition and post-transition metals, which are comparatively less frequently observed in sulfate form (Yu et al., 2021) (Table 3.1).

Table 3.1. Transition Metals (Alkaline Earth Metals are in Bold Type)

H				
Li	**Be**			
Na	**Mg**			
K	**Ca**	Sc	Ti	V
Rb	**Sr**	Y	Zr	Nb
Cs	**Ba**	La	Hf	Ta
Fr	**Ra**	Ac	Th	Pa

3.2. BERYLLIUM

Beryllium, classified as element number 4 on the periodic table, is a metallic substance with a silvery appearance. It possesses a density of approximately 1.85 grams per cubic centimeter. In addition to radium, beryllium is the alkaline earth element with the lowest occurrence in Earth's crust, occupying the 47th position in terms of relative abundance. Lithium, the lightest metal, is surpassed in terms of atomic mass by beryllium, the second lightest metal. Moreover, beryllium exhibits lower chemical reactivity than lithium, rendering it more appropriate for structural engineering applications. Due to its lightweight nature and comparable strength to steel, Beryllium is employed in various applications involving lightweight structural materials. The properties of beryllium exhibit a greater resemblance to aluminum rather than magnesium and other alkaline earth elements. As a result, certain authors discuss beryllium in conjunction with aluminum, as opposed to its association with the alkaline earth elements (Reshak et al., 2015) (Figure 3.1).

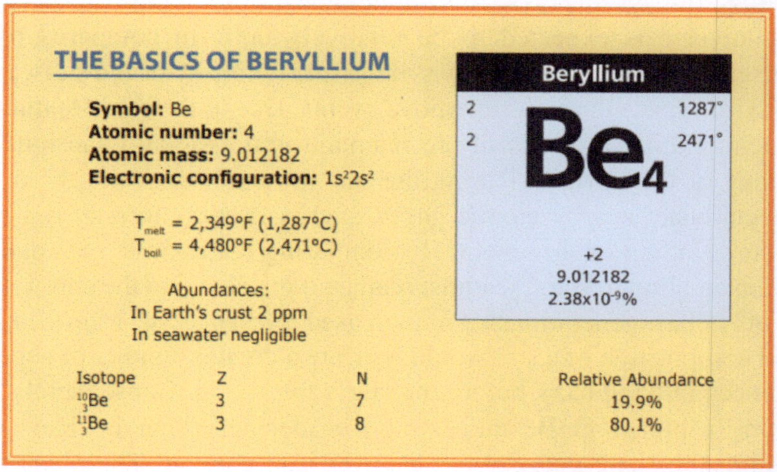

Figure 3.1. Basics of beryllium.

Source: Robin Garrell, creative commons license.

3.2.1. THE ASTROPHYSICS OF BERYLLIUM

The production of the beryllium-8 isotope occurs during the collapse of massive red-giant stars through a nuclear reaction known as the triple-alpha process. At the onset, two helium nuclei, commonly referred to as alpha particles, undergo fusion to generate an inherently unstable and exceedingly transitory state of beryllium, as denoted by the subsequent reaction (Parker et al., 1971):

$$^{4}_{2}He + ^{4}_{2}He \rightarrow ^{8}_{4}Be$$

As demonstrated in the subsequent reaction, the generated beryllium undergoes rapid destruction through fission or is transformed into carbon by capturing an alpha particle.

$$^{8}_{4}Be + ^{4}_{2}He \rightarrow ^{12}_{6}C + \gamma$$

where γ represents a high-energy photon, commonly called a gamma ray. The helium fusion process in stars is not considered the plausible origin of the beryllium currently detected in stellar spectra or the interstellar medium (ISM) (Saha & Fischer, 1987).

Beryllium-9, the sole stable isotope of beryllium, is formed through the collision of ambient carbon, nitrogen, and oxygen molecules with protons originating from cosmic rays. This implies that the distribution of beryllium-9 in the universe is expected to be relatively uniform compared to iron. Conversely, iron is predominantly synthesized within stars and subsequently expelled into space through supernova events. Due to their substantial mass, iron nuclei exhibit a propensity to maintain gravitational cohesion within the galaxy of their origin. During the star formation process, the budding stars accumulate nearby atomic gases, such as beryllium and iron nuclei, in the surrounding environment. By comparing the relative abundances of beryllium and iron in stars, scientists can gain insights into the chronological order of stellar generations. An increased presence of iron within the spectral composition of a star would indicate a greater number of supernova occurrences in the galaxy before the star's formation. Consequently, a star exhibiting a higher Fe/Be ratio would be deemed younger than another star within the same galaxy with a lower iron content. The phenomenon above has bestowed upon beryllium the esteemed recognition as a "Cosmo chronometer," thereby endowing it with the ability to serve as a temporal archive of cosmological events (Lolur et al., 2014).

Astronomers have detected a significantly higher abundance of beryllium, up to 1,000 times greater than anticipated, in the spectra of certain ancient stars. The observed phenomenon is unlikely to be attributed to cosmic ray activity. Still, it potentially provides valuable insights into the characteristics and dynamics of primordial matter in the early stages of the universe's expansion, commonly called the big bang. The prevailing hypothesis regarding the big bang, which posits a uniform explosion, allows for a maximum beryllium production of 1 percent. If the distribution of matter during the big bang were uneven, it would have facilitated the production of additional beryllium-9 in regions with lower density. According to the proposal by Professor Alan Guth from the Massachusetts Institute of Technology, it is suggested that the alteration in the dispersion of primordial matter may have arisen from a peculiar subatomic phase transition occurring in the initial stages of the universe (Yip et al., 2010). The hypothesis in question could potentially undergo direct testing by scientists by utilizing the Large Hadronic Collider, which is recognized as the world's largest particle accelerator and is located close to Geneva, Switzerland. The experiments conducted at this facility are specifically designed to replicate the conditions that prevailed mere microseconds following the big bang. Dr. Steven Goldfarb, a senior research fellow involved in the project, states that the objective is to identify the fundamental constituents of the cosmos. We seek to ascertain the fundamental constituents that comprise our physical composition. Furthermore, we also need to keep track of the governing principles that determine the manner in which these particles engage with one another (Dieterich et al., 2012).

3.2.2. The Chemistry of Beryllium

The oxidation state of an element is a representation of its chemical bonding. Metals have a tendency to undergo the process of ionization, resulting in the formation of cations with a positive charge. The electrical charge carried by a basic metal ion corresponds to the oxidation state of the respective element. The oxidation state of an element can be determined by the number of covalent or ionic bonds it has formed with atoms of other elements. The '+1' oxidation state indicates that an atom is either involved in a single bond with another atom or exists as a '+1' ion. A '+2' state refers to either the presence of two chemical bonds or the existence of a '+2' ion. Like other alkaline earth metals, beryllium exclusively exists in compounds with a '+2' oxidation state (Schmidbaur, 2001). However, it exhibits notable distinctions from the other alkaline earth elements. Ionic bonding is observed in compounds that contain elements such as magnesium, calcium, strontium, barium, and radium. In chemical compounds that incorporate beryllium, the nature of bonding can be described as covalent. Covalent bonding is among various similarities between beryllium and aluminum in terms of their properties. Another commonality between these two metals is their lightweight nature and their tendency to exhibit resistance against corrosion. Beryllium exhibits a significantly lower reactivity towards water, including boiling water, than its adjacent alkali metal, lithium. At lower temperatures, a protective oxide coating is formed on the surface of beryllium metal. Beryllium exhibits solubility in both acidic and basic solutions. Both reactions lead to the production of hydrogen gas. In the presence of strong acids, such as hydrochloric acid, the ensuing chemical reaction occurs (Hazen & Finger, 1986).

$$Be\ (s) + 2HCl\ (aq) \rightarrow BeCl_2\ (aq) + H_2\ (g)$$

In strong bases, the following reaction occurs:

$$Be\ (s) + OH^-\ (aq) + H_2O \rightarrow Na^+\ (aq) + HbeO_2^-\ (aq) + H_2\ (g)$$

Beryllium oxide (BeO) exhibits exceptional hardness, surpassing the oxides of other alkaline earth elements. Beryllium hydroxide ($Be(OH)_2$)

exhibits amphoteric properties, indicating its solubility in both acidic and basic solutions. The reaction described occurs in hydrochloric acid:

$$Be(OH)_2\ (s) + 2HCl\ (aq) \rightarrow BeCl_2\ (aq) + 2H_2O\ (l)$$

In sodium hydroxide, the following reaction occurs (Iversen et al., 2015):

$$Be(OH)_2\ (s) + 2OH^-\ (aq) \rightarrow BeO_2^{2-}\ (aq) + 2H_2O\ (l)$$

The beryllium ion exhibits instability when forming compounds with the carbonate ion (CO_3^{2-}), sulfate (SO_3^{2-}), and hydride (H^-). In contrast, it does demonstrate the ability to form halides such as beryllium chloride ($BeCl_2$) and beryllium bromide ($BeBr_2$). In contrast to most metal halides, which exhibit high electrical conductivity when in a molten state, beryllium halides do not possess this characteristic. The disparity can be elucidated by the variations in bonding within the two categories of compounds. Most metal halides exhibit ionic bonding, wherein the ions can freely conduct electricity when in a molten state. In the case of beryllium halides, the bonding mechanism is covalent, thereby lacking the presence of ions necessary for electrical conductivity. The distinctive characteristics of the beryllium ion can be ascribed to its significantly smaller size in comparison to the ions of the other alkaline earth elements. The compact nature of a beryllium atom results in a strong attraction between the valence electrons and the nucleus, thereby impeding the formation of a cation (Buchner, 2019).

3.3. MAGNESIUM

Magnesium, denoted by its atomic number 12, is a metallic substance characterized by its firmness, sheen, and silvery appearance. Possessing a density of 1.74 g/cm^3, this element holds a prominent position in terms of its prevalence within the Earth's composition, ranking seventh in relative abundance. Like the other alkaline earth elements, the elemental form of magnesium is not naturally occurring, and all magnesium compounds exist in the form of the Mg^{2+} ion. Two primary sources from which magnesium can be obtained: seawater and the mineral known as magnesite ($MgCO_3$). During the process of sodium chloride (NaCl) production, the evaporation of seawater enables the recovery of magnesium salts in addition to sodium salts. Magnesite, predominantly located in Colorado within the United States, is a prevalent magnesium resource. Magnesium plays a crucial role in the biological processes of both plants and animals, serving as a constituent of chlorophyll in photosynthetic organisms, contributing to the structural integrity of animal bones, and participating in the catalytic activities of diverse enzymes found in various living organisms. Furthermore, magnesium plays a crucial role in protein synthesis and facilitates the process of DNA replication (Elkins et al., 2016) (Figure 3.2).

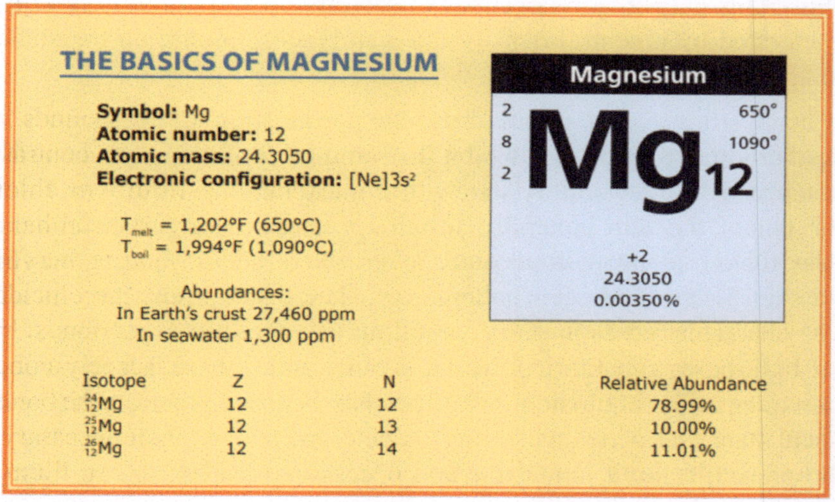

Figure 3.2. Basics of magnesium.

Source: Nicole Leonard, creative commons license.

3.3.1. The Astrophysics of Magnesium

The synthesis of magnesium in stars is contingent upon various stellar characteristics, including mass, temperature, and density. In stars of sufficient mass, where the core attains a temperature conducive to carbon burning, the fusion of two carbon atoms can lead to the production of two distinct isotopes of magnesium, namely ^{23}Mg and ^{24}Mg, through the following reactions (Lambert & Warner, 1968):

$$^{12}_{6}C + ^{12}_{6}C \rightarrow ^{23}_{12}Mg + ^{1}_{0}n$$

$$^{12}_{6}C + ^{12}_{6}C \rightarrow ^{24}_{12}Mg + \gamma$$

In the given context, the symbol "n" denotes a neutron, while the symbol "γ" represents a high-energy photon. The process of carbon combustion can also result in the generation of oxygen nuclei. In the event that the star depletes its reserves of lighter elements within its core, these oxygen nuclei can subsequently undergo fusion reactions, leading to the formation of magnesium-24:

$^{16}_{8}O + ^{16}_{8}O \rightarrow ^{24}_{12}Mg + 2\ ^{4}_{2}á$

The symbol α denotes an alpha particle, equivalent to a helium nucleus. This phenomenon occurs exclusively within highly massive stars, characterized by a central density that can reach a minimum threshold of 1010 kg/m³ (Kaur et al., 2021).

Cataclysmic events characterize the termination of these stars' life cycles, although they do not exhibit the same level of complete devastation as a supernova explosion. A star with a mass ranging from four to eight times that of the Sun generally sustains a core where carbon undergoes nuclear fusion, along with an outer layer where helium undergoes fusion and a shell where hydrogen undergoes fusion. The current state of affairs can be characterized as unstable, resulting in a forceful expulsion of most of the light gases constituting the star's outer atmosphere, accompanied by a subsequent core implosion. What persists is an enigmatic entity with a physical state that defies feasibility on celestial bodies, including the Sun. In the present state, it is feasible and necessary for electrons to exist at a greater density than what can be achieved within an atom. These celestial entities of an exotic nature are referred to as white dwarfs. The celestial bodies in question can no longer be classified as stars due to the cessation of fusion processes. The luminosity of these objects is solely attributed to the emission of infrared radiation, which is predominantly generated as heat, and their gradual deterioration persists throughout the passage of centuries (Chen, 1998).

In contrast, a supernova event results in the complete destruction of a star, causing the expulsion of all its constituent matter into the surrounding space. This ejected material is subsequently acquired by other stars, facilitating the ongoing nucleosynthesis process. The detonation results in the emission of alpha particles, which have the potential to engage in various interactions leading to the formation of elements characterized by nuclear multiples of four. One example of a nucleus is magnesium-24, which is formed through the collision of alpha particles with neon nuclei during an explosive event (Newsom, 1971):

$^{20}_{10}Ne + \alpha \rightarrow ^{24}_{12}Mg + \gamma$

$^{20}_{10}Ne + 24\alpha \rightarrow ^{24}_{12}Mg + ^{1}_{0}n$

One particular isotope of magnesium does not depend on the aforementioned unconventional mechanisms but instead undergoes a gradual accumulation process over several millennia within stellar atmospheres through alpha capture, as outlined below:

$${}^{20}_{10}Ne + 24\alpha \rightarrow {}^{25}_{12}Mg + {}^{1}_{0}n$$

Primarily observed in aged high-mass stars, commonly called AGB stars. These stars release their constituent elements into the atmosphere through thermal pulsations, a phenomenon that remains inadequately comprehended. However, these pulsations play a significant role in enriching the ISM with the high atomic weight elements that are presently observed on Earth (Sainz et al., 2014).

3.3.2. The Chemistry of Magnesium

The production of magnesium metal involves the process of electrolysis, specifically in the molten state of magnesium chloride ($MgCl_2$). During this procedure, the compound $MgCl_2$ is subjected to heat until it reaches its melting point; at this point, an electric current is applied. The collection of molten magnesium metal occurs at the cathode, while the generation of chlorine gas takes place at the anode, as depicted in the subsequent reaction (Westerhausen, 1991):

$$MgCl_2\ (l) \rightarrow Mg\ (l) + Cl_2\ (g)$$

Similar to the electrolysis of other salts, the chlorine gas generated during this process serves as a significant chlorine source for water purification systems. The combustion of magnesium in the presence of oxygen results in a luminous white illumination emission. The chemical reaction under consideration yields magnesium oxide as the resulting product, as evidenced by the following equation (Liu et al., 2016):

$$2Mg\ (s) + O_k\ (g) \rightarrow 2MgO\ (s)$$

Due to its pronounced exothermic nature, magnesium has the potential to be employed in the construction of incendiary bombs. Magnesium metal finds application in signal flares and fireworks due to its combustion properties, which result in heightened luminosity upon burning. During a previous era of flash photography, characterized by disposable flashbulbs, batteries generated an electrical current that traversed a minute magnesium component, thereby generating the desired flash. Magnesium oxide, also known as "magnesia," exhibits a remarkably elevated melting point of 4,788 °F (2,642 °C). This characteristic renders it suitable for producing magnesium carbon bricks, which are employed in furnace walls requiring exceptional resistance to elevated temperatures. The magnesium ion (Mg^{2+}) can form stable compounds with commonly encountered anions (Shalomeev et al., 2022).

The compound magnesium sulfate is obtained from the mineral epsomite in natural settings. This mineral is commonly utilized for producing Epsom salt, denoted as $MgSO_4$ Hawthorne × $7H_2O$. Epsom salt is commercially available in pharmacies and is commonly utilized as a laxative. Magnesium is a fundamental element that plays a crucial role in the biological processes of both plant and animal organisms. Green plants are capable of synthesizing their nourishment through a process known as photosynthesis, which occurs in the presence of sunlight. This ability is attributed to chlorophyll, a plant pigment that contains the essential magnesium ion. The presence of magnesium ions is essential for the replication of DNA and RNA molecules within the cellular structures of all living organisms (Kosar et al., 2020). Enzymes serve as catalysts for all biochemical reactions occurring within living organisms. Magnesium ions activate the enzyme molecules. Magnesium, alongside other metals such as zinc, is employed in the cathodic steel protection process. Steel structures situated below ground or utilized in the fabrication of marine vessels tend to undergo corrosion when exposed to moisture and dissolved oxygen. Due to the higher reactivity of magnesium compared to iron, introducing a magnesium block in contact with steel results in the preferential corrosion of magnesium. The application of this protective measure serves to prevent corrosion on steel surfaces, thereby enhancing the durability of underground tanks and ship hulls. The phenomenon can be described as an electrochemical process known as galvanic action, which resembles the processes observed in commercial batteries. In a galvanic cell, oxidation occurs, which is the negatively charged electrode (Lyu et al., 2020).

In contrast, the reduction process occurs at the cathode, which is the positively charged electrode. Iron would function as the anode and undergo oxidation if a protective metal such as magnesium or zinc is absent. Nevertheless, when a metal that is more reactive than iron is brought into contact with iron, the more reactive metal assumes the role of the anode and undergoes oxidation. This phenomenon is commonly referred to as a sacrificial anode.

In contrast, iron (or steel) assumes the role of the cathode. In the context of neutral metals or alloys, it is observed that reduction does not occur. Consequently, the iron component of these materials remains inert, while the process of reduction is observed in the dissolved oxygen or other substances present in the surrounding environment. The process involves sacrificing either magnesium or zinc, wherein replacing a block of either of these metals is comparatively more convenient and cost-effective than replacing the structure they are safeguarding (Giamello et al., 1991).

3.4. CALCIUM

Calcium, denoted by its atomic number 20, is a metallic element characterized by its silvery-gray appearance and a 1.55 g/cm³ density. Seawater is considered to be the primary reservoir of calcium. Various sedimentary materials that possess calcium content encompass calcite, limestone, marble, and chalk, all classified as calcium carbonate ($CaCO_3$) derivatives (Wang et al., 2022). Additionally, gypsum ($CaSO_4 \times 2H_2O$) and dolomite ($CaMg(CO_3)_2$) are also notable examples of such materials. Calcium-based compounds are widely distributed throughout the natural environment. Therefore, it is imperative for students enrolled in introductory chemistry and physical geology courses to have a thorough understanding of these concepts. For instance, $CaCO_3$ plays a significant role as a constituent in the formation of corals and shells. Sedimentary rocks frequently exhibit the preserved remains of marine organisms, with the identification of $CaCO_3$ in these rocks being readily discernible. Geologists typically transport dropper bottles containing hydrochloric acid (HCl) that has been diluted for fieldwork purposes. If there is suspicion regarding the presence of ($CaCO_3$ within a rock specimen, a small quantity of hydrochloric acid (HCl) will be introduced to the sample. If $CaCO_3$ is detected, the rock will exhibit enthusiasm as a result of the liberation of carbon dioxide gas, as illustrated by the subsequent chemical equation:

$$CaCO_3\ (s) + 2HCl\ (aq) \rightarrow CaCl_2\ (aq) + CO_2\ (g) + H_2O\ (l).$$

The general population widely recognizes calcium due to its indispensable role as a fundamental constituent of dental and skeletal structures. Osteoporosis, a condition characterized by reduced bone density and heightened susceptibility to bone fractures, can arise from a calcium deficiency (He et al., 2004) (Figure 3.3).

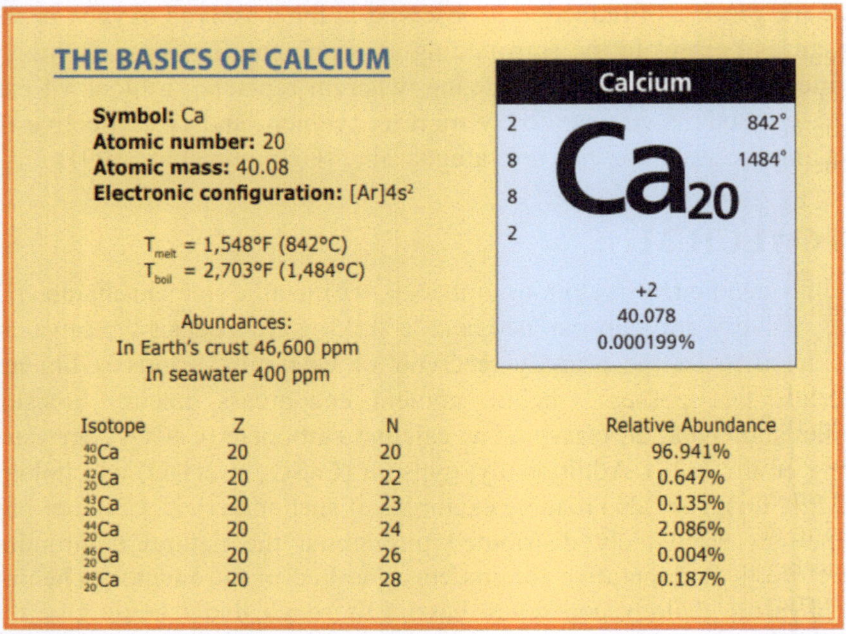

Figure 3.3. Basics of calcium.

Source: Nicole Leonard, creative commons license.

3.4.1. The Astrophysics of Calcium

Calcium's atomic nucleus consists of a multiple of four nuclei, rendering it an element that undergoes alpha decay. This implies that the production of calcium in stars primarily occurs through the process of collisional capture, where a helium nucleus (α) combines with an argon 36 atom, as depicted in the subsequent illustration (Derevianko et al., 2010):

$${}^{36}_{18}Ar + 24\alpha \rightarrow {}^{40}_{20}Ca$$

One intriguing observation is that calcium exhibits a relatively lower abundance than other alpha-process elements within galaxies. The precise causes remain uncertain; however, empirical evidence suggests a positive correlation between the calcium abundance in stars and the stellar mass and the velocity fluctuations during the formation of calcium nuclei. Additional investigations into supernovae, characterized by intricate velocity distributions, can potentially enhance our understanding of calcium nucleosynthesis theories. Type I supernovae are important in investigating calcium synthesis. This particular class of supernova transpires within binary star systems, wherein one constituent is a white dwarf, and the other is a typical main sequence star. Within this system, the celestial bodies known as stars exhibit a rotational motion around a shared central point, referred to as the center of mass. The compact white dwarf exerts a gravitational pull on the gaseous constituents present in the outer atmosphere of its companion star. The gases undergoing acceleration experience a significant increase in temperature as they approach the surface of the white dwarf. Eventually, these gases undergo ignition in a highly energetic thermonuclear explosion, resulting in the simultaneous destruction of both stars (Mitroy & Bromley, 2003).

Consequently, the matter from the stars is expelled into the surrounding space. In 2001, an observed supernova exhibited an uncommon characteristic in its emission spectrum, namely the presence of an ionized calcium jet with an exceptionally high velocity. This jet displayed a distinct behavior, moving at twice the speed and possessing a distinct polarization compared to another jet within the same explosion, which exhibited more conventional properties. The observed Sne I explosion is postulated to lack spherical symmetry, as evidenced by the varying polarization angles of the emitted light. Scientists hypothesize that the phenomenon potentially involved the ejection of condensed matter rather than complete vaporization. It is of utmost significance to meticulously analyze a multitude of supernovae spectra in order to establish with greater certainty the underlying mechanisms that take place within different types of supernovae. Astrophysicists persist in examining the spectra of solar prominences within our galaxy, specifically emphasizing the Sun. Their objective is to quantify the calcium/hydrogen ratio, a crucial parameter that indicates gas pressure within the Sun's atmosphere. The emission of visible light with a wavelength of 854 nanometers by doubly ionized calcium was initially observed and captured in a photograph by the American astronomer George Hale (1868–1938) in 1889. This significant event took place using Hale's newly developed spectroheliograph. The researcher observed non-uniform distribution of calcium around the Sun, characterized by localized concentrations resembling "clouds" in specific regions. Subsequently, it has been noted that these regions exhibit mobility and undergo alterations (Dzuba et al., 1991).

Furthermore, it is worth noting that the meteor spectra exhibit a significantly elevated relative density of Ca II, estimated to be approximately 1014 particles per cubic centimeter. The presence of ionized calcium gas in the Sun's atmosphere is believed to be primarily caused by the passage of meteors and comets through the region surrounding the Sun. As celestial bodies traverse their orbits, minute particles emanate from their tails and undergo sublimation due to the radiant energy emitted by the Sun. According to existing models investigating the temperature dependence of calcium ionization, it is inferred that the ionization of calcium atoms is most favorable at temperatures below 6,000 K. These temperatures are typically observed in the vicinity of the Sun's surface (Okumura & Kitano, 1986).

3.4.2. The Chemistry of Calcium

Under typical conditions, calcium metal exhibits stability in the air, provided that moisture is absent. Upon being subjected to heat, the metal undergoes a chemical reaction with oxygen present in the atmosphere, forming calcium oxide, which manifests as a white substance. The following equation can represent this reaction:

$2Ca\ (s) + O_2\ (g) + heat \rightarrow 2CaO\ (s).$

The metal exhibits a pronounced incandescence, emitting a luminous intensity of significant magnitude. Additionally, the vapors released can induce irritation upon contact with the skin and eyes. When exposed to moisture, the metal undergoes a rapid reaction resulting in the formation of calcium hydroxide and hydrogen gas, as depicted in the subsequent chemical equation:

$Ca\ (s) + 2H_2O\ (l) \rightarrow Ca(OH)_2\ (s) + H_2\ (g).$

The bonding observed in calcium compounds is consistently characterized as ionic. The calcium ion, denoted as Ca^{2+}, is known to generate a substantial array of compounds, which encompass the subsequent examples (Miron et al., 2022):

- Calcium nitrate ($Ca(NO_3)_2$)
- Calcium halides (CaF_2, $CaCl_2$, $CaBr_2$, and CaI_2)
- Calcium carbonate ($CaCO_3$)
- Various calcium phosphates ($Ca_3(PO_4)_2$, $Ca\ (H_2\ PO_4)_2$, $CaHPO_4$, and other more complex compounds)
- Calcium oxide (CaO)
- Calcium hydroxide ($Ca\ (OH_2)$)
- Calcium nitride ($Ca_3\ N_2$)
- Calcium phosphide ($Ca_3\ P_2$)
- Calcium sulfite ($CaSO_3$), and

- Calcium sulfate ($CaSO_4$).

Both carbonates and phosphates exhibit low solubility in aqueous solutions. Both nitrates and halides exhibit solubility. The compounds that remain exhibit solubilities that are generally in the intermediate range. Limestone can be generated through two distinct mechanisms: one involving the involvement of living organisms and the other occurring independently of any biological influence. The initiation of calcium carbonate formation in inorganic mechanisms involves the dissolution of atmospheric CO_2 in aqueous solution. When carbon dioxide (CO_2) is dissolved in water, it undergoes a reaction to produce carbonic acid (H_2CO_3), which is commonly found in carbonated beverages. Subsequently, carbonic acid reacts with water to generate bicarbonate ions ($HCO{-}_3$). The subsequent equations denote the reactions above (Platero Prats et al., 2010):

1. $CO_2\ (g) \leftrightarrow CO_2\ (aq)$
2. $CO_2\ (aq) + H_2O\ (l) \leftrightarrow H_2CO_3\ (aq)$
3. $H_2CO_3\ (aq) + H_2O\ (l) \leftrightarrow HCO_3^-\ (aq) + H_3O^+\ (aq)$,

   The double-arrow symbol (↔) denotes the bidirectional nature of the reactions. The subsequent reaction involves the conversion of bicarbonate ions into carbonate ions, as illustrated by the subsequent fourth equation:

4. $2\,HCO_3^-\ (aq) \leftrightarrow CO_3^{2-}\ (aq) + CO_2\ (aq) + H_2O\ (l)$

   The proportion of bicarbonate ions (HCO_3^-) to carbonate ions (CO_3^{2-}) is contingent upon the acidity level of the solution. Specifically, as the solution becomes more alkaline, there is an increase in the formation of carbonate ions (CO_3^{2-}). The presence of Ca^{2+} ions in water results from the natural dissolution of minerals. Once the carbonate ion (CO_3^{2-}) is present, it reacts with the calcium ion (Ca^{2+}) to form a precipitate of calcium carbonate ($CaCO_3$), as depicted in the fifth equation:

5. $Ca^{2+}\ (aq) + CO_3^{2-}\ (aq) \leftrightarrow CaCO_3\ (s)$

Various mechanisms facilitate the precipitation of calcium carbonate ($CaCO_3$). For instance, when water undergoes evaporation, the ions in the solution progressively rise until saturation is reached. Beyond this threshold, precipitation takes place. In the event of a temperature rise, the solubility of CO_2 in water decreases, leading to a modification in the equilibrium described by equation (4). Another mechanism that can be considered is wave action. Wave action along the coastlines of lakes and oceans leads to releasing CO_2 into the atmosphere. This process subsequently alters the equilibrium described in equation (4), increasing the concentration of carbonate ions (CO_3^{2-}) (Mukherjee et al., 2018).

3.5. STRONTIUM AND BARIUM

Strontium, denoted by its atomic number 38, exhibits a density of 2.63 grams per cubic centimeter and is recognized as the sixteenth most prevalent element within the Earth's composition. Barium, denoted by its atomic number 56, exhibits a density of 3.51 g/cm^3 and occupies the 14th position in terms of its relative abundance. Both of the elements possess a silvery hue. Due to the highly reactive nature of alkaline earth metals with water in their surroundings, the occurrence of these elements in their native metal form is highly unlikely. In the context of all mineral deposits, these elements are found predominantly in the form of divalent cations. The primary sources of these two elements are the minerals celestite ($SrSO_4$), strontianite ($SrCO_3$), and barite ($BaSO_4$) (Gärtner et al., 2007). In its natural state, neither of these elements possesses any significant biological relevance. In certain instances, however, strontium may potentially risk human health. Strontium is primarily recognized for its radioactive isotope Sr-90, which exhibits a half-life of 26 years. Strontium-90 is among the various byproducts resulting from the process of uranium fission. It constitutes a portion of the waste materials generated by nuclear reactors and comprises a component of the radioactive residue resulting from atomic detonations. Barium sulfate ($BaSO_4$) is widely recognized as the most prominent form of barium. Barium sulfate is a solid substance of white coloration, characterized by its inability to dissolve within the gastrointestinal (GI) tract of humans. The contrast agent is administered in powdered form to facilitate the visualization of the upper and lower gastrointestinal tracts in X-ray imaging. The presence of any atypical formations, such as malignant tumors, will be readily apparent as areas of increased darkness in X-ray images, contrasting against the white background provided by $BaSO_4$ (Crimmin et al., 2008) (Figure 3.4 and 3.5).

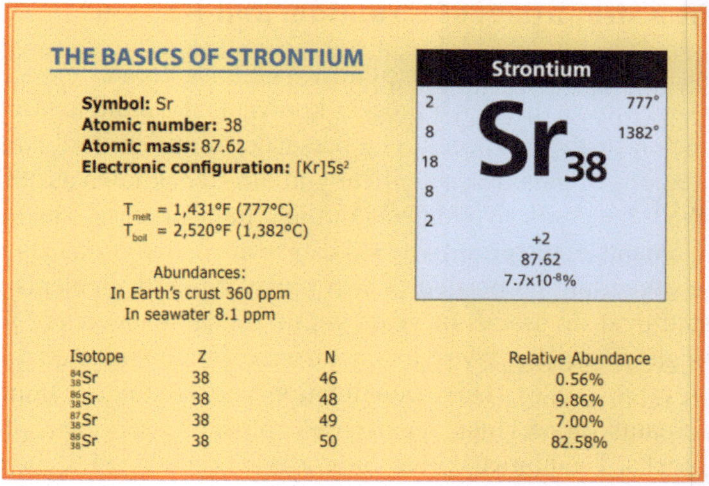

Figure 3.4. Basics of strontium.

Source: Greg Nobles, creative commons license.

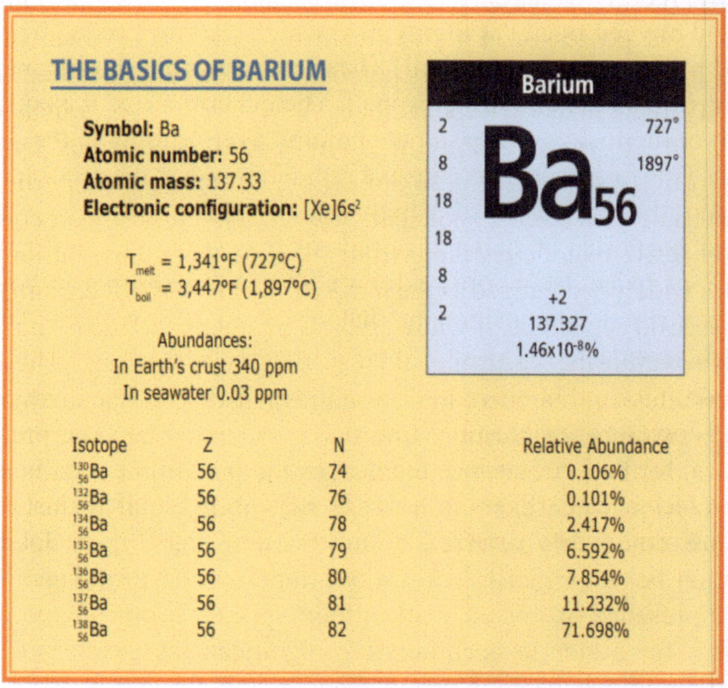

Figure 3.5. Basics of barium.

Source: Greg Nobles, creative commons license.

3.5.1. The Astrophysics of Strontium and Barium

The synthesis of strontium and barium, which have greater mass than iron, occurs in supernovae through a process known as the rapid neutron capture by iron nuclei, or the r-process. The presence of barium and strontium in the outer regions of galaxies, specifically in old stars known as "thick-disk" stars, is likely the result of synthesis through the r-process. The abundance of heavy elements in the primitive ISM can be attributed primarily to Type II supernovae. A certain proportion of heavy alkaline-earth elements gradually accumulated over thousands of years within the atmospheres of high-mass stars through processes involving neutron capture and electron release (Parkinson et al., 1976). This accumulation was contingent upon iron 56 nuclei, remnants of previous supernova explosions, serving as the initial building blocks. The nomenclature "s-process" is attributed to the relatively sluggish progression of this synthesis, which can be attributed to the scarcity of neutrons. The s-process can also take place within the inner shells of low-mass AGB stars during their helium-burning phase as they progress through the red-giant stage. AGB stars are characterized by the presence of carbon and oxygen nuclei at their cores, which are surrounded by a helium shell and an outer hydrogen shell. These stars exhibit significant thermal pulsations in their outer hydrogen shell. The occurrence of these pulsations is a result of the fusion process in the helium layer, leading to the formation of carbon. This fusion process causes a rapid increase in the star's luminosity and triggers the generation of a high-velocity stellar wind. Consequently, significant quantities of stellar material, such as strontium and barium, are expelled into space. Meteorites exhibit a comparable proportion of barium in their elemental composition to that observed in low-mass AGB stars (Russell & Saunders, 1925).

Nevertheless, researchers have documented certain stars exhibiting an atypical prevalence of barium, strontium, and other elements of greater atomic weight. This occurrence challenges the prevailing understanding of stellar evolution when examining these stars individually. These peculiar entities are commonly referred to as "barium stars," not due to their composition being primarily barium but rather owing to the unanticipated excessive presence of this element in their spectral profiles. One plausible mechanism for a star to accumulate a significant abundance of barium, strontium, and other heavy s-process elements is through the accretion of gases from a neighboring star in close proximity. The occurrence of binary star systems, wherein the majority of stars in the universe are found, renders this situation unsurprising. The majority of well-known barium stars are classified as red-giant stars. These stars have undergone a process in which they gravitationally captured the material that was expelled by a companion AGB star during its thermal pulsations (Oufqir et al., 2014). The production of strontium in astrophysical contexts has garnered significant attention, particularly in relation to a specific stellar spectrum associated with the highly luminous star Eta Carinae, which underwent a spectacular explosion. The event was hypothesized to be a supernova of sufficient magnitude to result in the complete destruction of the star. However, it did not. The post-explosion nebula has the potential to serve as a subject of extensive scholarly investigation for an extended duration. The feature known as the strontium filament, which possesses a high strontium concentration, can be described as a source of curiosity. The mere presence of supernovae has engendered a novel perspective on the subject. Supernovae are recognized for their ability to generate significant quantities of neutrinos and their corresponding antiparticles, referred to as antineutrinos. Antineutrinos can induce a conversion of a proton to a neutron, thereby potentially exerting a significant influence on the formation of uncommon isotopes in the context of stellar explosions (Roederer, 2012).

3.5.2. The Chemistry of Strontium and Barium

Similar to the remaining alkaline earth elements, strontium, and barium are not naturally occurring in a pure state within the Earth's crust. When exposed to water, halogens, and oxygen, these elements exhibit a high degree of reactivity. The system's chemistry is primarily influenced by the presence of the '+2' ions, specifically Sr^{2+} and Ba^{2+}. The chemical compounds formed by these ions exhibit significant similarities. Moreover, these compounds exhibit notable similarities to those formed by calcium and, to a lesser degree, magnesium compounds.

The majority of compounds containing alkaline earth ions exhibit a white solid state and are frequently encountered in chemistry laboratories at both high school and college levels. It is worth noting that these compounds are more prevalent than those of beryllium or radium. This discrepancy can be attributed to beryllium's toxicity and radium's scarcity and radioactivity. A compilation of various compounds is presented in the subsequent table 3.2 (Stegner et al., 2020):

Table 3.2. Compounds of the Alkaline Earth Ions

	NO_3^-	Cl^-	OH^-	O^{2-}	CO_3^{2-}	SO_4^{2-}	PO_4^{3-}
Mg^{2+}	$Mg(NO_3)_2$	$MgCl_2$	$Mg(OH)_2$	MgO	$MgCO_3$	$MgSO_4$	$Mg_3 (PO_4)_2$
Ca^{2+}	$Ca(NO_3)_2$	$CaCl_2$	$Ca(OH)_2$	CaO	$CaCO_3$	$CaSO_4$	$Ca_3 (PO_4)_2$
Sr^{2+}	$Sr(NO_3)_2$	$SrCl_2$	$Sr(OH)_2$	SrO	$SrCO_3$	$SrSO_4$	$Sr_3 (PO_4)_2$
Ba^{2+}	$Ba(NO_3)_2$	$BaCl_2$	$Ba(OH)_2$	BaO	$BaCO_3$	$BaSO_4$	$Ba_3 (PO_4)_2$

All compounds within a given column exhibit a strikingly similar appearance in their solid state. Moreover, it is noteworthy that they exhibit comparable characteristics. As an illustration, it is worth noting that nitrates and chlorides exhibit high solubility in aqueous solutions. Nevertheless, it is important to acknowledge that there exist certain distinctions. An instance of this phenomenon is observed in the ascending column of hydroxides and oxides, where the solubility in water tends to increase. Magnesium hydroxide ($Mg(OH)_2$) and magnesium oxide (MgO) are classified as weak bases due to their limited ability to accept protons in aqueous solutions. Additionally, these compounds exhibit low solubility when dissolved in dilute aqueous solutions (Suarez, 1996).

As one moves down the periodic table, the strength of bases increases, resulting in $Ba(OH)_2$ and BaO being classified as strong bases. The carbonates and phosphates of all four cations exhibit limited solubility in aqueous solutions, with the solubility decreasing as one moves down a group. In conclusion, magnesium sulfate ($MgSO_4$) exhibits solubility in water, calcium sulfate ($CaSO_4$) demonstrates only limited solubility, while strontium sulfate ($SrSO_4$) and barium sulfate ($BaSO_4$) are both characterized by complete insolubility. The precipitation behaviors of oxalate ($C_2O_4^{2-}$) and chromate (CrO_4^{2-}) anions exhibit significant differences when interacting with different alkaline earth ions. The solid forms of the four oxalates exhibit a white coloration, although it is noteworthy that only magnesium oxalate (MgC_2O_4) demonstrates solubility. The chromate ion imparts a yellow color to the solids of the four chromates. Both $MgCrO_4$ and $CaCrO_4$ exhibit high solubility in aqueous solutions. Barium chromate ($BaCrO_4$) exhibits a high degree of insolubility. The solubility of strontium chromate ($SrCrO_4$) exhibits an intermediate level.

When K_2CrO_4 is added to a solution containing Ba^{2+} and Sr^{2+}, a precipitation reaction occurs, resulting in the formation of $BaCrO_4$ as the primary precipitate. The disparities in solubility are employed in the qualitative evaluation of these ions. Consider a hypothetical scenario where a solution is initially comprised of one or more of the following ions: magnesium (Mg^{2+}), calcium (Ca^{2+}), strontium (Sr^{2+}), and barium (Ba^{2+}) (Green et al., 1999). Furthermore, it is worth considering the possibility that the solution may comprise one or more alkali metal ions, with particular emphasis on Na^+ and/or K^+. In practical applications, it is customary to eliminate all other metal ions, including transition and post-transition metals, prior to conducting tests for alkaline earth or alkali metal elements. The introduction of Na_2CO_3 into the solution results in the precipitation of alkaline earth ions, thereby causing the retention of alkali metal ions in the solution. The liquid portion of the solution, known as the supernatant, can be carefully poured off, thereby facilitating the separation of the alkaline earth metals from the alkali metals. The detection of alkali metals can be achieved through the utilization of flame tests if desired. The solid carbonates can undergo a process of washing and subsequent redissolution when exposed to a dilute acetic acid solution. The introduction of potassium chromate (K_2CrO_4) to the resultant solution will induce the formation of a precipitate consisting of barium chromate ($BaCrO_4$), thereby causing the retention of magnesium ions (Mg^{2+}), calcium ions (Ca^{2+}), and strontium ions (Sr^{2+}) within the solution (Murugavel et al., 2001). The liquid portion can be separated by pouring it off, a process known as decantation. The introduction of Na_2SO_4 into the supernatant solution will result in the formation of a precipitate composed of $SrSO_4$, while the remaining Mg^{2+} and Ca^{2+} ions will remain in the solution. The liquid portion can be separated by pouring it off. The addition of potassium oxalate ($K_2C_2O_4$) to the liquid remaining after a precipitation process will cause calcium oxalate (CaC_2O_4) to form as a solid. In contrast, magnesium ions ($Mg^{2+)}$ will remain in the liquid phase. This confirms the existence of calcium ions ($Ca^{2+)}$ in the initial solution. The liquid portion can once again be separated by pouring it off. The addition of sodium hydroxide (NaOH) will precipitate any magnesium ions ($Mg^{2+)}$ that may be present, thereby confirming the presence of magnesium ions in the initial solution. Alternatively, on certain occasions, the precipitation of $Mg_3(PO_4)_2$ may occur. These reactions exemplify the nature of the work conducted in mineral assays.

Did You Know?

Strontium titanate possesses a significantly elevated refractive index, thereby allowing it to exhibit a sparkling appearance akin to that of a genuine diamond. It is commonly employed in jewelry of lower cost.

In the mining sector, unprocessed mineral ores are extracted from beneath the Earth's surface. Chemists employed by mining corporations are required to possess the capability to disintegrate ores into their constituent elements. During this process, the minerals are subjected to assay or analysis in order to ascertain two key aspects: the presence of elements within them, which is referred to as qualitative analysis, and the proportion of a specific element in an ore, which is known as quantitative analysis. The flowchart depicted in Figure 3.6 illustrates the process of separating alkaline earth and alkali metals, exemplifying qualitative analysis (Lorenz et al., 2013).

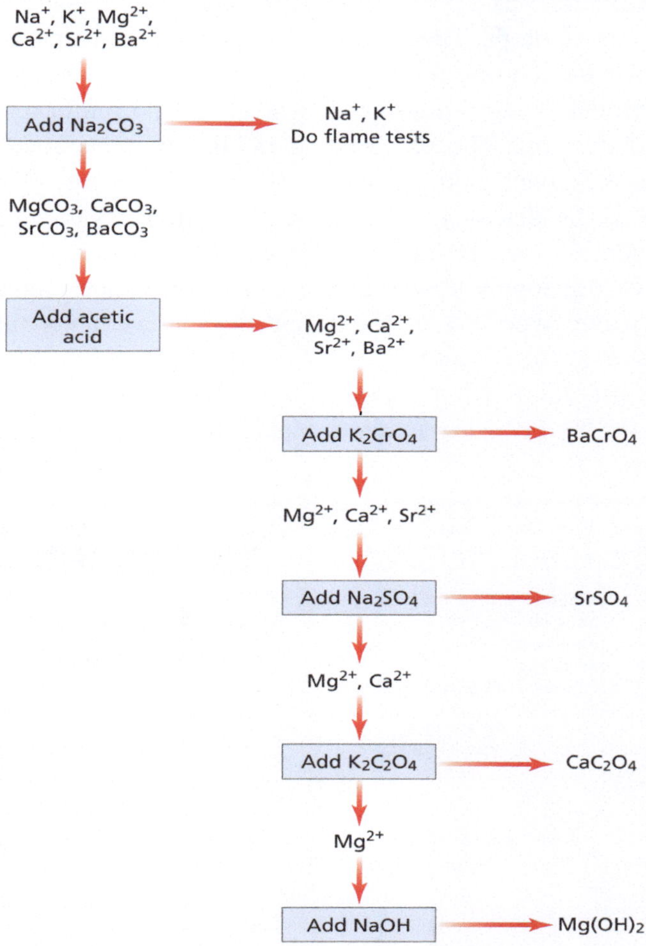

Figure 3.6. A flowchart illustrating the separation processes of alkaline earth and alkali metals.

Source: Brian Nordstrom, creative commons license.

3.6. RADIUM

Radium, designated as element 88 in the periodic table, demonstrates the property of radioactivity in all of its isotopes. Therefore, the occurrence of radium on Earth is primarily detected in small amounts, mainly as a minor component within uranium deposits. The most commonly observed isotope possesses a mass number of 226 and demonstrates a half-life of 1,604 years. Radium-228, which has a half-life of 5.77 years, is considered the second most long-lasting isotope. The remaining isotopes demonstrate considerably reduced half-lives, ranging from microseconds to days (Chen et al., 1999). Radium is produced continuously as a result of the radioactive decay of uranium and thorium. The presence of radium is constrained to minute quantities at any given moment due to its rapid decay rate. The historical significance of the discovery of radium stems from its connection to the esteemed scientific partnership of Pierre and Marie Curie, who conducted thorough investigations to reveal its presence. This chapter aims to present a comprehensive overview of the scientific achievements made by the Curies. This text offers a comprehensive examination of the chemistry and practical uses of radium while also shedding light on the natural mechanisms responsible for the genesis of radium inside the Earth's crust (Dietz et al., 1997) (Figure 3.7).

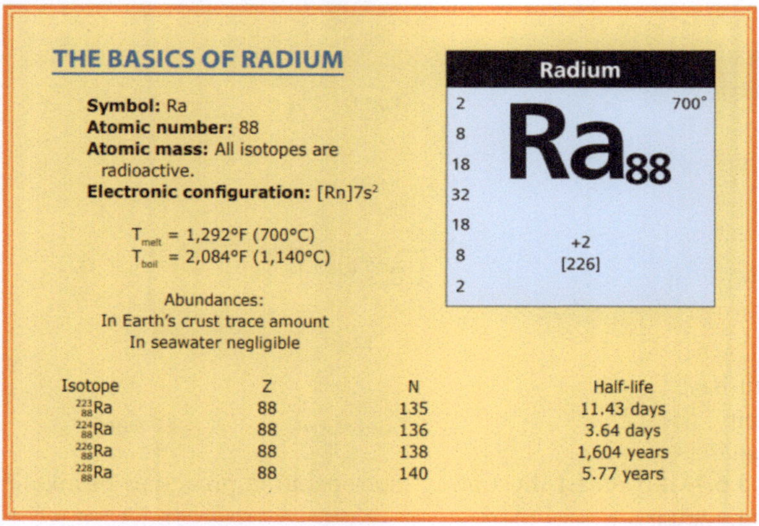

Figure 3.7. Basics of radium.

Source: Greg Nobles, creative commons license.

In the periodic table, radium is just below barium, and its chemistry is the same as that of barium. In a well-known story, misunderstanding between the two elements was a big part of what happened (Molinari & Snodgrass, 1990). In the 1930s, Enrico Fermi, an Italian scientist who lived from 1901 to 1954, and his colleagues studied what happened when neutrons hit uranium samples. (It wasn't until 1930 that the neutron was found. It wasn't used much in physics at the time.) They thought that when neutrons hit uranium, it would create trans-uranium elements, which are elements in the periodic table that come after uranium. This is shown by the following equation (Guseva et al., 2004):

92NU + ü → 92N + 1U → −10e + 93N + 1X,

In this context, N represents the mass number of a specific uranium isotope. The symbol $^{1}{}_{0}n$ is used to denote a neutron, while $-10e$ is the symbol for a beta particle, specifically an electron. Additionally, ü $N+\ X$ signifies the formation of element 93. Upon analyzing the outcomes of the neutron bombardment, Fermi's team observed the presence of radium, which was unexpected given their lack of prior knowledge suggesting the possibility of barium as a product (McIntyre et al., 1997).

Given that radium is produced as a result of two consecutive alpha decays of a uranium atom, the researchers concluded that their pursuit of a transuranic element was not fruitful. Following the initial findings, a group of German scientists consisting of Otto Hahn (1879–1968), Fritz Strassmann (1902–80), and Lise Meitner (1878–1968) proceeded to reevaluate the outcomes, ultimately demonstrating that the products generated were not radium atoms, but rather barium atoms resulting from the process of nuclear fission in uranium. Therefore, Fermi and his research team narrowly failed to uncover the phenomenon of nuclear fission (Shaposhnikova & Rachkova, 2022).

Did You Know?

Radium-based luminescent paint was employed to illuminate objects without electricity, particularly in the context of aircraft and maritime instruments, emergency signage, and timepieces.

3.7. SUMMARY

Alkaline earth metals are a group of chemical elements. The second leftmost group of the periodic table consists of six chemical elements, namely beryllium, magnesium, calcium, strontium, barium, and radium. The nomenclature of the individual in question can be traced back to the era of medieval alchemy. The atoms possess a valence shell consisting of two electrons, thereby exhibiting a high reactivity, leading to the formation of a multitude of compounds. Consequently, these atoms are not encountered in a free state within the natural environment.

The alkaline earth metals exhibit a silver hue and possess a soft texture, along with comparatively low densities, melting points, and boiling points. From a chemical perspective, it can be observed that the alkaline earth metals exhibit reactivity with the halogens, resulting in the formation of alkaline earth metal halides. These compounds are predominantly characterized as ionic crystalline substances, except beryllium chloride, beryllium bromide, and beryllium iodide, which display covalent bonding. Except beryllium, all alkaline earth metals exhibit a reaction with water, resulting in the formation of highly alkaline hydroxides. Consequently, it is imperative to exercise caution when handling these metals. The reactivity of alkaline earth metals increases with increasing atomic mass. The alkaline earth metals exhibit relatively low first ionization energies within their respective periods on the periodic table due to their comparatively modest effective nuclear charges and their capacity to achieve a complete outer shell configuration by relinquishing a mere two electrons. The second ionization energy of all alkaline metals exhibits a relatively low value.

REFERENCES

1. Arzamendi, G., Arguinarena, E., Campo, I., Zabala, S., & Gandía, L. M., (2008). Alkaline and alkaline-earth metals compounds as catalysts for the methanolysis of sunflower oil. *Catalysis Today, 133*, 305–313.
2. Buchner, M. R., (2019). Recent contributions to the coordination chemistry of beryllium. *Chemistry–A European Journal, 25*(52), 12018–12036.
3. He, X., Noll, B. C., Beatty, A., Mulvey, R. E., & Henderson, K. W. (2004). Ketone Deprotonation Mediated by Mono-and Heterobimetallic Alkali and Alkaline Earth Metal Amide Bases: Structural Characterization of Potassium, Calcium, and Mixed Potassium− Calcium Enolates. *Journal of the American Chemical Society, 126*(24), 7444-7445.
4. Chen, P. D., (1998). Spectroscopy and chemical dynamics of weakly bound alkaline-earth metal ion-H and alkaline-earth metal ion-hydrocarbon complexes 2. *International Reviews in Physical Chemistry, 17*(1), 1–34.
5. Chen, X., Ji, M., Fisher, D. R., & Wai, C. M., (1999). Ionizable calixarene-crown ethers with high selectivity for radium over light alkaline earth metal ions. *Inorganic Chemistry, 38*(23), 5449–5452.
6. Crimmin, M. R., Barrett, A. G., Hill, M. S., MacDougall, D. J., Mahon, M. F., & Procopiou, P. A., (2008). Bis (trimethylsilyl) methyl derivatives of calcium, strontium, and barium: Potentially useful dialkyls of the heavy alkaline earth elements. *Chemistry–A European Journal, 14*(36), 11292–11295.
7. Derevianko, A., Porsev, S. G., & Babb, J. F., (2010). Electric dipole polarizabilities at imaginary frequencies for hydrogen, the alkali–metal, alkaline–earth, and noble gas atoms. *Atomic Data and Nuclear Data Tables, 96*(3), 323–331.
8. Dieterich, J. M., Gerke, S., & Mata, R. A., (2012). A first-principles-based potential for the description of alkaline earth metals. *Journal of Atomic and Molecular Physics, 201, 331–335.*
9. Dietz, M. L., Chiarizia, R., Horwitz, E. P., Bartsch, R. A., & Talanov, V., (1997). Effect of crown ethers on the ion-exchange behavior of alkaline earth metals. Toward improved ion-exchange methods for the separation and preconcentration of radium. *Analytical Chemistry, 69*(15), 3028–3037.
10. Dzuba, V. A., Flambaum, V. V., Gribakin, G. F., & Sushkov, D. P.,

(1991). Fine structure of negative ions of alkaline-earth-metal atoms. *Physical Review A*, *44*(5), 2823.

11. Elkins, T. W., Roberts, S. J., & Hagelin-Weaver, H. E., (2016). Effects of alkali and alkaline-earth metal dopants on magnesium oxide supported rare-earth oxide catalysts in the oxidative coupling of methane. *Applied Catalysis A: General, 528*, 175–190.
12. Gärtner, M., Görls, H., & Westerhausen, M., (2007). Arylphosphanide complexes of the heavy alkaline earth metals calcium, strontium, and barium of the formula (thf) nM [P (R) Aryl] 2. *Zeitschrift für Anorganische und Allgemeine Chemie, 633*(11, 12), 2025–2031.
13. Giamello, E., Ferrero, A., Coluccia, S., & Zecchina, A., (1991). Defect centers induced by evaporation of alkali and alkaline-earth metals on magnesium oxide: An EPR study. *The Journal of Physical Chemistry, 95*(23), 9385–9391.
14. Green, D. C., Englich, U., & Ruhlandt-Senge, K., (1999). Calcium, strontium, and barium acetylides—new evidence for bending in the structures of heavy alkaline earth metal derivatives. *Angewandte Chemie International Edition, 38*(3), 354–357.
15. Guseva, L. I., Tikhomirova, G. S., & Dogadkin, N. N., (2004). Anion-exchange separation of radium from alkaline-earth metals and actinides in aqueous-methanol solutions of HNO_3. 227 Ac-223 Ra generator. *Radiochemistry, 46*, 58–62.
16. Hazen, R. M., & Finger, L. W., (1986). High-pressure and high-temperature crystal chemistry of beryllium oxide. *Journal of Applied Physics, 59*(11), 3728–3733.
17. He, X., Noll, B. C., Beatty, A., Mulvey, R. E., & Henderson, K. W. (2004). Ketone Deprotonation Mediated by Mono-and Heterobimetallic Alkali and Alkaline Earth Metal Amide Bases: Structural Characterization of Potassium, Calcium, and Mixed Potassium− Calcium Enolates. *Journal of the American Chemical Society*, *126*(24), 7444-7445.
18. Iversen, K. J., Couchman, S. A., Wilson, D. J., & Dutton, J. L., (2015). Modern organometallic and coordination chemistry of beryllium. *Coordination Chemistry Reviews, 297*, 40–48.
19. Kaur, M., Dar, D. F., Sahoo, B. K., & Arora, B., (2021). Radiative transition properties of singly charged magnesium, calcium, strontium, and barium ions. *Atomic Data and Nuclear Data Tables, 137*, 101381.
20. Kosar, N., Gul, S., Ayub, K., Bahader, A., Gilani, M. A., Arshad, M.,

& Mahmood, T., (2020). Significant nonlinear optical response of alkaline earth metals doped beryllium and magnesium oxide nanocages. *Materials Chemistry and Physics, 242*, 122507.

21. Lambert, D. L., & Warner, B., (1968). The abundances of the elements in the solar photosphere—V: The alkaline earths Mg, Ca, Sr, Ba. *Monthly Notices of the Royal Astronomical Society, 140*(2), 197–221.
22. Liu, Y., Merinov, B. V., & Goddard, III. W. A., (2016). Origin of low sodium capacity in graphite and generally weak substrate binding of Na and Mg among alkali and alkaline earth metals. *Proceedings of the National Academy of Sciences, 113*(14), 3735–3739.
23. Lolur, P., Dawes, R., & Heaven, M., (2014). Vibronic perturbations in the electronic spectrum of Mgc. In: *69th International Symposium on Molecular Spectroscopy* (Vol. 2, pp. 132–135).
24. Lorenz, V., Hrib, C. G., Grote, D., Hilfert, L., Krasnopolski, M., & Edelmann, F. T., (2013). Diazadiene complexes of the heavy alkaline-earth metals strontium and barium: Structures and reactivity. *Organometallics, 32*(16), 4636–4642.
25. Lyu, M., Lee, D. K., & Park, N. G., (2020). Effect of alkaline earth metal chloride additives BCl_2 (B= Mg, Ca, Sr and Ba) on the photovoltaic performance of $FAPbI_3$ based perovskite solar cells. *Nanoscale Horizons, 5*(9), 1332–1343.
26. McIntyre, R. S. C., Gregoire, D. C., & Chakrabarti, C. L., (1997). Vaporization of radium and other alkaline earth elements in electrothermal vaporization inductively coupled plasma mass spectrometry. *Journal of Analytical Atomic Spectrometry, 12*(5), 547–551.
27. Miron, G. D., Kulik, D. A., Yan, Y., Tits, J., & Lothenbach, B., (2022). Extensions of CASH+ thermodynamic solid solution model for the uptake of alkali metals and alkaline earth metals in CSH. *Cement and Concrete Research, 152*, 106667.
28. Mitroy, J., & Bromley, M. W., (2003). Dispersion coefficients for H and He interactions with alkali-metal and alkaline-earth-metal atoms. *Physical Review A, 68*(6), 062710.
29. Molinari, J., & Snodgrass, W. J., (1990). The chemistry and radiochemistry of radium and the other elements of the uranium and thorium natural decay series. *The Environmental Behavior of Radium, 1*, 11–56.

30. Mukherjee, D., Schuhknecht, D., & Okuda, J., (2018). Hydrido complexes of calcium: A new family of molecular alkaline-earth-metal compounds. *Angewandte Chemie International Edition, 57*(31), 9590–9602.
31. Murugavel, R., Baheti, K., & Anantharaman, G., (2001). Reactions of 2-mercaptobenzoic acid with divalent alkaline earth metal ions: Synthesis, spectral studies, and single-crystal x-ray structures of calcium, strontium, and barium complexes of 2, 2 '-dithiobis (benzoic acid). *Inorganic Chemistry, 40*(27), 6870–6878.
32. Newsom, G. H., (1971). Inner-shell absorption in the spectra of the alkaline earths. I. Magnesium (Mg I). *Astrophysical Journal* (Vol. 166, pp. 243–245).
33. Okumura, M., & Kitano, Y., (1986). Coprecipitation of alkali metal ions with calcium carbonate. *Geochimica et Cosmochimica Acta, 50*(1), 49–58.
34. Oufqir, S., Bloom, P. R., & Torner, B. M., (2014). The retention of calcium, barium, and strontium ions by a mollisol humic acid: Spectroscopic investigation. In: *EGU General Assembly Conference Abstracts* (Vol. 2, pp. 16985).
35. Parker, V. B., Wagman, D. D., & Evans, W. H., (1971). *Selected Values of Chemical Thermodynamic Properties: Tables for the Alkaline Earth Elements (Elements 92 Through 97 in the Standard Order of Arrangement)* (Vol. 270, pp. 254–256). National Bureau of Standards. National Bureau of Standards.
36. Parkinson, W. H., Reeves, E. M., & Tomkins, F. S., (1976). Neutral calcium, strontium, and barium: Determination of f values of the principal series by the hook method. *Journal of Physics B: Atomic and Molecular Physics, 9*(2), 157.
37. Platero, P. A. E., De La Peña-O'Shea, V. A., Iglesias, M., Snejko, N., Monge, A., & Gutiérrez-Puebla, E., (2010). Heterogeneous catalysis with alkaline-earth metal-based MOFs: A green calcium catalyst. *ChemCatChem, 2*(2), 147–149.
38. Reshak, A. H., Huang, H., Kamarudin, H., & Auluck, S., (2015). Alkali-metal/alkaline-earth-metal fluorine beryllium borate $NaSr_3Be_3B_3O_9F_4$ with large nonlinear optical properties in the deep-ultraviolet region. *Journal of Applied Physics, 117*(8), 9871–9875.
39. Roederer, I. U., (2012). Are there any stars lacking neutron-capture

elements? Evidence from strontium and barium. *The Astronomical Journal, 145*(1), 26.

40. Russell, H. N., & Saunders, F. A., (1925). New regularities in the spectra of the alkaline earths. *Astrophysical Journal, 61*, 38–40.
41. Saha, H. P., & Fischer, C. F., (1987). Multiconfiguration Hartree-Fock calculations of the oscillator strength for the 2s2p p 2 1, 2snd D 1 (n= 3–6) transitions in beryllium. *Physical Review A, 35*(12), 5240.
42. Sainz, R. M., Roncero, O., Sanz-Sanz, C., Aguado, A., Ramos, A. A., & Bueno, J. T., (2014). Depolarizing collisions with hydrogen: Neutral and singly ionized alkaline earths. *The Astrophysical Journal, 788*(2), 118.
43. Schmidbaur, H., (2001). Recent contributions to the aqueous coordination chemistry of beryllium. *Coordination Chemistry Reviews, 215*(1), 223–242.
44. Shalomeev, V., Tabunshchyk, G., Greshta, V., Nykiel, M., & Korniejenko, K., (2022). Influence of alkaline earth metals on structure formation and magnesium alloy properties. *Materials, 15*(12), 4341.
45. Shaposhnikova, L. M., & Rachkova, N. G., (2022). The role of bryophytes in the migration of uranium, radium-226, and non-radioactive alkaline-earth elements in the taiga aquatic and terrestrial ecosystems. *Geochemistry International, 60*(3), 298–310.
46. Stegner, P., Färber, C., Oetzel, J., Siemeling, U., Wiesinger, M., Langer, J., & Harder, S., (2020). d–d dative bonding between iron and the alkaline-earth metals calcium, strontium, and barium. *Angewandte Chemie International Edition, 59*(34), 14615–14620.
47. Suarez, D. L., (1996). Beryllium, magnesium, calcium, strontium, and barium. *Methods of Soil Analysis: Part 3 Chemical Methods, 5*, 575–601.
48. Wang, W., Lemaire, R., Bensakhria, A., & Luart, D., (2022). Review on the catalytic effects of alkali and alkaline earth metals (AAEMs) including sodium, potassium, calcium, and magnesium on the pyrolysis of lignocellulosic biomass and on the co-pyrolysis of coal with biomass. *Journal of Analytical and Applied Pyrolysis, 163*, 105479.
49. Westerhausen, M., (1991). Synthesis and spectroscopic properties of bis (trimethylsilyl) amides of the alkaline-earth metals magnesium, calcium, strontium, and barium. *Inorganic Chemistry, 30*(1), 96–101.

50. Yip, F. L., McCurdy, C. W., & Rescigno, T. N., (2010). Double photoionization of atomic beryllium. In: *APS Division of Atomic, Molecular and Optical Physics Meeting Abstracts* (Vol. 55, pp. 45–50).
51. Yu, J., Guo, Q., Gong, Y., Ding, L., Wang, J., & Yu, G., (2021). A review of the effects of alkali and alkaline earth metal species on biomass gasification. *Fuel Processing Technology, 214*, 106723.

CHAPTER 4

Transition Metals

Contents

4.1. INTRODUCTION

Transition metals encompass a selection of highly consequential substances with significant industrial applications. Iron, copper, silver, gold, and mercury are well-known metallic elements that have significantly influenced human civilization over an extensive period of time. Titanium, chromium, tungsten, and platinum exemplify significant elements in the contemporary economy (Ogawa, 2000).

The characteristic that sets transition metals apart from other elements in the periodic table is the presence of a partially occupied d subshell containing electrons. Given that a d subshell can accommodate a maximum of 10 electrons, it follows that there are precisely 10 columns dedicated to transition metals. The transition metals, which are positioned in the central region of the periodic table, are organized into four rows that align with the principal energy levels of the d subshells. Electrons gradually occupy these subshells. The elements from scandium to zinc are classified as belonging to the third row of transition metals. This categorization is based on the fact that electrons progressively occupy the d subshell of the third principal energy level. The elements yttrium through cadmium are located in the 4d period of the periodic table. The elements lanthanum and hafnium up to mercury are in the fifth row of the periodic table, specifically in the 5d block. The synthetic elements actinium and rutherfordium up to copernicium are located in the 6d period of the periodic table. The observed discontinuities in atomic number between lanthanum and hafnium, as well as between actinium and rutherfordium, can be attributed to the presence of the lanthanide and actinide series of elements. The elements above are categorized as a distinct group of elements, commonly known as the inner transition elements (Xia et al., 2021).

The observed resemblances in characteristics among the transition metals can be attributed to the inherent attributes of the valence electrons residing within the d subshells. Transition metals generally exhibit elevated melting and boiling points. The majority of metals located in the 3d period of the periodic table exhibit a notable propensity to dissolve readily when exposed to hydrochloric acid, except copper. In contrast, metals with higher atomic masses generally exhibit greater chemical stability when exposed to acidic environments (Zeng et al., 2020).

One significant implication of possessing valence d electrons is that transition metals typically demonstrate a significantly higher range of oxidation states than main group metals. The alkali metals exclusively exhibit a '+1' oxidation state when forming ions. The alkaline earth elements exhibit a valence state of '+2' when forming ions. Compounds and ions involving aluminum and gallium exclusively exhibit an oxidation state of '+3.' Compounds and ions of tin and lead exclusively exhibit oxidation states of '+2' and '+4.' Bismuth exclusively demonstrates oxidation states of '+3' and '+5.' On the contrary, it is frequently observed that transition metals tend to generate compounds or ions that can manifest four or five distinct oxidation states. Consequently, the inclusion of transition metals results in a significantly broader range of compounds that can be synthesized, accompanied by more intricate chemical bonding patterns than other elements in the periodic table (Wang & Jiang, 2021).

4.2. THE SCANDIUM GROUP

The scandium group is comprised of two elements, namely scandium and yttrium. Apart from aluminum, scandium is the initial metallic element that predominantly occurs in compounds with an oxidation state of '+3.' From the first element, hydrogen, to the twentieth element, calcium, all electrons in atoms in their lowest energy state, known as the ground state, are situated exclusively within the s or p subshells. A significant transition occurs in the element scandium, marking the initial instance where an electron occupies a d subshell. Scandium, yttrium, and lanthanum, respectively, are the initial members of distinct transition element series. The outermost electron of scandium occupies the 3d subshell, the outermost electron of yttrium occupies the 4d subshell, and the outermost electron of lanthanum occupies the 5d subshell. The summation of two valence electrons within a s subshell and the presence of one electron within a d subshell results in a collective count of three valence electrons for each of these elements (Lu et al., 2018) (Figures 4.1 and 4.2).

The properties of elements undergo a significant change due to the transition in valency from the alkaline earth elements in group IIA to the scandium family elements in group IIIA. The elements of the scandium family exhibit notable distinctions in terms of their melting and boiling points, which are considerably higher than the alkaline earth elements (Tolbert & Beauchamp, 1984).

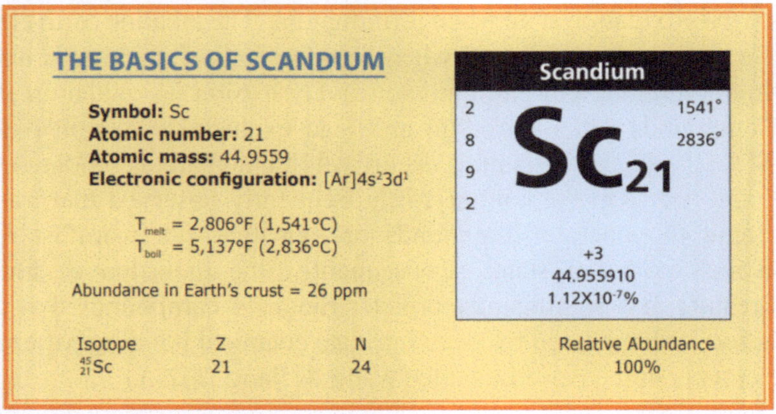

Figure 4.1. Basics of scandium.

Source: Monica Halka, creative commons license.

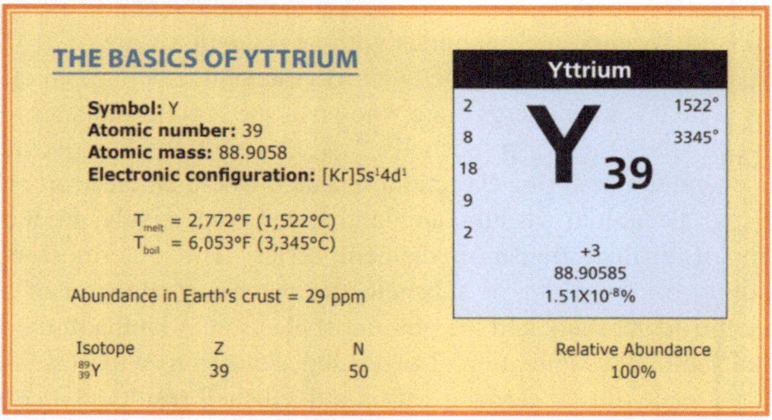

Figure 4.2. Basics of yttrium.

Source: Kenneth street, creative commons license.

The nuclei of scandium and yttrium are not synthesized within the central regions of typical stars such as the Sun. Conversely, the accumulation of these constituents occurs gradually over extended periods of time within the atmospheres of significantly larger celestial bodies through the process of neutron capture. Neutrons are generated through the process of alpha particle capture by carbon-13 (^{13}C) and neon-22 (^{22}Ne), as outlined below (Corbett, 1978):

$$^{13}_{6}C + ^{4}_{2}\alpha \rightarrow ^{16}_{8}O + ^{1}_{0}n$$

$${}^{22}_{10}\mathrm{Ne} \aleph {}^{4}_{2}a \qquad {}^{25}_{12}\mathrm{Mg} \quad {}^{1}_{0}\mathrm{n}$$

The nomenclature "slow process" or "s-process" is attributed to the gradual advancement of synthesis, which can be attributed to the scarcity of neutrons. The synthesis of both of these elements can also occur in supernovae through the process known as the rapid neutron capture by iron nuclei, commonly referred to as the r-process (Watanabe & Kleppa, 1984).

The variability of scandium abundance in different stellar classifications presents a perplexing phenomenon. Chemically anomalous stars have garnered significant attention in this context. Stars classified as spectral type A exhibit a mass ranging from approximately one and a half to three times that of the Sun, accompanied by significantly higher surface temperatures. The spectral composition of their spectrum is predominantly characterized by hydrogen lines, accompanied by relatively faint iron and calcium lines. Additionally, it exhibits a relatively high rotational velocity. Certain type A stars, particularly those found in binary systems, exhibit a significantly reduced rotational velocity, potentially attributed to the gravitational effects exerted by their companion star. The stars under consideration exhibit a higher abundance of metallic elements, although the observed levels of scandium are significantly lower than anticipated. One potential explanation for the observed discrepancy could be attributed to the star's slow rotation, which may impede the efficient mixing of elements on a large scale. Consequently, it is plausible that scandium could be more abundantly present in the deeper regions of the star (Geoffrey N áCloke, 1993).

The star known as "φ Herculis" is classified as a manganese-scandium star and exhibits an intriguing characteristic. While its magnesium and silicon abundances are within the expected range, it demonstrates significantly elevated levels of scandium, surpassing the normal abundance by a factor of 25. Additionally, the star displays an extraordinary overabundance of yttrium, exceeding the typical levels by thousands of times. Yttrium exhibits a notable deficiency in stars that possess a scarcity of metals (Barsukova et al., 2018).

Scandium, denoted by its atomic number 21, occupies the 35th position in terms of its relative abundance. It is commonly observed in conjunction with yttrium and lanthanides. Scandium is a malleable and lustrous metallic element, exhibiting a 3.0 g/cm^3 density. While it is possible to detect trace amounts of scandium in human blood and tissues, typically measured in parts per billion, the element has no established biological role. Scandium exhibits a moderate level of reactivity. The substance exhibits solubility in water and various acids, resulting in the formation of the Sc^{3+} ion. The scandium ion exhibits a propensity to engage in compound formation with a wide range of commonly encountered univalent anions characterized by a charge of -1. Hence, it can be observed that $Sc(NO_3)_3$, ScF_3, $ScCl_3$, $ScBr_3$, and ScI_3 can be readily synthesized. All of these compounds exhibit a white coloration and demonstrate solubility in a diluted aqueous solution (Blasse & Dirksen, 1988).

Yttrium, with an atomic number of 39, is ranked as Earth's 43rd most prevalent element. Despite yttrium being relatively unknown to the general population, it is worth noting that its abundance is twice that of lead. The majority of yttrium in the contemporary era is sourced from Australia. The mineral xenotime, also known as yttrium phosphate (YPO_4), is the principal source of yttrium. An additional source of yttrium can be found in uranium mines located in Canada. Yttrium possesses a density of 4.5 grams per cubic centimeter. Metallic yttrium, in conjunction with the rare earth elements (REEs) it is typically associated with, is commonly synthesized through the process of electrolysis of the chloride salt, as depicted in the subsequent chemical reaction (Holleck, 1977):

$$2YCl_3\ (l) \rightarrow 2Y\ (s) + 3Cl_2\ (g).$$

The elements discovered in conjunction with yttrium exhibit a remarkable degree of similarity, resulting in the formation of yttrium metal through electrolysis that is typically impure to some extent. The most effective approach for isolating yttrium ions from closely related ions is through the process of ion exchange. This technique involves passing a solution containing a mixture of different ions through a resin material. The velocity at which various ions traverse the resin is contingent upon the respective charges and sizes of said ions. During the passage of ions through the resin, it is possible to collect the individual ions. The procedure above is commonly known as elution. The chemistry of yttrium is primarily characterized by its predominant oxidation state of '+3.' At elevated temperatures, the metallic element yttrium undergoes oxidation, forming yttrium oxide (Y_2O_3). The only compound that exhibits solubility in water is nitrate, specifically $Y(NO_3)_3$. Additional yttrium compounds that are frequently encountered include yttrium hydroxide ($Y(OH)_3$), yttrium carbonate ($Y_2(CO_3)_3$), yttrium oxalate ($Y_2(C_2O_4)_3$), yttrium fluoride (YF_3), and yttrium phosphate (YPO_4). Yttrium hydroxide exhibits a yellow hue and possesses limited solubility in aqueous solutions. The remaining compounds exhibit a white hue and demonstrate insolubility (Moslemzadeh et al., 2006).

4.3. THE TITANIUM GROUP

The group of elements known as the titanium group comprises titanium, zirconium, and hafnium. The vanadium group comprises three elements: vanadium, niobium, and tantalum. Titanium metals are typically located within group IVA of the periodic table. These elements exhibit a tetravalent nature, indicating that their chemical behavior is primarily characterized by the oxidation state of +4. Titanium is widely recognized as the most well-known element within its respective group. Titanium dioxide (TiO_2) serves as a pigment in the form of white paint. Titanium alloys are renowned for their remarkable strength, making them highly desirable materials in various applications, including but not limited to golf clubs, surgical instruments, and prosthetics. Zircons are well-known gemstones resembling diamonds and are primarily composed of zirconium silicate, denoted as $ZrSiO_4$. Although relatively unfamiliar to the general population, Hafnium holds significant importance as a constituent of control rods within nuclear reactor fuel assemblies and plays a crucial role in the semiconductor industry (Aubert et al., 2002).

The vanadium metals are located within Group VA of the periodic table. All of these elements exhibit a pentavalent nature, indicating the significance of the '+5' oxidation state for each element. Conversely, Vanadium demonstrates the ability to form compounds that display a considerable range of oxidation states. In addition to the state denoted as '+5,' there exist three other states, namely '+2,' '+3,' and '+4.' Similarly, niobium demonstrates a valence state of '+3' alongside its more commonly observed valence state of '+5' (Dehnicke et al., 1999).

Titanium-44, classified as an α-process element, is produced through nucleosynthesis in exceptionally massive supernovae. Alpha-process elements are produced by successive incorporation of alpha particles, comprising four particles each. Consequently, these elements exhibit mass numbers that are integer multiples of four. The synthesis of titanium-44 occurs within the atmospheres of expanding supernovae, leading to the dispersion of this radioactive isotope into the surrounding space (Abdel-Baki & El-Diasty, 2006).

$${}^{40}_{20}\mathrm{Ca} + {}^{4}_{2}\alpha \rightarrow {}^{40}_{22}\mathrm{Ti}$$

The radioactive isotope titanium-44 undergoes radioactive decay with a half-life of 59 days, forming the isotope scandium-44. The element Scandium-44 exhibits inherent instability and undergoes decay with a half-life of 1.8 days, during which it emits a photon with an energy of 1.156 million electron volts (MeV). This decay process results in the formation of the stable isotope Calcium-44 (44Ca). The emission above the line, observed in a stellar spectrum, is a distinctive marker indicating the previous existence of ^{44}Ti, as recognized by astronomers. One intriguing celestial entity is the titanium star, also known as Cas A, situated approximately 9,000 light years distant. This stellar object represents a supernova remnant featuring a neutron star positioned at its core (Cotton, 1955) (Figures 4.3, 4.4, and 4.5).

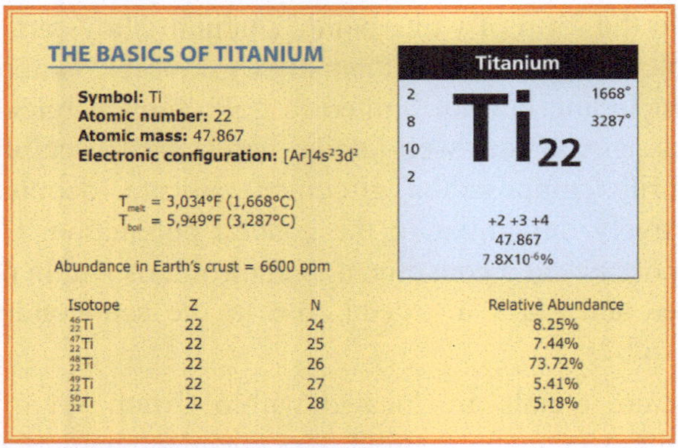

Figure 4.3. Basics of titanium.

Source: George trilling, creative commons license.

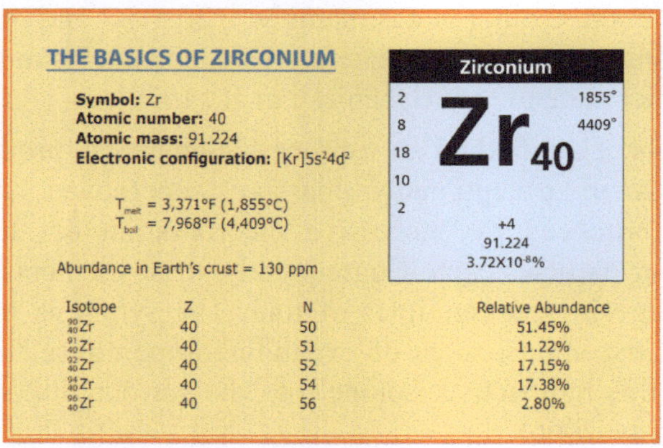

Figure 4.4. Basics of zirconium.

Source: John Phillips, creative commons license.

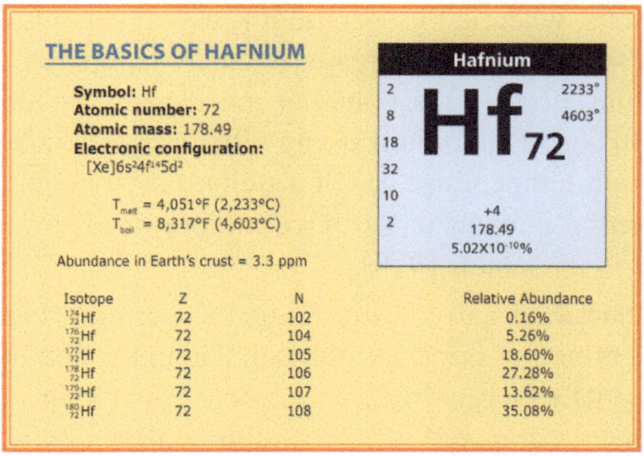

Figure 4.5. Basics of hafnium.

Source: Nicole Leonard, creative commons license.

The explosion likely possessed the potential for terrestrial observation during the mid-17th century, although no historical documentation exists to substantiate any actual sightings. This particular supernova is distinguished by its remarkable production of an unprecedented surplus of ^{44}Ti in comparison to ^{56}Ni, the latter being a byproduct observed in all supernovae. Given that the synthesis of titanium necessitates temperatures in the range of approximately 5 billion Kelvin, it follows that the temperatures resulting from gravitational contraction are directly proportional to the mass of a star (Weber, 1983). Consequently, it can be inferred that the progenitor star must have possessed an extraordinarily substantial mass before its eventual demise. The prevailing isotope of titanium found on Earth, namely Ti-48, is considered stable and is not generated through stellar nucleosynthesis. Instead, it is formed due to the beta decay of terrestrial scandium 48 or vanadium 48. Like yttrium and other elements with higher atomic numbers than iron, the syntheses of zirconium and hafnium take place within stars through neutron capture onto lighter elements. The presence of zirconium in stellar atmospheres yields unexpected findings, nonetheless. Like yttrium, it seems to exhibit a scarcity in stars characterized by low metallicity. However, the relative ratios of these two elements in comparison to iron display variations across different stars, even when said stars share similar classifications (Narath, 1967).

Nevertheless, the abundance of these elements exhibits less variability than titanium, a phenomenon that remains perplexing. An additional subject of inquiry pertains to the zirconium conflict, a phenomenon observed within certain HgMn stars, wherein the spectral line associated with doubly ionized zirconium exhibits significantly greater intensity compared to that of singly ionized zirconium atoms. One potential mechanism for this phenomenon involves the collision of zirconium atoms with highly abundant free electrons present in the high-temperature gas of a stellar environment. The impact of electron broadening is expected to have a more pronounced effect on Zr^{+} compared to Zr^{2+} (Jeitschko, 1970).

Titanium, classified as element 22 in the periodic table, possesses a density of 4.5 grams per cubic centimeter. Titanium is a metallic element characterized by its silvery-white appearance, possessing superior strength and a lower density compared to steel. Additionally, it exhibits exceptional resistance to corrosion. Titanium, which ranks as the ninth most abundant element in the Earth's crust and the fourth most abundant metal, exhibits a wide distribution. The primary titanium ores of significance encompass the igneous minerals known as ilmenite ($FeTiO_3$) and rutile (TiO_2). Ilmenite exhibits opaque crystals ranging in color from black to brownish-black, while rutile can manifest in hues of reddish-brown, red, yellow, or black. Brookite and anatase exhibit compositional similarities to rutile yet are predominantly found in metamorphic rocks and possess distinct crystalline structures in comparison to rutile. The majority of ilmenite extracted in the United States is sourced from the state of New York; however, the volume of imported titanium surpasses the amount produced domestically (Bradley & Chisholm, 1976).

Zirconium, denoted by the atomic number 40, has a 6.5 grams per cubic centimeter density. Zirconium demonstrates comparable mechanical strength and corrosion resistance to titanium. Moreover, it possesses the advantage of being resistant to the corrosive effects of concentrated hydrochloric, nitric, and sulfuric acids. Zirconium is utilized in nuclear reactors due to its exceptional corrosion resistance. Zirconium, in its pure form exhibits characteristics of being soft, ductile, and malleable. However, the addition of a mere 1 percent impurity to zirconium results in a significant alteration of its properties, rendering it hard and brittle. Zirconium, the 18th most prevalent element in the Earth's crust, exhibits a limited distribution. The primary sources of this substance are primarily derived from two minerals, baddeleyite (ZrO_2) and zircon. Zircon, a gemstone with a wide range of color variations, including transparent, opaque, colorless, red, brown, yellow, green, and gray, is one of the sources for this substance. The crystals of zircon exhibit dimensions of at least one inch and occasionally exceed this size, particularly in deposits located in Canada and Australia. Approximately

10% of the zirconium utilized within the US is sourced from zircon mines located in North Carolina, New Jersey, New York, and Pennsylvania. The remaining portion of the zircon supply is primarily sourced from Brazil and Australia, where the extraction of zircon from coastal sand deposits takes place (Murray & Yoffe, 1972).

Hafnium, classified as element 72 in the periodic table, exhibits a notable density of 13.1 g/cm^3. The element in question ranks as the 45th most prevalent constituent within the Earth's crust. Hafnium is characterized by a limited number of minerals that are exclusively associated with it. However, it typically comprises approximately 1 to 2 percent of zirconium minerals. Furthermore, less common minerals exist wherein the proportion of hafnium surpasses that of zirconium. Illustrative instances encompass alvite, which is constituted by oxides of beryllium, hafnium, thorium, yttrium, zirconium, and silicon, as well as cyrtolite, which is constituted by oxides of hafnium, zirconium, and silicon (Vayssilov, 1997).

The extraction of all three metals poses challenges due to their limited availability in their respective ores. One method of obtaining titanium involves the combination of TiO_2 and carbon within a furnace. This procedure produces both titanium metal and titanium carbide (TiC), necessitating the subsequent separation of the pure metal. Both titanium and zirconium can be acquired by reducing their respective chlorides, namely $TiCl_4$ and $ZrCl_4$, utilizing magnesium metal at elevated temperatures. This process is exemplified by the chemical equation provided for titanium (Chen et al., 2013):

$$TiCl_2\ (l) + 2Mg\ (l) \rightarrow Ti\ (l) + 2MgCl_2\ (l).$$

Once the mixture has reached a suitable temperature for the solidification of titanium, the water solubility of magnesium chloride enables its removal through washing, thereby leaving behind solid titanium metal. The chemical compounds predominantly contain these elements in the '+4' oxidation state. Among the three elements, titanium exhibits a restricted range of chemical behavior solely in the oxidation states of '+2' and '+3.' Titanium exhibits ionic bonding in its '+2' oxidation state. In the states denoted as '+3" and '+4,' the chemical bonding displayed by these elements predominantly manifests as covalent in nature. The following table presents a selection of compounds wherein these elements exhibit an oxidation state of '+4,' highlighting both their shared characteristics and distinguishing features (Samuel et al., 1977). The majority of these compounds exhibit a white coloration. The chlorides, nitrates, and sulfates demonstrate solubility in diluted aqueous solutions when they are present. The oxides and sulfides exhibit insolubility. Titanium exhibits the oxidation state of '+2' and consequently gives rise to the formation of titanium oxide (TiO) and titanium dichloride ($TiCl_2$). Titanium(II) oxide (TiO) is an ionic compound that exhibits properties akin to the oxides of alkaline earth metals, such as calcium oxide (CaO) or magnesium oxide (MgO). Both titanium oxide (TiO) and titanium dichloride ($TiCl_2$) exhibit strong reducing properties and readily undergo reactions with water. These reactions involve the reduction of hydrogen in water, resulting in the formation of hydrogen gas. The following chemical equations can represent this (Hillman & Shelton, 2003):

$$TiO\ (s) + H_2O\ (l) \rightarrow TiO_2(s) + H_2\ (g)$$

$$TiCl_2\ (aq) + 2H_2O\ (l) \rightarrow TiO_2\ (s) + 2HCl\ (aq) + H_2\ (g)$$

Due to its high reactivity with water, Ti^{2+} is deemed unstable in aqueous environments, thereby limiting its chemical reactivity in aqueous solutions. In the oxidation state of "+3," titanium is present as the titanous ion, denoted as Ti^{3+}, exhibiting a distinct violet hue. The trivalent titanium ion is known to react with various compounds, resulting in the formation of titanium sesquioxide, titanous hydroxide, and titanous trichloride ($TiCl_3$). Titanium(III) oxide (Ti_2O_3) exhibits complete insolubility in water and can be synthesized through the reduction of titanium dioxide (TiO_2) using hydrogen gas (H_2), as demonstrated by the subsequent chemical equation (Kohal et al., 2009).:

$$2TiO_2\ (s) + H_2\ (g) \rightarrow Ti_2O_3\ (s) + H_2O\ (l)$$

Titanous hydroxide is a strong reducing agent because it tends to change to the "+4" state when it is exposed to oxygen. As demonstrated by the following process, it will also break down into TiO_2 and hydrogen gas:

$$2Ti(OH)_3(aq) \rightarrow 2TiO_2\ (s) + 2H_2O\ (l) + H_2\ (g)$$

Titanium dioxide (TiO_2), commonly referred to as pure titanic oxide, exhibits a white coloration and has been employed as a substitute for white lead pigments in paint formulations due to the toxic nature of lead. Titanium tetrachloride ($TiCl_4$), commonly referred to as Titanic chloride, finds application in the production of smoke screens, wherein the resulting smoke predominantly comprises $TiCl_4 \cdot 5H_2O$. Titanium is capable of forming oxyanions known as titanates, which carry a negative charge (Han et al., 1998).

The titanium chlorides, namely $TiCl_2$, $TiCl_3$, and $TiCl_4$, exhibit a consistent pattern in bonding and properties that is generally observed in transition metal halides with higher oxidation states of the metal. Titanium dichloride ($TiCl_2$) and titanium trichloride ($TiCl_3$) are ionic compounds exhibiting crystalline structures at standard room temperature. These compounds possess relatively elevated melting points (Lindner et al., 2017).

The chemical properties of zirconium and hafnium exhibit such a high degree of similarity that, historically, their mixtures were erroneously perceived as a singular elemental entity. Similar to titanium, Zirconium exhibits the formation of zirconates, predominantly in the ZrO_{3-}^{2} composition. Illustrative examples encompass Na_2 ZrO_3, $CaZrO_3$, and $PbZrO_3$. Moreover, similar to titanium, zirconium exhibits the formation of a tetrachloride compound, $ZrCl_4$, and zirconyl chloride, $ZrOCl_2$, in analogy. There exist a number of fluoro zirconates that have been identified (Baan, 2007). Illustrative instances encompass K_2 ZrF_6 and $BaZrF_6$. Zircon, a mineral containing the silicate compound $ZrSiO_4$, possesses gem-like qualities and has the potential to serve as a substitute for diamonds. In each of these instances, zirconium exists in the oxidation state of +4. The chemical properties of hafnium closely resemble those of zirconium, thus accounting for the presence of hafnium in all zirconium minerals. The separation of hafnium and zirconium poses significant challenges due to their high similarity in chemical properties. The most effective method for achieving separation is through the utilization of ion exchange chromatography (Stimmelmayr et al., 2012).

4.4. THE VANADIUM GROUP

The elements within the vanadium group possess the potential to offer valuable insights to astrophysicists engaged in the examination of phenomena such as supernovae or the genesis of Earth's Moon and the terrestrial planets (Esayed & Northwood, 1992).

The luminosity of a supernova is significantly influenced by the quantity of vanadium 47 generated during the explosive event. As elucidated in the preceding section titled "The Astrophysics of the Titanium Group," the decay process of titanium 44 nuclei, which are synthesized in supernovae, involves the emission of high-energy photons that can be effectively detected by specialized satellites. However, the interaction between alpha-particles and Ti-44 can destroy the latter before it undergoes spontaneous decay, resulting in the formation of the unstable isotope vanadium 47 (Whittingham et al., 2005).

$$^{44}_{22}\mathrm{Ti} + {}^{4}_{2}\alpha \rightarrow {}^{47}_{23}\mathrm{V} + {}^{1}_{1}\mathrm{p}$$

In the context of this reaction, the absence of photon emission results in a diminished luminosity of the supernova compared to a scenario where all Ti-44 undergoes radioactive decay. Recent experimental findings have indicated a higher incidence of alpha collisions than initially anticipated, resulting in a reduction of approximately 25% in the intensity of the gamma-ray signal emitted by a supernova. The aspect above necessitates contemplation by instrument designers who aim to develop devices capable of detecting gamma rays emitted by supernovae (Huang et al., 2021) (Figure 4.6, 4.7 and 4.8).

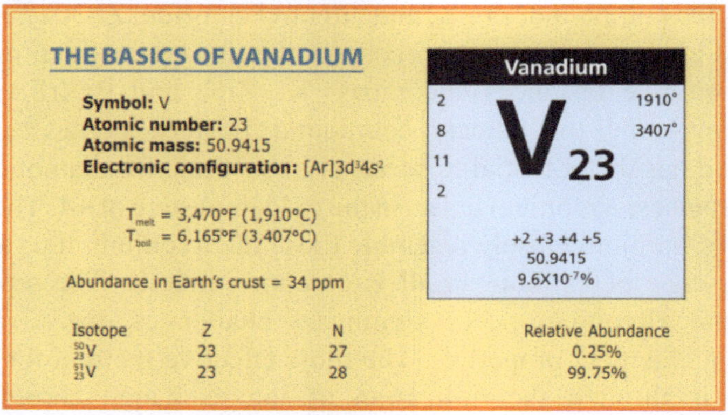

Figure 4.6. Basics of vanadium.

Source: John Phillips, creative commons license.

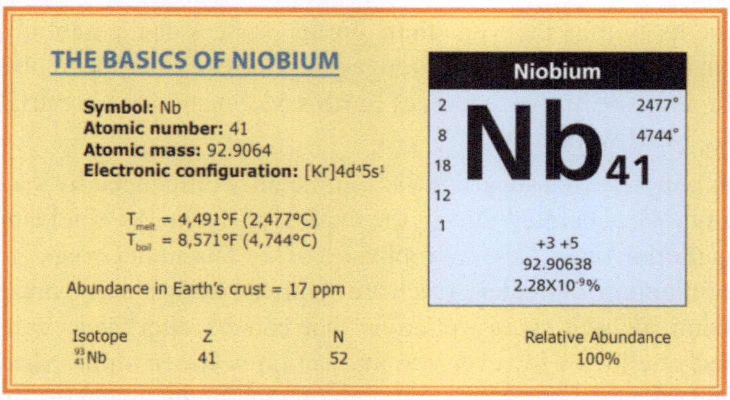

Figure 4.7. Basics of niobium.

Source: John Phillips, creative commons license.

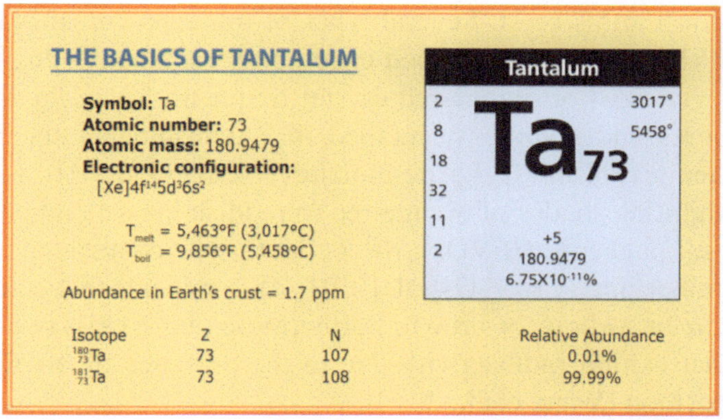

Figure 4.8. Basics of tantalum.

Source: Robert brown, creative commons license.

Using spectral analysis of vanadium-51 and tantalum-180 in supernovae holds significant potential to enhance our comprehension of neutrino oscillations. The neutrino process facilitates the production of vanadium group nuclides, which occurs when a substantial flux of neutrinos, generated during a supernova event, traverses the outer layer of the exploding star. In this process, the neutrinos interact with heavy nuclei, causing the removal of protons or neutrons from these nuclei. The production method described may yield Tantalum 180 with a heightened capacity to serve as a sensitive gauge for the electron neutrino temperature during the detonation. The discovery of neutrino mass has been a recent development in the par field. To date, experiments have provided valuable insights into the potential mass range of neutrinos.

Investigating extinct radionuclides, such as niobium 92, which is synthesized in supernovae and undergoes decay to form ^{92}Zr over 36 million years, can potentially provide insights into the temporal aspects of early and pre-solar system events and processes. The presence of residual quantities of this particular isotope during the solar system's formation can be determined by observing the levels of its Zr-92 daughter, which is found in varying abundances among different meteorites. Planetary geologists find this type of selective formation highly intriguing. Further investigation in this field can enhance our comprehension of the factors contributing to the diverse elemental compositions observed among planets (Brec, 1986).

Vanadium, denoted by its atomic number 23, has a 5.8 grams per cubic centimeter density. The element in question ranks as the 19th most abundant constituent in Earth's crust. Several types of minerals contain the element vanadium. The minerals mentioned encompass vanadinite ($Pb_5[VO_4]_3Cl$), which is a vivid red or orange-red crystal frequently found in conjunction with lead ores and commonly retailed in establishments specializing in rocks; dechenite ($PbZn[VO_3]_2$); descloizite ($PbZn[VO_4][OH]$), a translucent mineral exhibiting shades of orange-red to reddish brown and possessing a greasy sheen; pucherite ($BiVO_4$); roscoelite (a type of mica containing V_2 O_3); and volborthite ($Cu_3V_2O_7[OH]_2 \cdot 2H_2O$), a mineral that can appear in shades of green, yellow, or brown. Furthermore, there exist two vanadium minerals that exhibit radioactivity due to the presence of uranium within their composition (Wang et al., 2019):

i. Tyuyamunite ($Ca\ [UO_2]_2\ V_2O_8 \cdot nH_2O$)
ii. Carnotite ($K_2\ [UO_2]_2V_2O_8 \cdot 3H_2O$)

Niobium, denoted by the atomic number 41, possesses a density of 8.55 g/cm³. Niobium is the 33rd most prevalent element within the Earth's crust. Columbite, a mineral primarily composed of niobium ($[Fe, Mn]$ $[Nb, Ta]_2O_6$), is characterized by its gray to brownish-black coloration and concurrent presence of tantalum. Columbite is commonly observed in conjunction with granite. Samarskite, a mineral with a black or brownish appearance, is an additional source of niobium (Urbankowski et al., 2017).

Tantalum, classified as element 73 in the periodic table, exhibits a notable density of 16.6 g/cm³. Tantalum is the 51st most abundant element in the Earth's crust and is primarily acquired through the by-product of tin extraction. The predominant mineral in question is tantalite ($FeTa_2O_6$), which exhibits colors ranging from black to red-brown or colorless. It is commonly found in conjunction with vanadium and niobium ores. As previously stated, columbite serves as a significant source of tantalum. Most tantalum utilized in the United States is sourced through imports from nations such as Australia, Canada, and China. Furthermore, due to its limited availability, a substantial quantity of tantalum undergoes recycling.

All three metals exhibit a gray or silvery-white coloration. These metals exhibit high hardness levels while simultaneously demonstrating malleability and ductility. The elements belonging to the vanadium group exhibit semi-noble characteristics, indicating their inherent oxidation resistance. Consequently, these materials resist tarnishing due to their propensity to engage solely in oxidation reactions with atmospheric oxygen under elevated temperatures. The resistance to oxidation of these substances is further evidenced by their inherent ability to resist corrosion when exposed to saline solutions. Both vanadium and niobium have the ability to undergo dissolution in nitric acid, while tantalum remains unaffected by this chemical reaction.

Vanadium is the initial metallic element listed in the periodic table that demonstrates a significant range of oxidation states, specifically including +2, +3, +4, and +5. Moreover, vanadium, as the initial element in the inaugural series of transition metals, exhibits the ability to generate a diverse array of highly vibrant compounds and intricate ions. The observed colors of first-row transition elements can be attributed to the manner in which light interacts with their 3d electrons. The distinctiveness of the colors associated with each oxidation state enables their utilization for the identification of the oxidation state of vanadium.

Vanadium is found in its vanadous ion form, characterized by an oxidation state of "+2." The vanadous ion gives rise to vanadous oxide (VO) formation, which exhibits a metallic appearance that can occasionally lead to its misidentification as the actual metal. The compound vanadous sulfate (VSO_4) exhibits a profound violet hue, similar to other salts containing the V^{2+} ion (Campen et al., 2001).

When in the oxidation state of +3, the vanadium element exists in the form of the vanadic ion. The chemical properties of vanadium (V^{3+}) closely resemble those of the ferric ion (Fe^{3+}), with the notable distinction that Fe^{3+} exhibits significant resistance to oxidation to higher oxidation states. In contrast, V^{3+} can readily undergo oxidation to form the "+4" and "+5" oxidation states. The compounds comprising salts, hydroxide, and oxide of V^{3+} exhibit a characteristic green coloration. Several examples of vanadium compounds are $V(OH)_3$, $V_2 O_3$, VCl_3, VOCl, $V_2 (SO_4)_2$, VN, and $V_2 S_3$. The vanadic ion is capable of forming numerous complex salts, such as $(NH_4)_3 VF_6$, $K_3 V(CN)_6$, and $K_3V(CNS)_6$.

Vanadium exhibits the formation of vanadium dioxide (VO_2) in its "+4" oxidation state. One illustrative instance is the potassium tetra-oxovanadate(V) heptahydrate compound, with the chemical formula $K_2V_4 O_9 \cdot 7H_2O$. Vanadites typically exhibit a deep black hue.

Vanadium, classified as a group VA element, exhibits its highest oxidation state as "+5." In the state denoted as "+5," the primary compound of utmost significance for vanadium is V_2O_5, an oxide. Furthermore, it is worth noting that vanadium in the oxidation state of +5 exists either as the metavanadate ion (VO^{3-}) or as the pervanadyl ion (VO^{2+}). The pervanadyl ion is generated through the dissolution of V_2O_5 in an acidic medium, as illustrated by the subsequent chemical equation:

V_2O_5 (s) + $2H^+$ (aq) → $2VO^{+2}$ (aq) + H_2O (l).

The pervanadyl ion has a tendency to polymerize, as shown in the following equation:

$10VO^{+2}$ (aq) + $8H_2O$ (l) → $H_2V_{10}O_{28}^{4-}$ (aq) + $14H^+$ (aq) (Carrick et al., 1960).

Metavanadic acid, chemically represented as HVO_3, exhibits a yellow coloration in its solid state. Several salts of metavandic acid can be synthesized, such as sodium metavanadate ($NaVO_3$), potassium metavanadate (KVO_3), ammonium metavanadate (NH_4VO_3), and barium metavandate (Ba $[VO_3]_2$). Furthermore, a diverse range of sodium salts can be found, such as sodium orthovanadate (Na_3VO_4), sodium pyrovanadate ($Na_4V_2O_7$), and sodium hexavanadate ($Na_2H_2V_6O_{17}$). The compound ammonium sulfide [$(NH_4)_2$ S] undergoes a chemical reaction with vanadates, resulting in the formation of the thiovanadate ion ($VS4^{-4}$). In chemical nomenclature, the prefix "thio" indicates substances where sulfur atoms have replaced oxygen atoms. Introducing acid to the solution results in the formation of vanadium (V) sulfide (V_2S_5) as a precipitate. The $(NH_4)_3VS_4$ compound exhibits a characteristic violet-red hue, which serves as a means to identify and confirm the existence of vanadium within a given solution.

Niobium and tantalum's chemical and physical characteristics exhibit a high degree of similarity. Niobium is characterized by its inherent passivity, exhibiting noble metal behavior. The chemistry of niobium is predominantly confined to the oxidation states of "+3" and "+5." The ion with a charge of "+3," known as the niobic ion (Nb^{3+}), is responsible for conferring a blue hue to solutions. Nevertheless, the ion with a charge of "+3" exhibits significant instability, leading to swift oxidation by oxygen present in the atmosphere, ultimately forming the "+5" oxidation state.

Several niobium compounds exist in which niobium exhibits an oxidation state of +5. These compounds include niobium pentoxide (Nb_2O_5), niobium pentachloride ($NbCl_5$), and niobium nitride (Nb_3N_5).

Tantalum's chemical reactivity is constrained to the oxidation state of +5 when it comes to the formation of stable compounds. Upon combustion in air, tantalum undergoes a chemical reaction resulting in the formation of tantalum pentoxide, as represented by the subsequent chemical equation (Wei et al., 2016):

$4Ta\ (s) + 5O_2(g) \rightarrow 2Ta_2O_5(s).$

4.5. THE CHROMIUM GROUP

The VIA group in the periodic table comprises three elements: chromium (Cr, atomic number 24), molybdenum (Mo, atomic number 42), and tungsten (W, atomic number 74). Molybdenum and tungsten exhibit oxidation states of +2, +3, +4, +5, and +6, while chromium is predominantly in oxidation states of +2, +3, and +6 (Lomer, 1962) (Figure 4.9, 4.10 an 4.10).

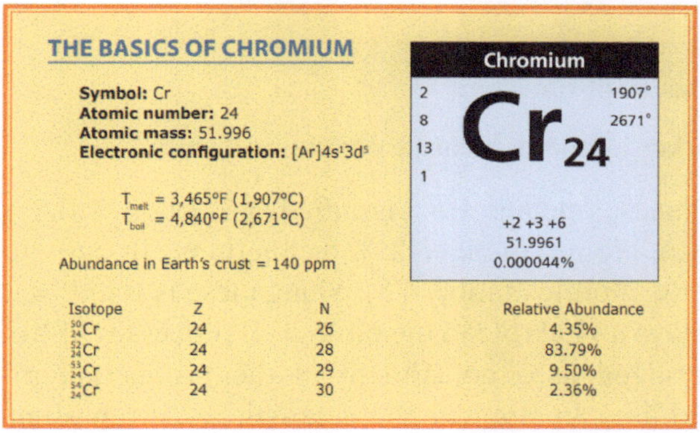

Figure 4.9. Basics of chromium.

Source: Samuel Markowitz, creative commons license.

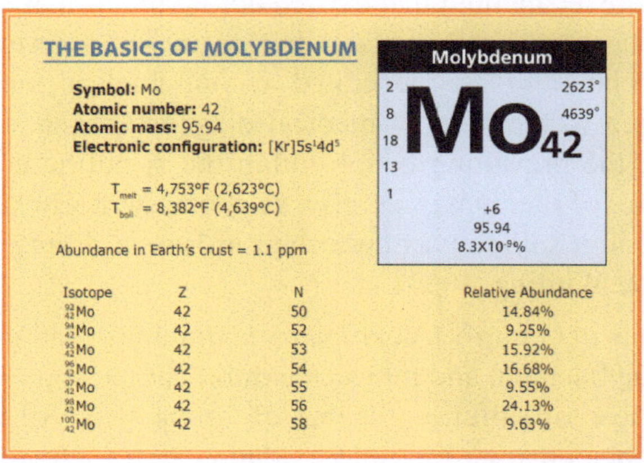

Figure 4.10. Basics of molybdenum.

Source: John Phillips, creative commons license.

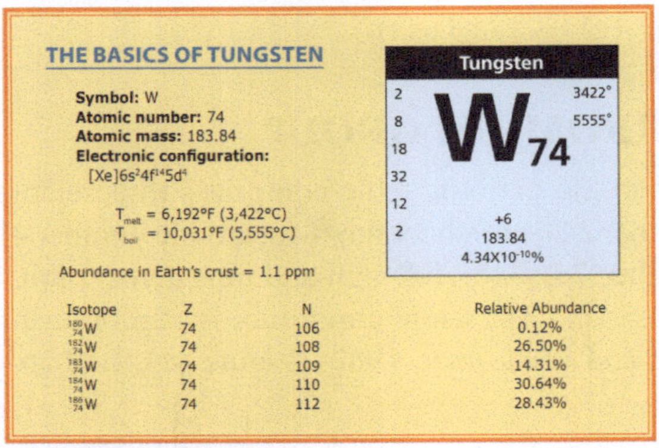

Figure 4.11. Basics of tungsten.

Source: A. Starker Leopold, creative commons license.

The manganese metals are classified as Group VIIA and comprise manganese (Mn, atomic number 25), technetium (Tc, atomic number 43), and rhenium (Re, atomic number 75). Manganese is a well-known chemical element that plays a vital role in the biological processes of living organisms and is also utilized as a constituent in the production of steel alloys. Technetium exhibits an atypical characteristic, as it represents the element with the lowest atomic number lacking any stable isotopes. Even though two isotopes of technetium exhibit half-lives in the range of one million years, it is noteworthy that the age of the Earth spans billions of years. Consequently, the entirety of the technetium that existed during the formation of Earth has undergone radioactive decay and is no longer present. However, significant amounts of technetium are generated within nuclear reactors and are routinely utilized in the field of medical diagnostics and therapy. Before the year 1994, the prevailing belief within the scientific community was that the existence of rhenium minerals was nonexistent. However, a mineral containing rhenium sulfide was unearthed in Russia during that particular year (Nicholls & Whiting, 1959).

The elements in Group VIA exhibit a maximum oxidation state of "+6." Chromium, molybdenum, and tungsten exhibit chemical properties that are analogous to those of sulfur, a member of Group VIB. All three elements in Group VIA demonstrate multiple oxidation states. However, the states of "plus two" and "plus three" are exclusively observed in the case of chromium.

The most stable oxidation state of chromium is +3, which is present as the chromic ion (Cr^{3+}) exhibiting a violet coloration. Hence, the lower oxidation states of chromium, specifically the chromous ion (Cr^{2+}), exhibit strong reducing properties due to their susceptibility to oxidation. Conversely, chromium in the "6+" state, such as chromate or dichromate, demonstrates notable oxidizing capabilities owing to its propensity for reduction. Both the chromous and chromic ions exhibit a wide range of compounds that possess vibrant hues, as illustrated in the subsequent table 4.1 (Zeiss & Tsutsui, 1957):

Table 4.1. Chromous and chronic Ions Forming Compounds with Bright Colors

	Cl^-	OH^-	NO_3^-	SO_4^{2-}	PO_4^{3-}	S^{2-}
Cr^{2+}	$CrCl_2$ (blue)	$Cr(OH)_2$ (br-yellow)	none	$CrSO_4$ (blue)	None	None
Cr^{3}+ (violet)	$CrCl_3$ (violet)	$Cr(OH)_3$ (green)	$Cr(NO_3)_3$ (purple)	$Cr_2(SO_4)_3$ (violet)	$CrPO_4$ (blue-green)	Cr_2S_3 (brown)

An additional significant chromium (III) ion is chromite, exhibiting a green hue. Ferrous chromite ($FeCr_2O_4$) finds application as a refractory material for furnace lining in the steelmaking sector. The acid-base characteristics of chromium oxides exhibit variability in relation to the oxidation state of chromium. Chromous oxide (CrO) exhibits basic properties as it undergoes dissolution in the presence of acids while remaining unaffected by bases. Chromium(III) oxide (Cr_2O_3) exhibits amphoteric properties as it can undergo dissolution when exposed to both acidic and basic solutions. Chromic trioxide, also known as chromium anhydride (CrO_3), exhibits acidic properties as it can be dissolved by bases while remaining unaffected by acids. The amphoteric nature of chromium(III) oxide (Cr_2O_3) is demonstrated by the subsequent reactions (Brandt & Rayne, 1963):

$$Cr_2O_3(s) + 6HCl\ (aq) \rightarrow 2CrCl_3(aq) + 3H_2O\ (l)$$

$$Cr_2O_3(s) + 2NaOH\ (aq) + 3H_2O\ (l) \rightarrow$$

$$2Na+\ (aq) + 2Cr(OH)_4^-\ (aq)$$

In the initial reaction, the dissolution of Cr_2O_3 is facilitated by an acid, while in the subsequent reaction, it is facilitated by a base.

In the oxidation state of "+6," chromium exhibits two significant ions, namely chromate (CrO_4^{2-}) and dichromate ($Cr_2O_7^{2-}$). In an aqueous solution, chromate exhibits basic properties, while dichromate demonstrates acidic properties. The conversion of chromate to dichromate can be achieved by introducing acid into a chromate solution, as demonstrated by the subsequent chemical equation (Nesmeyanov et al., 1978):

$$2\ CrO_4^{2-}\ (aq) + 2H^+\ (aq) \rightarrow Cr_2O_7^{2-}\ (aq) + H_2O\ (l).$$

The dichromate ion exhibits remarkable oxidizing capabilities in an acidic environment. Potassium dichromate ($K_2Cr_2O_7$) is frequently employed in the field of quantitative analytical chemistry. One method for determining the quantity of iron in a compound or mixture involves the process of titration, wherein the iron sample is reacted with $K_2Cr_2O_7$ (Eikey & Abu-Omar, 2003):

$$6Fe^{2+}\ (aq) + Cr_2O_7^{2-}\ (aq) + 14H^+(aq) \rightarrow$$

$$6Fe^{3+}(aq) + 2Cr^{3+}(aq) + 7H_2O\ (l).$$

Regrettably, chromate ions, dichromate ions, and other manifestations of chromium in its hexavalent state (+6 oxidation state) is associated with potential health risks. According to the National Institute for Occupational

Safety and Health, chromium (VI) is regarded as a plausible occupational carcinogen that is linked to an elevated likelihood of developing lung cancer. Additional potential negative health consequences linked to exposure to hexavalent chromium (Cr (VI)) encompass dermatological issues, bronchial asthma, nasal inflammation, rhinitis, epistaxis, and various respiratory ailments. Furthermore, exposure to Cr (VI) may lead to the development of nasal and sinus malignancies, ocular irritation, as well as renal and hepatic impairments. The optimal approach for handling hexavalent chromium compounds involves minimizing skin contact and preventing inhalation of dust particles (Preuss & Anderson, 1998).

The identification of chromous or chromic ions in qualitative analysis methods can be achieved through their oxidation to the chromate ion in a basic solution. The oxidizing agent commonly employed is a 3 percent solution of hydrogen peroxide (H_2O_2), as demonstrated by the subsequent equation;

$$2\,Cr\left(OH\right)_4^- (aq) + 3H_2O_2\,(aq) + 2OH^-\,(aq) \rightarrow$$

$$2\,CrO\quad (aq) + 8H_2O\,(l)$$

The manifestation of a yellow hue, which is commonly linked to the chromate ion, serves as visual confirmation that the reaction has taken place. The concentration of hydrogen peroxide, specifically 3 percent H_2O_2, is equivalent to the strength of hydrogen peroxide commonly available for purchase at pharmacies (Mertz, 1993).

If other species potentially display similar yellow hues or colors that could conceal the color of chromate, further examinations can be conducted. As an illustration, introducing a solution containing Ba^{2+} may be considered.

When the anion CrO_4^{2-} is present, the compound $BaCrO_2$, which is the sole yellow barium compound, will form a solid precipitate. This reaction can be written as (Bailey, 2014):

$$Ba^{2+}\,(aq) + CrO_4^{2-}\,(aq) \rightarrow BaCrO_4\,(s).$$

An alternative approach involves the introduction of supplementary hydrogen peroxide into a solution containing nitric acid. The acid facilitates the conversion of chromate to dichromate. Subsequently, dichromate undergoes oxidation by H_2O_2 to form chromium peroxide (CrO_5), as depicted in the following chemical equation.:

$$Cr_2O_7^{2-}\ (aq) + 4H_2O_2\ (aq) + 2H^+\ (aq) \rightarrow 2CrO_5\ (aq) + 5H_2O\ (l).$$

The compound CrO_5 is known for its inherent instability, yet it possesses a distinct blue hue distinguishing it from any other coexisting species. In certain cases, liquid ether is introduced before hydrogen peroxide to mitigate the rapid fading of the blue color. The compound CrO_5 exhibits greater stability in an ether environment, leading to its preferential extraction into the ether phase upon shaking the solution. Consequently, the blue color associated with CrO_5 remains observable for an extended duration in the ether phase (Tate et al., 1962).

Compounds in the oxidation states of "+2," "+3," "+4," "+5," and "+6" can be formed by molybdenum and tungsten. In the state denoted as "+2," both elements tend to form chlorides, namely $MoCl_2$ and WCl_2, which display an inherent instability when dissolved in aqueous solutions. In the state denoted as "+3," molybdenum is present in a solution as a species exhibiting an olive-green coloration. The compound exhibits the formation of a hydroxide ($Mo[OH]_3$), an oxide (Mo_2O_3), a sulfide (Mo_2S_3), and a phosphate ($Mo_3[PO_4]_2$). Tungsten exhibits a lack of reactivity in forming simple compounds while in the oxidation state of +3 (Trent & Thieding-Cancel, 1995).

Molybdenite is the primary source of molybdenum, wherein molybdenum exists in the oxidation state of "+4" as molybdenum sulfide (MoS_2). The ion with a charge of "+4" is inherently unstable. However, it is possible to synthesize stable cyanide complexes such as $K_4Mo(CN)_8$. Compounds of tungsten in the oxidation state of +4 encompass tungsten dioxide (WO_2) and tungsten tetrachloride (WCl_4). Tungsten (IV) compounds, on the other hand, tend to readily undergo conversion to the "plus six" oxidation state (Anderson et al., 1997).

In the state denoted as "+5," it is observed that molybdenum pentachloride ($MoCl_5$) is the sole compound lacking the presence of oxygen. Several species that possess oxygen atoms are molybdenum hydroxide ($MoO[OH]_3$), molybdenum pentoxide (Mo_2O_5), and a solid substance known as "molybdenum blue" ($[MoO]_3[MoO_4]_2$), which exhibits a deep-blue coloration. Tungsten (V) can also form pentachloride, denoted as WCl_5. Tungsten has been observed to form a compound known as "tungsten blue" which exhibits similarities to the compound molybdenum blue (Clancy et al., 1994). The compound CrO_5 is highly unstable, yet it displays a distinct blue hue that distinguishes it from any other coexisting species. In order to mitigate the rapid fading of the blue color, it is occasionally deemed necessary to introduce liquid ether before adding hydrogen peroxide. The compound CrO_5 exhibits greater stability in an ether medium. Thus, agitation of the solution results in the extraction of CrO_5 into the ether phase. Consequently, the blue coloration of the solution is retained for an extended duration (Yazaki et al., 2010).

Transition metals such as iron, copper, chromium, and manganese exhibit different oxidation states, including +2, +3, +4, +5, and +6. For example, iron forms $FeCl_2$ in its +2 oxidation state, copper forms $CuCl_2$, chromium forms CrO_2, manganese forms MnO_3, and vanadium forms VO_2. Each of these compounds demonstrates unique chemical properties, making them valuable subjects of study in inorganic chemistry. (Uyanik, 2001).

4.6. THE MANGANESE GROUP

Ongoing research is being conducted on the nucleosynthesis of members belonging to the manganese group due to variations in astrophysical abundances, despite their chemical similarities. Technetium and rhenium, which possess greater atomic masses than iron, are produced through the process known as rapid neutron capture, or the r-process, within supernovae. This process involves the successive capture of neutrons by iron nuclei. The heavy atoms are rapidly expelled into the vacuum of space, where a portion of them is subsequently accreted during the process of stellar formation. Manganese, on the other hand, has the capacity to accumulate gradually over extended periods of time, spanning thousands of years, within the atmospheres of massive stars through a process known as neutron capture. This process necessitates the presence of iron-56 nuclei as initial particles, which are remnants of previous supernova events. The nomenclature "s-process" is attributed to the relatively sluggish progression of this synthesis, which can be attributed to a diminished neutron density (Hock et al., 2013) (Figures 4.12, 4.13 and 4.14).

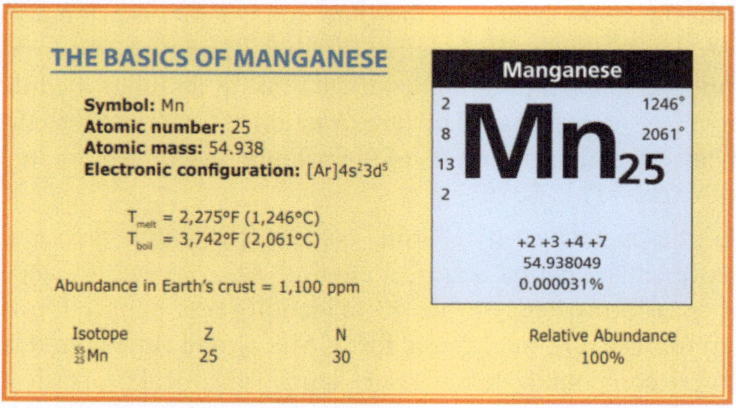

Figure 4.12. Basics of manganese.

Source: George Trilling, creative commons license.

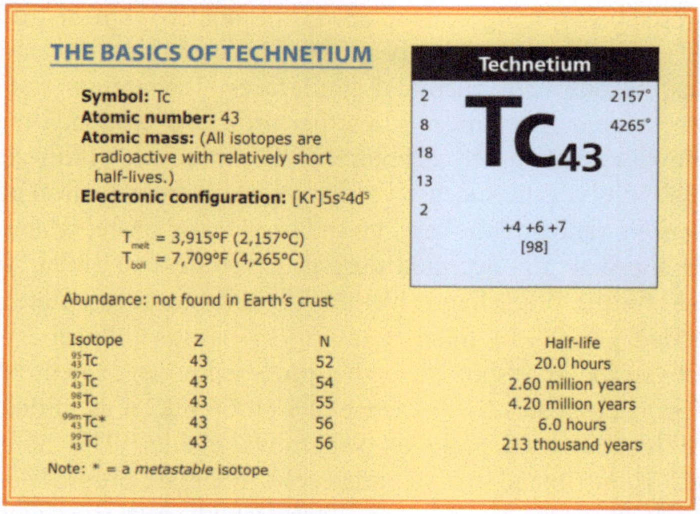

Figure 4.13. Basics of technetium.

Source: Robert Brown, creative commons license.

Manganese, denoted by its atomic number 25, exhibits a density of 7.4 g/cm^3 and holds the 12th position in terms of its abundance on the planet Earth. Due to the relatively low manganese contents found in ores within the United States, it is deemed more cost-effective to import manganese rather than pursue domestic recovery methods. Typically, manganese is commonly encountered in the form of the mineral manganite, which exhibits a black or gray coloration and has the chemical formula MnO[OH)]. The ocean floor has been identified as a potential reservoir of manganese, as it contains black nodules composed of pyrolusite, a mineral consisting of manganese dioxide (MnO_2). However, the depth of these nodules often renders the economic feasibility of recovery unattainable (Irving & Williams, 1953).

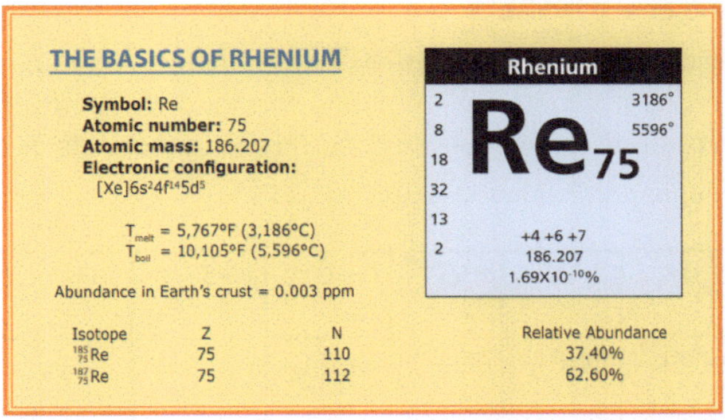

Figure 4.14. Basics of rhenium.

Source: Glenn Seaborg, creative commons license.

Technetium, which is identified as element 43 on the periodic table, does not occur naturally on Earth. All isotopes of technetium exhibit radioactive properties, characterized by half-lives that are considerably shorter than the estimated age of Earth. Consequently, any technetium that might have originally existed during the formation of Earth has undergone complete decay over an extended period. Nevertheless, significant amounts of technetium are produced within nuclear reactors. The element technetium exhibits a density of 11.5 grams per cubic centimeter. The comprehensive investigation of technetium's chemistry remains limited; however, its compound formation resembles rhenium's. The utilization of the metastable isotope Tc-99, which serves as a gamma-ray emitter in nuclear medicine, holds significant significance.

Rhenium, denoted by its atomic number 75, exhibits a significantly elevated 21.0 g/cm^3 density. Regarding its occurrence on Earth, rhenium is positioned at the 77th rank in terms of abundance.

The significance of the " +7" oxidation state can be attributed to the presence of manganese, technetium, and rhenium in Group VIIA. Various compounds containing manganese and rhenium have been identified, wherein these elements exhibit a range of positive oxidation states spanning from "+2" to "+7." In the state denoted as "+7," the chemical elements exhibit the formation of compounds such as Mn_2O_7, characterized by a red hue, and Re_2O_7, distinguished by a yellow coloration. Furthermore, it has been observed that rhenium exhibits an oxidation state of "-1," a remarkably uncommon occurrence for a metallic element. The investigation of technetium compounds has predominantly focused on the oxidation state of "+7" (Florio et al., 1952).

Manganese exhibits a range of oxidation states, spanning from +2 to +7, encompassing a variety of compound formations. The following series provides illustrations of manganese occurrences in these states (Table 4.2):

Table 4.2. Manganese Oxidation States

Oxidation state	0	"+2"	"+3"	"+4"	"+5"	"+6"	"+7"
Chemical species	Mn	Mn^{2+}	Mn^{3+}	MnO_2	MnO_3^-	ü $^{2-}_{4}$	MnO_4^-

The manganic ion (Mn^{3+}) exhibits a propensity for instability when present in an aqueous solution, leading to its disproportionate transformation into Mn (II) and Mn(IV), as evidenced by the subsequent reaction (Krämer et al., 2007):

$$2Mn^{3+} (aq) + 2H_2O (l) \rightarrow Mn^{2+} (aq) + MnO_2(s) + 4H^+ (aq).$$

Disproportionation refers to the phenomenon wherein an element, which is initially present in a singular oxidation state, undergoes a transformation resulting in the formation of two distinct products. One of these products exhibits a lower oxidation state, while the other demonstrates a higher oxidation state (Edwards & Hayward, 1968).

The oxides that are produced in the states of "plus two" and "plus three" are referred to as manganous oxide (MnO) and manganic oxide (Mn_2O_3), respectively. These oxides exhibit basic properties as a result of their ability to dissolve in acidic solutions. This can be demonstrated through the following example (Vincent et al., 1987):

$$MnO (s) + 2HCl (aq) \rightarrow MnCl_2(aq) + H_2O (l).$$

In the state denoted as "+4," manganese undergoes a chemical reaction resulting in the formation of manganese dioxide (MnO_2). This black-colored compound exclusively represents the presence of manganese in its oxidation state of +4. Manganese dioxide exhibits amphoteric properties due to its solubility in acidic and alkaline solutions. Manganese oxides exhibit acidic properties in higher oxidation states due to their solubility in basic solutions (Butler, 1977).

Limited research has been conducted on the chemical properties of technetium. There exist a minimum of three distinct oxides, namely TcO_2, TcO_3, and Tc_2O_7, each exhibiting distinct coloration properties, with TcO_2 appearing black, TcO_3 displaying a purple hue, and Tc_2O_7 manifesting a yellow color. Technetium exists in these compounds' oxidation states of +4, +6, and +7. Technetium can also be found in the oxidation state of +7 in the pertechnetate ion (TcO_4^-). This ion can be easily reduced by copper metal to form metallic technetium. The reduction reaction occurs in an acidic solution (Ruther et al., 2014):

$$2\,TcO_4^-\ (aq) + 7Cu\ (s) + 16H^+\ (aq) \rightarrow$$
$$2Tc\ (s) + 7Cu^{2+}\ (aq) + 8H_2O\ (l).$$

In addition, it should be noted that technetium can form halides. The compound $TeCl_4$ exhibits a deep red color, while $TeCl_6$ displays a green hue, and TeF_6 appears yellow in color.

The chemical properties of rhenium closely resemble those of manganese. Rhenium exhibits the formation of various colored oxides across multiple oxidation states, as illustrated in the subsequent tables 4.3 and 4.4 (Roels et al., 1992):

Table 4.3. Rhenium Forming Colored Oxides in Several Oxidation States

Oxidation State	Formula of Oxide	Color of Oxide
"+3"	Re_2O_3	Black
"+4"	ReO_2	Brown
"+5"	Re_2O_5	Blue
"+6"	ReO_3	Red
"+7"	Re_2O_7	Yellow

Table 4.4. Rhenium Forming Colored Halides

Oxidation State	Formula of Halide	Color of Halide
"+2"	ReI_2	Black
"+3"	Re_3Cl_9 Re_3Br_9 Re_3I_9	Red Red-Brown Black

“+4”	ReF_4 $ReCl_4$ $ReBr_4$ ReI_4	Blue Black Red Black
“+5”	ReF_5 $ReCl_5$ $ReBr_5$	Green-Yellow Red-Brown Green-Blue
“+6”	ReF_6 $ReCl_6$	Yellow Green-Brown
“+7”	ReF_7	Yellow

The perrhenate ion (ReO_4^-) is formed by rhenium, similar to how the permanganate ion (MnO_4^-) and pertechnetate ion (TcO_4^-) are formed by manganese and technetium, respectively. In an acidic solution, it can be observed that all three ions exhibit strong oxidizing properties (Zhan et al., 2016).

4.7. THE IRON, COBALT, AND NICKEL GROUPS

The iron group is comprised of three elements: iron (Fe, atomic number 26; known as Ferrum in Latin), ruthenium (Ru, atomic number 44), and osmium (Os, atomic number 76). The cobalt group is comprised of three elements: cobalt (Co, atomic number 27), rhodium (Rh, atomic number 45), and iridium (Ir, atomic number 77). The nickel group is comprised of three elements: nickel (Ni, atomic number 28), palladium (Pd, atomic number 46), and platinum (Pt, atomic number 78) (Burstall & Nyholm, 1952) (Figures 4.15, 4.16 and 4.17).

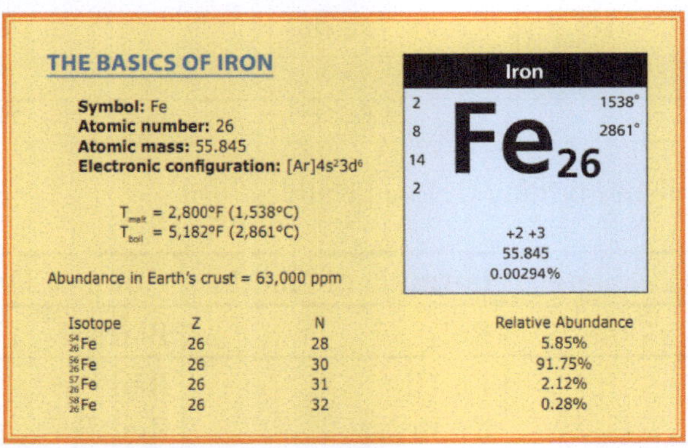

Figure 4.15. Basics of iron.

Source: John Phillips, creative commons license.

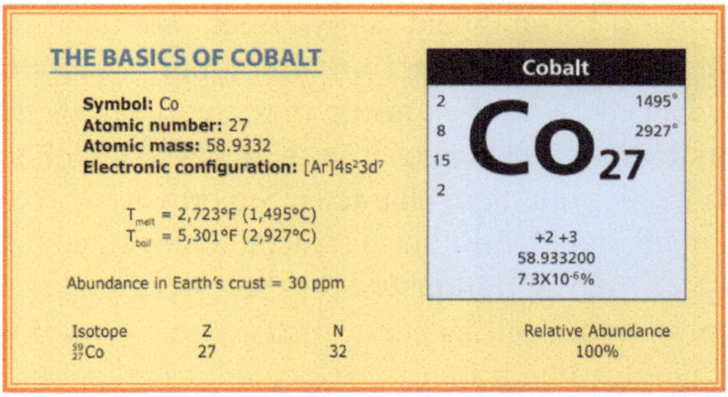

Figure 4.16. Basics of cobalt.

Source: Samuel Markowitz, creative commons license.

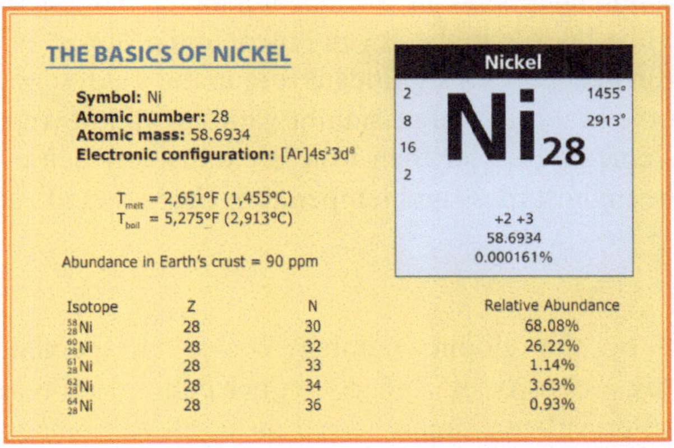

Figure 4.17. Basics of nickel.

Source: Samuel Markowitz, creative commons license.

Iron, cobalt, and nickel exhibit numerous similarities. Initially, it is noteworthy that all of them exhibit a propensity to undergo reactions with commonly encountered acids readily. On the contrary, platinum metals demonstrate noble characteristics, indicating their tendency to resist chemical reactions with potent reagents such as hydrochloric or nitric acid (Holland et al., 2002).

Furthermore, iron, cobalt, and nickel exhibit exceptional hardness and elevated melting and boiling points. Moreover, it is worth noting that all three metals demonstrate magnetic properties. The phenomenon of enduring magnetic properties in substances is scientifically referred to as ferromagnetism, with the prefix "ferro" denoting its association with iron (Jutzi et al., 1998).

Iron and its most significant alloy, steel, are ubiquitously present in various contexts. Iron and steel are commonly recognized by young children, as evidenced by their exposure to these materials through popular culture, such as cartoons featuring iron as a key component in Popeye's spinach and the iconic superhero character, Superman, who is often referred to as the "man of steel." Iron is a crucial constituent of hemoglobin, facilitating oxygen transportation within the circulatory system. Various forms of transportation, such as cars, trucks, and locomotives, as well as architectural structures like bridges and skyscrapers, and maritime vessels, such as oceangoing ships, rely on steel as a fundamental component (Li et al., 2015). Additionally, steel finds application in producing essential items such as nails, bolts, tanks, armor, and weaponry. The Iron Age is widely regarded as a pivotal period in the development of modern civilization. Steel played a pivotal role in the development and progress of the Industrial Revolution. Iron also plays a significant role in the field of cosmology. Iron nuclei exhibit the highest level of stability among all known elements. The synthesis of elements heavier than iron is exclusively achievable by stars that possess the highest mass and temperature (Daniel et al., 2021).

4.7.1. Iron

Iron, denoted by the atomic number 26, is an elemental substance characterized by a density of 7.86 grams per cubic centimeter. Oxygen is the fourth most prevalent element found on Earth and ranks as the second most abundant metal, surpassed only by aluminum. Iron is widely utilized in various applications, and its historical importance and crucial role in animal biology contribute to its widespread recognition as one of the most familiar elements. Numerous significant iron ore deposits are distributed extensively across the globe. Hematite, also known as ferric oxide (Fe_2O_3), is widely regarded as one of the most significant ores. It exhibits a range of colors, including a recognizable rusty red, steel-gray, and even a metallic black hue. Another significant type of iron ore is magnetite (Fe_3O_4), a magnetic form of black iron oxide commonly referred to as lodestone. Goethite and limonite, which are FeO(OH) variants, exhibit characteristic yellow, brown, or black hues, representing iron oxide compounds. Iron pyrite, scientifically known as FeS_2, is a sulfide mineral composed of iron and sulfur (Rana et al., 2021). It is commonly referred to as fool's gold due to its resemblance to the precious metal. This mineral exhibits a pale-yellow coloration. Meteorites typically exhibit a significant abundance of metallic iron, accounting for their elevated density. Due to the widespread presence of iron in the Earth's crust, it is also found as a component in ores containing a metal mixture. Iron is a crucial element that serves as the primary constituent of hemoglobin, the compound responsible for transporting oxygen throughout the human body. Iron is a crucial nutrient that plays a vital role in supporting the growth and development of plants. The chemical behavior of iron is primarily characterized by its oxidation states of "+2" and "+3." The oxidation states mentioned are exemplified by the reactions occurring within a blast furnace during the reduction of iron ore to produce iron metal for steel production. The blast furnace exhibits variations in temperature across its distinct regions. In a geographical area characterized by a comparatively lower temperature of approximately 400°F (200°C), the chemical reaction between ferric oxide and carbon monoxide (CO) occurs as follows (Dolg et al., 1987):

$$3Fe_2O_3(s) + CO\ (g) \rightarrow 2Fe_3O_4(s) + CO_2(g).$$

The compound Fe_3O_4 exhibits iron in both the "+2" and "+3" oxidation states, which can be elucidated by reconfiguring the formula as $FeO{\cdot}Fe_2O_3$. In the compound FeO, the oxidation state of iron is +2, whereas in Fe_2O_3, it is +3. At elevated temperatures, specifically around 660°F (350°C), the reduction of Fe_3O_4 to ferrous oxide (FeO) occurs according to the reaction presented below (Chábera et al., 2021):

$$Fe_3O_4\ (s) + CO\ (g) \rightarrow 3FeO\ (s) + CO_2(g).$$

At still higher temperatures, the ferrous oxide is finally reduced to iron metal, as shown in the following reaction:

$$FeO\ (s) + CO\ (g) \rightarrow Fe\ (s) + CO_2\ (g).$$

After production of metallic iron, it is possible to create steel by alloying it. Ferric oxide, commonly known as rust, is the most widely recognized iron oxide compound. In a humid environment, the corrosion in metallic iron, leads to the formation of ferrous oxide. This reaction is facilitated by the presence of atmospheric oxygen, which acts as the oxidizing agent (Sun et al., 2019):

$$2Fe\ (s) + O_2(g) \rightarrow 2FeO\ (s).$$

It should be noted that the oxidizing agent may consist of water, environmentally present weak acids, or a combination of these substances along with oxygen gas. Ferrous oxide exhibits a propensity to undergo further oxidation, resulting in the formation of ferric oxide, as evidenced by the subsequent reactions:

$$4FeO\ (s) + O_2(g) + 6H_2O\ (l) \rightarrow 4Fe(OH)_3\ (s);$$

$$2Fe(OH)_3\ (s) \rightarrow Fe_2O_3\ (s) + 3H_2O\ (l)$$ (Clogston et al., 1962).

Both iron (III) hydroxide ($Fe(OH)_3$) and iron(III) oxide (Fe_2O_3) exhibit the characteristic reddish-brown hue commonly associated with rust. The presence of different compounds is primarily contingent upon the quantity of moisture available. The ferrous ion exhibits a high propensity for undergoing disproportionation reactions, wherein one ferrous ion is subject to reduction, resulting in the formation of a neutral iron atom. In contrast, ferrous ions undergo oxidation, leading to the generation of a ferric ion.

$3Fe_2 + (aq) \rightarrow Fe (s) + 2Fe_3+ (aq)$.

Aqueous solutions containing the ferrous ion exhibit instability due to the susceptibility of the ferrous ion to oxidation by atmospheric oxygen. In laboratory settings, it is advisable to prepare fresh solutions of ferrous chloride ($FeCl_2$) or ferrous sulfate ($FeSO_4$) and utilize them promptly. Both the ferrous ion (Fe^{2+}) and the ferric ion (Fe^{3+}) can form complex ions. The complex ions most commonly encountered in high school or introductory college chemistry laboratories are the cyanides of these ions. The given table 4.5 presents a compilation of various ferrous and ferric compounds, accompanied by their respective hues (Macdonald & Geerligs, 2004):

Table 4.5. Numerous Common Ferrous and Ferric Compounds, along with Colors

	Cl^-	NO_3^-	O^{2-}	CO_3^{2-}	SO_4^{2-}	S^{2-}
Fe^{2+}	$Fe Cl_2$ (gray)	$Fe (NO_3)_2$ (gray)	FeO (white)	$Fe CO_3$ (gray)	$Fe SO_4$ (blue-gray)	Fe S (black)
Fe^{3+}	$FeCl_3$ (yellow)	$Fe (NO_3)_3$ (gray)	Fe_2O_3 (red-brown)	$Fe_2(CO_3)_3$ (red)	$Fe_2(SO4)_3$ (yellow)	Fe_2S_3 (green)

4.7.2. Cobalt

Cobalt, denoted by its atomic number 27, has a density of 8.9 g/cm^3 and is positioned as Earth's 32nd most abundant element. Cobaltite, characterized by its chemical formula CoAsS, stands out as the primary mineral of utmost significance in cobalt content. The United States possesses approximately 1 million tons of cobalt reserves, primarily concentrated in Minnesota (Colonna et al., 2020). Cobaltite is a mineral of relatively low abundance, primarily found in limited quantities within Norway, Sweden, and Ontario, Canada. Cobalt is present in meteorites that consist of iron and nickel, as well as in manganese nodules located on the ocean floor. Additionally, it is commonly associated with other transition metals, including nickel. Cobalt, a metal of moderate reactivity, undergoes dissolution in hydrochloric acid resulting in the formation of the cobalt ion with a charge of +2, as represented by the following chemical equation (Ferguson et al., 2009):

$Co\ (s) + 2HCl\ (aq) \rightarrow Co^{2+}\ (aq) + 2Cl^{-}\ (aq) + H_2\ (g)$.

The Co^{2+} ion exhibits a pink hue and is referred to as the cobaltous ion. Notably, this particular cobalt ion exclusively manifests in uncomplicated chemical compounds. Several examples of Co^{2+} compounds can be observed, such as cobaltous chloride ($CoCl_2$), cobaltous nitrate ($Co[NO_3]_2$), cobaltous sulfate ($CoSO_4$), cobaltous hydroxide ($Co[OH]_2$), cobaltous oxide (CoO), cobaltous carbonate ($CoCO_3$), cobaltous phosphate ($Co_3[PO_4]_3$), and cobaltous sulfide (CoS). The initial three substances exhibit solubility in diluted aqueous solutions, while the subsequent five demonstrate insolubility. Each of these compounds demonstrates vibrant hues, typically in red, blue, or black shades. The pigments used in the painting encompass cobalt blue, identified as cobalt aluminate ($Co[AlO_2]_2$), and cobalt green, known as cobalt zincate ($CoZnO_2$) (Cadar et al., 2020). The cobaltic ion, denoted as Co^{3+}, represents the ion with a charge of +3. The cobaltic ion exhibits limited stability in aqueous solution; however, it can be found in various compounds such as cobaltic fluoride (CoF_3) and cobaltic hydroxide ($Co[OH]_3$). The Co^{3+} ion exhibits the ability to generate a variety of vividly pigmented complex compounds. These compounds typically consist of six additional molecules or ions, known as ligands, which encircle the cobaltic ion in an octahedral configuration. Some examples of compounds are as follows: reddish-violet chloro pentamidine cobalt chloride, denoted as ($Co[NH_3]_6\ Cl)Cl_2$; orange hexamine cobalt chloride, represented by $Co(NH_3)_6Cl_3$; triamine cobaltinitrite, written as $Co(NH_3)_3\ (NO_2)_3$; aqua-pentamidine cobalti-hydroxide, indicated as ($Co[NH_3]_5\ [H_2O])(OH)_3$; aqua-pentamidine cobalt chloride, denoted as ($Co[NH_3\ ]_5\ H_2O)Cl_3$; and potassium cobaltinitrite, known as $K_3Co(NO_2)_6$, which is commonly used as a paint pigment referred to as cobalt yellow (Xiao et al., 2008). Cobalt, as a Group VIIIA member, exhibits oxidation states that surpass those of "+2" or "+3." Due to their inherent instability and susceptibility to reduction, compounds featuring cobalt in higher oxidation states exhibit pronounced oxidizing capabilities. Several examples of cobalt compounds can be cited, such as cobalt (IV) oxide (CoO_2), cobaltous cobaltite ($Co(CoO_3)$), and tricobalt tetroxide (Co_3O_4). In cobaltous cobaltite, the first cobalt atom is in the "+2" oxidation state, while the second is in the "+4" oxidation state. Tricobalt tetroxide, alternatively represented as $2CoO{\cdot}CoO_2$, consists of two cobalt atoms in the "+2" oxidation state and one cobalt atom in the "+4" oxidation state (Sanpo et al., 2014).

4.7.3. Nickel

Nickel, denoted by its atomic number 28, has a density of 8.9 g/cm^3 and occupies the 23rd position in terms of its relative abundance on Earth. Nickel-iron is a mineral characterized by the coexistence of nickel and iron in their neutral metallic states. The presence of nickel-iron is infrequent in terrestrial sources. However, it is relatively prevalent in meteorites. The prevailing belief is that Earth's core's composition predominantly consists of nickel and iron. Nickel, iron, and cobalt constitute one of the three metallic elements with magnetic properties (Zaanen et al., 1986). Various minerals are known to contain nickel. One such mineral is pentlandite, which exhibits a bronze color and is composed of a combination of iron and nickel sulfides. Another mineral, garnierite, displays a green hue and is composed of a mixture of nickel and magnesium silicates. Lastly, niccolite, characterized by its copper-red color, is composed of nickel arsenide (NiAs). The predominant origin of nickel in North America is derived from the mines located in Sudbury, Ontario, Canada. New Caledonia is globally recognized as a significant supplier of nickel. The chemistry of nickel exhibits significant similarities to that of cobalt. The nickelous ion (Ni^{2+}) exhibits a vibrant green hue when dissolved in an aqueous medium, and it represents the sole prevalent form of nickel ion (Zheng et al., 2017).

4.8. THE PLATINUM METALS

The platinum group metals consist of ruthenium (Ru), rhodium (Rh), palladium (Pd), osmium (Os), iridium (Ir), and platinum (Pt). These elements are considered to be among the scarcest in the Earth's crust. Each of these substances exhibits relatively elevated melting and boiling points as well as high densities. Palladium, with a density of 12.0 g/cm^3, ranks 71st in terms of abundance within the Earth's crust. Platinum ranks 72nd in terms of its abundance and possesses a density of 21.5 g/cm^3. Osmium, with its atomic number 76, ranks 74th in terms of abundance among the known elements. It holds the distinction of being the element with the highest density, measuring 22.6 grams per cubic centimeter. Iridium is ranked 77th in terms of its position within the periodic table and has a density of 22.4 g/cm^3 (Levison & Robinson, 1970).

Ruthenium is ranked as the 80th element in the periodic table and has a 12.4 g/cm^3 density. Rhodium, with a 12.4 g/cm^3 density, is positioned at the 81st place. Noble gases and radioactive elements necessitating artificial production are among the few elements that exhibit scarcity in Earth's crust. Platinum exhibits the highest degree of ductility among metals within its group, thereby possessing the ability to be elongated into wire-like structures. All of these metallic elements exhibit the shared characteristic of being comparatively chemically unreactive. Collectively, these entities are extensively employed in various industrial sectors as catalysts, with the catalytic converters commonly observed in automobiles serving as the most recognizable example (Ma et al., 2022). Due to their limited availability and significant demand within various industries, these metals are actively traded within commodity markets. Metals such as platinum, palladium, and rhodium frequently exhibit greater value and rarity in comparison to gold. During the latter part of 2009, the market value of gold reached approximately $1,100 per ounce, while platinum and rhodium were valued at $1,400 and $2,000 per ounce, respectively. The prices of these commodities exhibit volatility in response to changes in demand. Specifically, during the latter part of 2009, the spot price for palladium was recorded to be less than $400 per ounce. Platinum exhibits a high degree of scarcity. A standard ore sample may possess a platinum metal content of less than 0.1 ounce per ton of ore. The additional metallic elements within the same group are frequently found as impurities in platinum deposits and can exist in the form of platinum alloys (Ishiyama & Miyaura, 2000) (Figures 4.18–4.23).

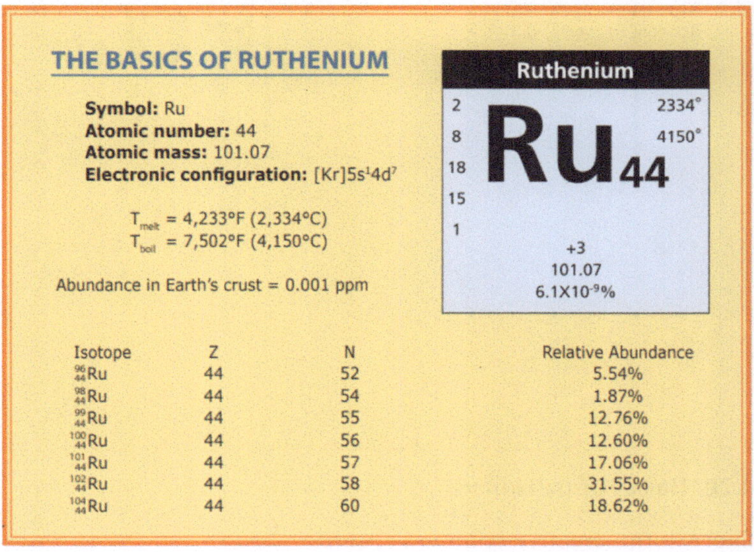

Figure 4.18. Basics of vanadium.

Monica Halka, creative commons license.

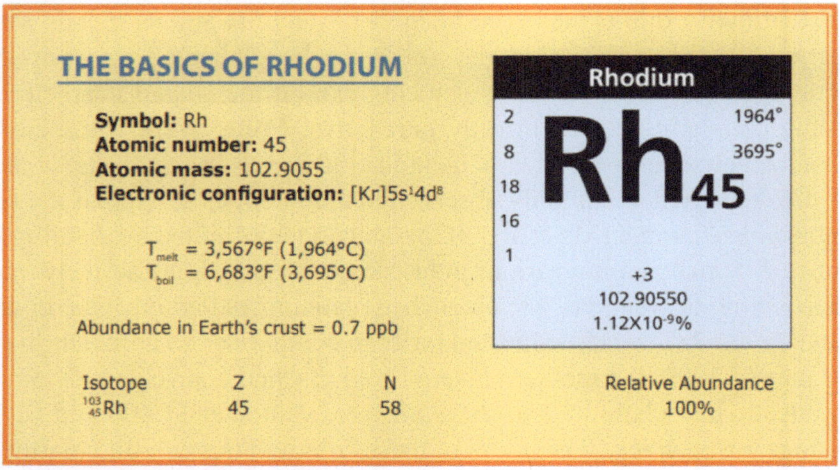

Figure 4.19. Basics of rhodium.

Source: Nordstorm, creative commons license.

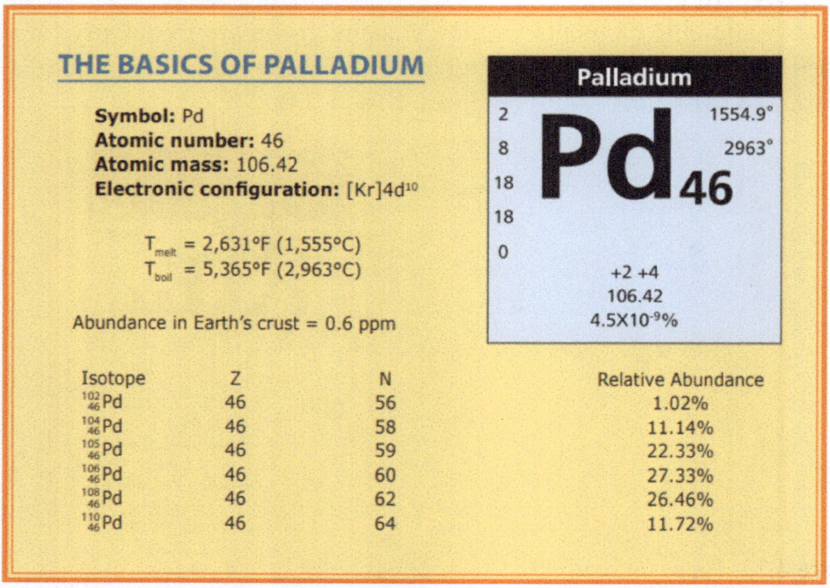

Figure 4.20. Basics of palladium.

Source: Nordstorm, creative commons license.

THE BASICS OF OSMIUM

Symbol: Os
Atomic number: 76
Atomic mass: 190.23
Electronic configuration:
[Xe]$6s^2 4f^{14} 5d^6$

T_{melt} = 5,491°F (3,033°C)
T_{boil} = 9,054°F (5,012°C)

Abundance in Earth's crust = 0.002 ppm

Osmium
Os 76
2, 8, 18, 32, 14, 2
3033°
5012°
+3 +4
190.23
2.20X10⁻⁹%

Isotope	Z	N	Relative Abundance
$^{184}_{76}Os$	76	108	0.02%
$^{186}_{76}Os$	76	110	1.59%
$^{187}_{76}Os$	76	111	1.96%
$^{188}_{76}Os$	76	112	13.24%
$^{189}_{76}Os$	76	113	16.15%
$^{190}_{76}Os$	76	114	26.26%
$^{192}_{76}Os$	76	116	40.78%

Figure 4.21. Basics of osmium.

Source: Harold Johnston, creative commons license.

THE BASICS OF IRIDIUM

Symbol: Ir
Atomic number: 77
Atomic mass: 192.217
Electronic configuration:
[Xe]$6s^2 4f^{14} 5d^7$

T_{melt} = 4,435°F (2,446°C)
T_{boil} = 8,002°F (4,428°C)

Abundance in Earth's crust = 1 ppb

Iridium
Ir 77
2, 8, 18, 32, 15, 2
2446°
4428°
+3 +4
192.217
2.16X10⁻⁹%

Isotope	Z	N	Relative Abundance
$^{191}_{77}Ir$	77	114	37.3%
$^{193}_{77}Ir$	77	116	62.7%

Figure 4.22. Basics of iridium.

Source: Harold Johnston, creative commons license.

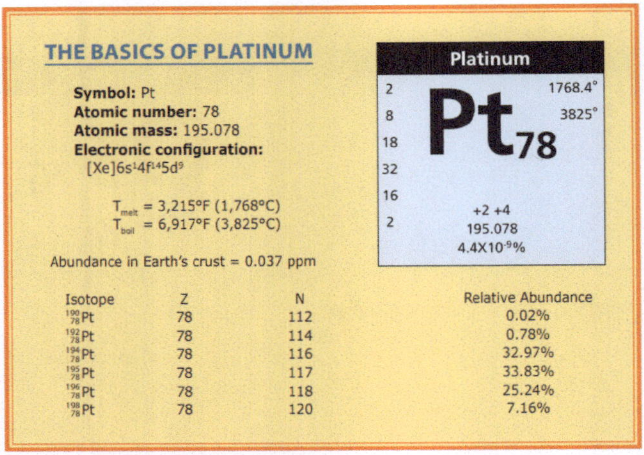

Figure 4.23. Basics of platinum.

Source: Nordstorm, creative commons license.

The platinum group metals exhibit a range of stable positive oxidation states, reaching a maximum of "+8." The states that exhibit the highest degree of stability are as follows: The oxidation states of the following transition metals are as follows: Ruthenium (Ru) can exhibit oxidation states of 2, 3, 4, 6, 7, and 8; Rhodium (Rh) can exhibit oxidation states of 3 and 4; Palladium (Pd) can exhibit oxidation states of 2 and 4; Osmium (Os) can exhibit oxidation states of 3, 4, 6, and 8; Iridium (Ir) can exhibit oxidation states of 3, 4, and 6; and Platinum (Pt) can exhibit oxidation states of 2 and 4. At elevated temperatures, it is observed that every metal tends to undergo a reaction with oxygen gas (O_2) (Cleare & Hoeschele, 1973). Furthermore, it is observed that every metal tends to undergo a chemical reaction with chlorine gas (Cl_2) upon exposure to heat. The platinum group metals exhibit a notable resistance to chemical reactions when exposed to hydrochloric acid (HCl) or nitric acid (HNO_3) individually. It is worth noting that among these metals, palladium is the sole member that undergoes dissolution upon exposure to nitric acid. They exhibit a higher propensity to dissolve, albeit at a relatively slow rate, in aqua regia. In this particular context, it is noteworthy that rhodium and iridium exhibit exceptional resistance to the corrosive effects of aqua regia. Platinum-group metals tend to avoid the formation of straightforward elemental compounds, instead favoring the creation of halide and oxide compounds (Rauchfuss, 1979). The catalytic properties of platinum metals are of greater significance than the chemical compounds they generate. Catalysts possess significant value within the industrial sector.

Catalysts possess the capacity to enhance the rate of chemical reactions without undergoing any significant consumption in the course of the reaction. There exist two broad categories of catalysts. Homogeneous catalysts are characterized by their presence in the identical phase as the chemical species undergoing catalytic reaction. Within the human body, enzymes, which are crucial proteins, are found in the cellular milieu. These enzymes facilitate the acceleration of chemical reactions occurring within cells, thereby playing a vital role in sustaining essential life processes. Conversely, heterogeneous catalysts are situated within a distinct phase from the phase occupied by the chemicals undergoing catalysis (Harding & Fielicke, 2014).

4.9. THE COPPER GROUP

The copper group is comprised of three metallic elements: copper (Cu, atomic number 29, known as cuprum in Latin), silver (Ag, atomic number 47, referred to as Argentum in Latin), and gold (Au, atomic number 79, known as aurum in Latin). All of these entities were discovered and given names during ancient times. Copper, silver, and gold are commonly encountered in their native metallic forms, although their occurrence in ores is more prevalent (Hathaway & Underhill, 1961). Copper, the primary metal in bronze and brass, played a pivotal role in the emergence of the Bronze Age as it facilitated the production of bronze tools and artifacts. Silver exhibits a higher degree of scarcity in comparison to copper while concurrently displaying diminished chemical reactivity. Silver is commonly utilized in the creation of jewelry, whereas its compounds are not commonly employed due to their relatively subdued coloration. Gold is widely recognized by the majority of individuals, albeit primarily through its reputation. Due to its comparatively low chemical reactivity, gold exhibits a higher propensity, relative to copper or silver, to occur in its elemental state, typically manifesting as nuggets or flakes (West et al., 1991) (Figures 4.24–4.26).

THE BASICS OF COPPER

Element: Copper
Symbol: Cu
Atomic number: 29
Atomic mass: 63.546
Electronic configuration: $[Ar]4s^1 3d^{10}$

T_{melt} = 1,984°F (1,085°C)
T_{boil} = 4,644°F (2,562°C)

Abundance in Earth's crust = 68 ppm

Copper

2
8
18
1

Cu 29

1084.62°
2562°

+1 +2
63.546
$1.70X10^{-6}\%$

Isotope	Z	N	Relative Abundance
$^{63}_{29}Cu$	29	34	69.17%
$^{65}_{29}Cu$	29	36	30.83%

Figure 4.24. Basics of copper.

Source: Monica Halka, creative commons license.

THE BASICS OF SILVER

Element: Silver
Symbol: Ag
Atomic number: 47
Atomic mass: 107.868
Electronic configuration: $[Kr]5s^1 4d^{10}$

T_{melt} = 1,763°F (962°C)
T_{boil} = 3,924°F (2,162°C)

Abundance in Earth's crust = 0.08 ppm

Silver

2
8
18
18
1

Ag 47

961.78°
2162°

+1
107.8682
$1.58X10^{-9}\%$

Isotope	Z	N	Relative Abundance
$^{107}_{47}Ag$	47	60	51.84%
$^{109}_{47}Ag$	47	62	48.16%

Note: $^{94}_{47}Ag$ has been synthesized at the National Superconducting Cyclotron Laboratory at Michigan State University. It is the heaviest odd N = Z isotope observed to date.

Figure 4.25. Basics of silver.

Source: Monica Halka, creative commons license.

THE BASICS OF GOLD

Element: Gold
Symbol: Au
Atomic number: 79
Atomic mass: 196.9665
Electronic configuration:
$[Xe]6s^1 4f^{14} 5d^{10}$

T_{melt} = 1,948°F (1,064°C)
T_{boil} = 5,173°F (2,856°C)

Abundance in Earth's crust = 0.003 ppm

Isotope	Z	N	Relative Abundance
$^{197}_{79}Au$	79	118	100%

Gold

	Au 79	
2		1064.18°
8		2856°
18		
32		
18	+1 +3	
1	196.96655	
	6.1×10^{-10}%	

Figure 4.26. Basics of gold.

Source: Nordstorm, creative commons license.

Copper is classified as the 29th element in the periodic table. The substance in question exhibits a reddish hue and possesses a density of 8.96 g/cm^3. Copper ranks as the 26th most prevalent element within the Earth's crust and is widely recognized as one of the most acquainted elements. Copper is occasionally encountered in its elemental form, known as native copper (Kuhl et al., 2012). The compounds and ions of copper exhibit a characteristic of vibrant coloration, typically manifesting in shades of green, blue, or purple. Minerals that possess copper, such as turquoise and malachite, are frequently employed in creating jewelry due to their distinctive hues. Copper possesses exceptional electrical conductivity, rendering it highly suitable for various applications, notably in electrical wiring. The prevalence of its thermal conductivity has led to its widespread use in cookware. Silver is classified as element 47 in the periodic table. The term "name" denotes the specific hue of an object or entity. Silver has a higher density than copper, with a value of 10.5 g/cm^3. The element in question is ranked as the 66th most prevalent element within the Earth's crust, making it a notable constituent (Natali & Giordani, 2012).

Additionally, it is widely recognized as one of the most familiar elements. Silver possesses exceptional electrical conductivity, surpassing that of copper. However, its cost renders its application in conventional electrical wiring unfeasible. Gold is classified as element 79 in the periodic table. With a density of 19.3 g/cm^3, this element exhibits a high level of compactness, surpassing that of lead. Gold is ranked 73rd in terms of its abundance on Earth. Gold exhibits a lackluster yellow hue and possesses a remarkable resistance to corrosion or tarnishing, surpassing that of copper or silver. Gold possesses the highest degree of malleability and ductility among all metallic elements. Hence, it is evident that diamonds possess significant applications in various industrial and technological domains, distinct from their conventional utilization in jewelry (Liu et al., 2020).

4.9.1. Copper

Copper is a vital element that is crucial in human physiology and plant development. In select species of mollusks and crustaceans, the presence of copper in hemocyanin is a replacement for the iron typically found in the hemoglobin of most animal species, enabling the transportation of oxygen within the bloodstream (Stadler et al., 2004). Iron is responsible for the reddish hue observed in the blood of the majority of animal species. Mollusks and crustaceans that utilize hemocyanin instead of hemoglobin exhibit a distinctive blue hue in their circulatory fluid, owing to the presence of copper. Copper is among the limited number of metallic elements that occur naturally in their elemental form (Piamonteze et al., 2009). An instance of the presence of concentrated deposits of copper metal can be observed in the Upper Peninsula of Michigan. The color of pure copper metal is red, a hue that is commonly recognized due to its association with copper coins, notably the Lincoln Cent in the United States. Primarily, copper is commonly encountered in mineral deposits (Gagne et al., 1981).

Copper in the "+1" oxidation state is present as the cuprous ion (Cu^+), which is characterized by its colorlessness. In the state known as "+2," the cupric ion (Cu^{2+}) is observed to have a sky-blue color. The cuprous ion exhibits instability and tends to undergo disproportionation, resulting in the formation of copper metal and the cupric ion, as demonstrated by the subsequent equation (Ye & Sanford, 2012):

$$2Cu^+ (aq) \rightarrow Cu (s) + Cu^{2+} (aq).$$

Copper(I) cation (Cu^+) does exhibit the ability to form a limited number of compounds, including cuprous oxide (Cu_2O). Cuprous oxide (Cu_2O) exhibits a characteristic red hue, occurs naturally, and is employed as a pigment for porcelain coloring purposes (Yasumatsu et al., 2007).

4.9.2. Silver

Before the onset of the 20th century, silver was primarily utilized for coin production. Silver has played a significant role in various industries for over a century, with particular emphasis on its application in the realm of photographic film. Experiments involving the utilization of silver compounds to create images were initiated as early as 1727. The phenomenon of darkening is observed in silver chloride (AgCl) when it is exposed to light. After the film undergoes processing, a photographic negative is produced, which can be used to create prints. Chemists employ the limited precipitation of certain metal ions with halide ions (Cl^-, Br^-, I^-), as a basis for analytical tests (Stork et al., 2007).

Consequently, when dilute HCl is applied to solutions containing metal ions, only the precipitation of AgCl, $PbCl_2$, and Hg_2Cl_2 will occur. All three compounds exhibit a white coloration. AgCl can be effectively isolated from the remaining solids by introducing a diluted aqueous solution of NH_3. This is due to the fact that AgCl exhibits solubility in the presence of ammonia, as evidenced by the following chemical equation (Jiang et al., 2018):

$$AgCl\ (s) + 2NH_3\ (aq) \rightarrow Ag(NH_3)_2^+(aq) + Cl^-\ (aq).$$

The reaction between $PbCl_2$ and aqueous ammonia yields $Pb(OH)_2$, which shares the same white color as the reactants. Consequently, the detection of the reaction may not be readily apparent. The compound $Hg_2\ Cl_2$ reacts with an aqueous solution of ammonia, resulting in a combination of metallic mercury and mercury ammono chloride ($HgNH_2Cl$). The separation of the silver complex ion from the mixture can be achieved through the process of filtration. Silver exhibits analogous reactions with bromide (Br^-) and iodide (I^-) ions (Anders et al., 2006). The solubility of each subsequent silver halide salt exhibits a downward trend as the atomic weight of the halide increases. Therefore, it can be observed that AgCl exhibits the highest solubility among the three compounds, followed by AgBr with a relatively lower solubility, and AgI displays the least solubility among all three substances. The colors of AgBr and AgI also exhibit variation, with both compounds appearing yellow in hue (Gulbiński et al., 2003).

Furthermore, it should be noted that AgBr and AgI exhibit lower reactivity when exposed to aqueous ammonia. While AgCl can dissolve in a dilute ammonia solution, AgBr necessitates a concentrated ammonia solution for dissolution. Conversely, AgI exhibits no solubility in ammonia. The silver ion (Ag+) tends to react with various negative ions, resulting in the formation of insoluble compounds. The following examples are listed in approximate order of solubility, starting from the most soluble: Ag_2CO_3, silver oxalate ($Ag_2C_2O_4$), Ag_2CrO_4, Ag_2O, the silver halides (AgCl, AgBr, and AgI), AgCN, and silver sulfide (Ag_2S). Most of these compounds exhibit a white coloration, while the most intense hues are observed in the reddish-brown silver chromate and black silver sulfide compounds. The formation of silver sulfide, commonly observed as tarnish on silverware, occurs as a consequence of the chemical reaction between silver metal and trace quantities of hydrogen sulfide gas (H_2S) present in the surrounding atmosphere (Janssens et al., 2005):

$Ag\ (s) + H_2S\ (g) \rightarrow Ag_2S\ (s) + H_2(g)$ (Garza-Cervantes et al., 2017).

4.9.3. Gold

Gold is primarily encountered naturally as the native metal, frequently occurring alongside native silver and platinum group metals. There are two notable instances of gold-containing compounds, namely tellurides such as $AuTe_2$ and a composite compound of gold and silver, denoted as $AuAgTe_4$. Seawater is found to possess trace amounts of gold, typically ranging from 0.1 to 0.2 milligrams per metric ton of seawater, which is believed to be a result of gold particles being carried by streams and subsequently deposited into the ocean. Gold compounds exhibit two oxidation states, namely "+1" and "+3," in which gold is present. The ion with a charge of "+1," denoted as Au+ is referred to as the aurous ion (Akita et al., 2013).

Similarly, the ion with a charge of "+3," denoted as Au^{3+} is known as the auric ion. The process of extracting gold from ore through the amalgamation of gold with liquid mercury was initiated around 500 B.C.E. Once the gold had been extracted from the ore, eliminating the mercury involved subjecting it to heat. Around 1890, the utilization of cyanide treatment in the extraction of gold significantly enhanced the efficacy of recovering gold from minor deposits. During this particular process, the chemical compound cyanide reacts with gold through the following chemical reaction (Packheiser et al., 2008):

$$4Au\ (s) + 8CN^-\ (aq) + O_2(g) + 2H_2O\ (l) \rightarrow$$

$$4Au\left(CN\right)_2^-\ (aq) + 4OH^-\ (aq).$$

The reaction rate is significantly sluggish, necessitating multiple days for completion. The dissolution of gold can occur when a new reagent, aqua regia, is formed by combining concentrated HCl and HNO_3, as demonstrated in the subsequent reaction (Kim et al., 2013):

$$Au\ (s) + 5H^+\ (aq) + 4Cl^-\ (aq) + NO_3^-\ (aq) \rightarrow$$
$$HAuCl_4\ (aq) + NO\ (g) + 2H_2O\ (l)$$

The nomenclature "aqua regia" aptly signifies its purpose. Given that gold was predominantly owned by the monarch, it was logical that the utilization of the "king's water" was necessary for its dissolution (Velicky et al., 2018).

The auric ion (Au^{3+}) exhibits greater stability in comparison to the aurous ion (Au^+). When subjected to heat in the absence of moisture, auric chloride undergoes decomposition and transforms into aurous chloride, as represented by the following chemical equation (Zhao et al., 2015):

$AuCl_3(g) \rightarrow AuCl\ (s) + Cl_2(g)$ (Campos, 2017).

4.10. THE ZINC GROUP

The zinc group is comprised of three elements: zinc (Zn, atomic number 30), cadmium (Cd, atomic number 48), and mercury (Hg, atomic number 80, known as hydragyrum in Greek). Zinc, cadmium, and mercury are the sole naturally transpiring elements within Group IIB, and they are situated in the terminal column of the d-block of elements. These three elements are classified differently by various authors. Authors who strictly adhere to the definition of a transition metal, which characterizes an element as having a d subshell that is only partially occupied, will exclude zinc, cadmium, and mercury due to the fact that their d subshells are fully filled. In addition to the transition metals, zinc, cadmium, and mercury are also classified as transition metals due to their properties, which exhibit greater similarity to the transition metals than the post-transition metals. The latter's properties are primarily influenced by the presence of partially filled p subshells. The decision was made to incorporate zinc, cadmium, and mercury into the transition metals category within this comprehensive multivolume collection dedicated to the periodic table (Thiel & Voityuk, 1996).

Zinc, denoted by its atomic number 30, has a 7.1 g/cm^3 density. It ranks 24th in terms of abundance within the Earth's crust. Numerous mineral deposits encompass zinc as a constituent element, such as ZnS found in the mineral sphalerite, commonly called zinc blende. Another example is ZnO present in the mineral zincite. Franklinite, a composite of zinc, manganese, and iron oxides, also contains zinc. Zinc silicate (Zn_2SiO_4) is found in the mineral willemite, and the mineral calamine ($Zn_2[OH]_2SiO_3$) also contains zinc. The recovery of metallic zinc can be achieved by reducing carbon as a reducing agent. Typically, zinc is generated via electrolytic methods. A newly exposed zinc surface exhibits a glossy metallic sheen.

Nevertheless, it is prone to tarnishing. Typically, zinc exhibits hardness and brittleness; however, when exposed to elevated temperatures surpassing 100°C, it transforms, rendering it malleable. In the context of zinc recovery from zinc sulfide, SO_2 production occurs as a secondary outcome (Molina et al., 1998) (Figures 4.27–4.29).

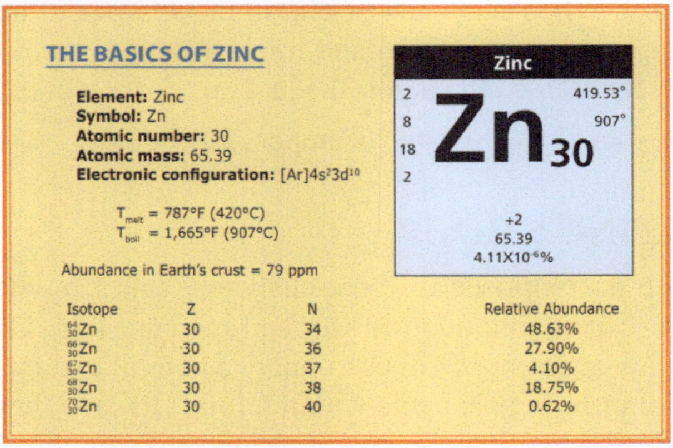

Figure 4.27. Basics of zinc.

Source: Harold Johnston, creative commons license.

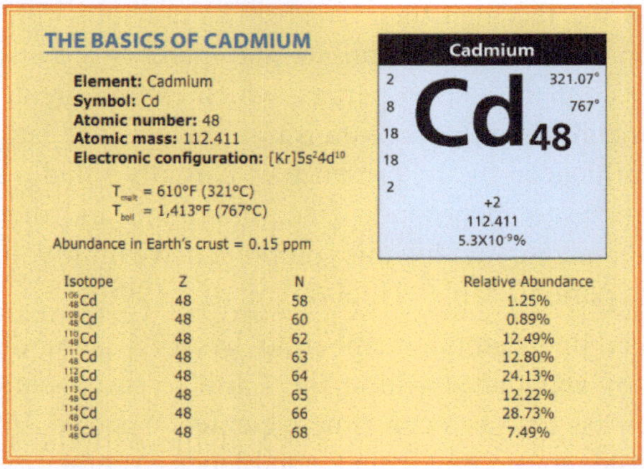

Figure 4.28. Basics of cadmium.

Source: A. Starker Leopold, creative commons license.

THE BASICS OF MERCURY

Element: Mercury
Symbol: Hg
Atomic number: 80
Atomic mass: 200.59
Electronic configuration: $[Xe]6s^2 4f^{14} 5d^{10}$

T_{melt} = -38°F (-39°C)
T_{boil} = 674°F (357°C)

Abundance in Earth's crust = 0.067 ppm

Isotope	Z	N	Relative Abundance
$^{196}_{80}Hg$	80	116	0.15%
$^{198}_{80}Hg$	80	118	9.97%
$^{199}_{80}Hg$	80	119	16.87%
$^{200}_{80}Hg$	80	120	23.10%
$^{201}_{80}Hg$	80	121	13.18%
$^{202}_{80}Hg$	80	122	29.86%
$^{204}_{80}Hg$	80	124	6.87%

Figure 4.29. Basics of mercury.

Source: Neil Bartlett, creative commons license.

Cadmium is classified as element 48 in the periodic table. The substance possesses a density of 8.65 grams per cubic centimeter and is ranked as the 65th most abundant element in the Earth's crust. Cadmium metal, in its pure form exhibits a white coloration with a subtle bluish undertone. Cadmium is characterized by its relatively low hardness, rendering it moderately ductile and malleable under ambient conditions. Despite its various technological applications, cadmium's toxicity raises significant health concerns. Cadmium exhibits carcinogenic and teratogenic properties, thereby inducing the development of cancerous cells and congenital abnormalities (DeLoach et al., 1996). These concerns have the potential to present health risks to individuals employed in the recycling industry who handle materials that may contain cadmium. The majority of cadmium is obtained through the extraction process from zinc ores, wherein the cadmium concentration typically accounts for approximately 0.5 percent of the zinc content. The initial step in the reduction process involves the production of metallic cadmium, which is subsequently succeeded by the formation of metallic zinc. The sole significant source of cadmium ore is greenockite, a yellow-colored mineral consisting of CdS.

Mercury is classified as element 80 in the periodic table. At room temperature, the liquid exhibits a relatively high density of 13.5 g/cm^3. The primary source of mercury extraction is the mineral known as cinnabar (HgS), which exhibits a reddish coloration. Mercury is naturally occurring in its elemental form and in association with gold and silver. The element above was recognized by individuals in ancient times. The observers were captivated by the liquid nature of the substance under typical circumstances, despite its resemblance to other metals of the era, characterized by a silvery hue. During the medieval era, individuals practicing alchemy and medicine called mercury "quicksilver" due to its distinctive liquid characteristics (Ekkert et al., 2016).

4.10.1. Zinc

Zinc is crucial in facilitating optimal growth and development in both animal and plant organisms. The absence of zinc in the human body results in impaired growth and hindered sexual maturation. Zinc has been found to mitigate prostate gland enlargement in elderly males. Zinc, being highly susceptible to oxidation, exhibits excellent reducing properties. Hence, it is widely employed as a prevalent metal in chemistry laboratories to illustrate the interaction between a reactive metal and hydrochloric acid. The chemical equation provided illustrates the reaction in which zinc acts as a reducing agent, leading to the reduction of hydrochloric acid (HCl) and the subsequent production of hydrogen gas (H_2) (Pan et al., 2019):

$$Zn\ (s) + 2\ HCl\ (aq) \rightarrow ZnCl_2(aq) + H_2(g).$$

The hydrogen gas vigorously bubbles (effervesces) and can be sparked with the heat from a blazing wood splint, as demonstrated by the following reaction:

$$2H_2\ (g) + O_2\ (g) \rightarrow 2H_2O\ (g) + \text{heat}.$$

Comparable reactions are observed when metallic zinc interacts with other commonly encountered acids, including sulfuric acid (H_2SO_4), dilute nitric acid (HNO_3), and acetic acid ($HC_2H_3O_2$). Furthermore, these acids lead to the production of hydrogen gas. In the presence of concentrated nitric acid, zinc exhibits a reduction reaction where it reduces the nitrate ion to form the ammonium ion:

$$4Zn\ (s) + 10H^+\ (aq) + NO_3^-\ (aq) \rightarrow$$

$$4Zn^{2+}\ (aq) + NH_4^+\ (aq) + 3H_2O\ (l)$$

In this particular instance, no gaseous byproducts are produced.

Zinc exclusively forms the Zn^{2+} ion, indicating that zinc is solely found in the "+2" oxidation state within compounds or complex ions. A white coloration characterizes the majority of zinc compounds that exhibit neutrality. Zinc compounds encompass a variety of chemical substances, such as $Zn(NO_3)_2$, $ZnCl_2$, $ZnBr_2$, ZnI_2, $ZnCO_3$, $ZnSO_4$, Zn_3P_2, $Zn_3(PO_4)_2$, ZnO, $Zn(OH)_2$, ZnS, and ZnC_2O_4. The solubility of nitrate and halide compounds is higher in dilute aqueous solutions, whereas the remaining compounds generally exhibit insolubility.

Zinc hydroxide ($Zn(OH)_2$) exhibits amphoteric behavior, as it is capable of dissolving in both acidic and basic solutions, as evidenced by the ensuing chemical reactions (Douka et al., 2021):

$$Zn(OH)_2\ (s) + 2HCl\ (aq) \rightarrow ZnCl_2\ (aq) + 2H_2O\ (l);$$

$$Zn(OH)_2\ (s) + 2NaOH\ (aq) \rightarrow Zn\left(OH\right)_4^{2-}\ (aq) + 2Na^{+}(aq).$$

The propensity of zinc to undergo oxidation makes it suitable for two significant applications. The initial category pertains to conventional batteries, such as those commonly employed in flashlight devices. Batteries function through the process of oxidation-reduction reactions. Zinc functions as the anode or cathode of negative polarity and serves as the location for the oxidation process. Various chemical species can function as cathodes, the positively charged electrodes responsible for facilitating reduction reactions. The movement of electrons occurs from the anode to the cathode, generating electrical energy that can be harnessed for purposes such as illumination or the operation of small electrical motors.

The second function of zinc pertains to its utilization as a sacrificial anode. Zinc exhibits greater reactivity compared to iron. Hence, it can be observed that zinc exhibits a higher susceptibility to corrosion compared to iron, thereby rendering it suitable for employment in safeguarding iron materials. Galvanized nails serve as a noteworthy illustration. Nails are primarily composed of steel, predominantly iron, which exhibits a high susceptibility to corrosion upon exposure to moisture. Applying a zinc coating, known as galvanization, on nails protects the iron by causing the zinc to undergo oxidation, acting as an anode. This process significantly prolongs the lifespan of the nail, surpassing what would be achievable without the presence of the protective zinc layer (Bresser et al., 2013).

Consequently, zinc has been utilized to preserve the iron. In the absence of a zinc coating, the iron would exhibit an anodic behavior. However, due to zinc's role as an anode, iron effectively assumes the role of a cathode, where reduction occurs. The concept of cathodic protection is employed to denote the process of safeguarding iron by designating it as the cathode. As a result, due to the inherent inability of a neutral metal to undergo reduction, the iron remains unaffected until the complete dissolution of the protective zinc coating. Additional instances of cathodic protection can be observed in scenarios where steel is submerged in soil or operates in aquatic environments. In the past, gasoline stations have traditionally employed underground steel tanks for the storage of gasoline. Due to the moisture absorption by the soil, the tanks were susceptible to corrosion. The utilization of a zinc block affixed to a tank can function as a sacrificial anode, thereby effectively prolonging the operational lifespan of said tank. In contemporary times, there has been a notable shift towards the substitution of steel gasoline tanks with robust plastic tanks. This transition has been primarily motivated by apprehensions surrounding the detrimental environmental impact resulting from the seepage of gasoline into the soil. In order to safeguard the hulls of boats and ships that navigate in salt water, a highly corrosive setting, the utilization of zinc (or occasionally magnesium) blocks as sacrificial anodes is employed, consider the substantial financial implications associated with the repair or replacement of the hull of a sizeable maritime vessel such as an oceangoing cruise ship, freighter, or oil tanker. Replacing a block of zinc or magnesium is a more cost-effective and convenient solution than replacing the entire hull (Nasch et al., 1997).

4.10.2. Cadmium

Like zinc, cadmium is found in compounds exclusively as a divalent cation with a charge of +2, denoted as Cd^{2+}. Soluble compounds of cadmium encompass $Cd(NO_3)_2$, $CdCl_2$, and $CdBr_2$. The list of insoluble compounds comprises $CdCO_3$, CdO, $Cd(OH)_2$, and CdS. Most cadmium compounds exhibit a white coloration, although there are a few exceptions. Specifically, $Cd(NO_3)_2$ and CdO display a brown hue, while $CdBr_2$ and CdS are yellow (Inoue & Negishi, 1986).

In contrast to $Zn(OH)_2$, $Cd(OH)_2$ does not exhibit amphoteric behavior. Consequently, $Cd(OH)_2$ exhibits solubility in acidic solutions while remaining insoluble in basic solutions (Mi et al., 2007).

4.10.3. Mercury

The acquisition of liquid mercury from cinnabar is a relatively straightforward process. The sole requirement entails subjecting the ore to heat through the process of roasting while ensuring that there is plenty of atmospheric oxygen. The subsequent equation depicts the reaction above.

$HgS (s) + O_2 (g) \rightarrow Hg (l) + SO_2(g)$.

In contrast to its counterparts, zinc, and cadmium, which are exclusively found in compounds and complex ions in the "2+" oxidation state, mercury is capable of existing in both the "1+" and "2+" oxidation states. The ion in the state of "+1" is not entirely unprecedented in chemistry; however, its occurrence is infrequent enough to warrant particular emphasis in introductory chemistry courses. The ionic species denoted as mercury "plus one" is referred to as the mercurous ion and is represented by the chemical formula Hg_2^{2+}. The peculiarity lies in forming a singular ion wherein two metal atoms are intricately bonded. The ion exhibits a positive charge of +2, whereas each mercury atom is solely in a +1 oxidation state. Therefore, when combined with a negatively charged ion such as Cl^-, the chemical formula for mercurous chloride is Hg_2Cl_2. The mercuric ion refers to the mercury ion in the oxidation state of "+2." The two ions display notable disparities in their chemical properties (Goodfellow, 1987). Metallic mercury exhibits a notably low reactivity as a reducing agent, indicating its inherent oxidation resistance. The mercuric ion exhibits notable oxidizing properties due to its relatively facile reduction to either the mercurous ion or liquid mercury. The mercurous ion can undergo a chemical reaction known as disproportionation. The following equation can represent such a reaction:

$$Hg_2^{2+} \text{ (aq)} \rightarrow Hg \text{ (l)} + Hg^{2+}$$

The chemical properties of the Hg_2^{2+} ion closely resemble those of the Ag+ ion. AgCl, Hg_2Cl_2, and $PbCl_2$ exhibit a white coloration and represent the exclusive set of metal chlorides that display insolubility in dilute aqueous solutions. The compound $PbCl_2$ exhibits solubility upon exposure to heat. In contrast, it should be noted that mercuric chloride ($HgCl_2$) exhibits solubility in aqueous solutions. One frequently employed qualitative method to detect the existence of Hg^{2+} in a solution involves the introduction of stannous chloride ($SnCl_2$) into the solution. The addition of a limited quantity of $SnCl_2$ results in the reduction of Hg^{2+} to Hg_2^{2+}, as demonstrated by the subsequent chemical reaction:

$$2Hg^{2+} \text{ (aq)} + SnCl_2 \text{ (aq)} \rightarrow Hg_2Cl_2 \text{ (s)} + Sn^{4+} \text{ (aq)}.$$

$$Hg_2Cl_2 \text{ (s)} + SnCl_2\text{(aq)} \rightarrow 2Hg \text{ (l)} + SnCl_4 \text{ (aq)}.$$

A white solid, specifically $Hg_2 Cl_2$, forms as a residue during the initial reaction. In the subsequent reaction, the element mercury exhibits a silvery hue. The typical outcome entails the formation of a gray amalgam, which can be inferred as a consequence of the existence of Hg^{2+} within the initial solution (Botana et al., 2015).

4.11. SUMMARY

Transition metals are a group of chemical elements characterized by valence electrons in two shells as opposed to just one. The term "transition" lacks specific chemical significance, but it serves as a useful designation to differentiate elements with similar atomic structures and resultant properties. The elements in question are situated within the intermediate regions of the extended periods of the periodic table, positioned between the leftmost groups and the rightmost groups. More precisely, they constitute elements belonging to Groups 3 (IIIb) to 12 (IIb).

The 24 elements under consideration exhibit notable resemblances, primarily in their classification as metals. Furthermore, a majority of these elements possess characteristics such as hardness, strength, luster, high melting, and boiling points, as well as proficiency in conducting heat and electricity. The properties of these elements exhibit a significant degree of variation, thus rendering the statements comparative in nature with respect to the overall properties of other elements.

Several elements, such as titanium, iron, nickel, and copper, hold significant technological importance. These elements find applications in both structural and electrical technology. Additionally, transition metals can create numerous valuable alloys, both among themselves and in combination with other metallic elements. Furthermore, the majority of these elements exhibit solubility in mineral acids, except a select few, namely platinum, silver, and gold, which are classified as "noble" due to their resistance to simple (nonoxidizing) acids.

REFERENCES

1. Abdel-Baki, M., & El-Diasty, F., (2006). Optical properties of oxide glasses containing transition metals: Case of titanium-and chromium-containing glasses. *Current Opinion in Solid State and Materials Science, 10*(5, 6), 217–229.
2. Akita, T., Kohyama, M., & Haruta, M., (2013). Electron microscopy study of gold nanoparticles deposited on transition metal oxides. *Accounts of Chemical Research, 46*(8), 1773–1782.
3. Anders, A., Byon, E., Kim, D. H., Fukuda, K., & Lim, S. H., (2006). Smoothing of ultrathin silver films by transition metal seeding. *Solid State Communications, 140*(5), 225–229.
4. Anderson, R. A., Cheng, N., Bryden, N. A., Polansky, M. M., Cheng, N., Chi, J., & Feng, J., (1997). Elevated intakes of supplemental chromium improve glucose and insulin variables in individuals with type 2 diabetes. *Diabetes, 46*(11), 1786–1791.
5. Aubert, C., Buisine, O., & Malacria, M., (2002). The behavior of 1, n-enynes in the presence of transition metals. *Chemical Reviews, 102*(3), 813–834.
6. Baan, R. A., (2007). Carcinogenic hazards from inhaled carbon black, titanium dioxide, and talc not containing asbestos or asbestiform fibers: Recent evaluations by an IARC monographs working group. *Inhalation Toxicology, 19*(1), 213–228.
7. Bailey, C. H., (2014). Improved meta-analytic methods show no effect of chromium supplements on fasting glucose. *Biological Trace Element Research, 157*, 1–8.
8. Barsukova, M. O., Sapchenko, S. A., Dybtsev, D. N., & Fedin, V. P., (2018). Scandium-organic frameworks: Progress and prospects. *Russian Chemical Reviews, 87*(11), 1139.
9. Blasse, G., & Dirksen, G. J., (1988). Scandium borate ($ScBO_3$) as a host lattice for luminescent lanthanide and transition metal ions. *Inorganica Chimica Acta, 145*(2), 303–308.
10. Botana, J., Wang, X., Hou, C., Yan, D., Lin, H., Ma, Y., & Miao, M. S., (2015). Mercury under pressure acts as a transition metal: Calculated from first principles. *Angewandte Chemie International Edition, 54*(32), 9280–9283.
11. Bradley, D. C., & Chisholm, M. H., (1976). Transition-metal

dialkylamides and disilylamides. *Accounts of Chemical Research, 9*(7), 273–280.

12. Brandt, G. B., & Rayne, J. A., (1963). Low-field de Haas-Van Alphen studies of chromium group transition elements. *Physical Review, 132*(5), 1945.
13. Brec, R., (1986). Review on structural and chemical properties of transition metal phosphorous trisulfides MPS3. *Solid State Ionics, 22*(1), 3–30.
14. Bresser, D., Mueller, F., Fiedler, M., Krueger, S., Kloepsch, R., Baither, D., & Passerini, S., (2013). Transition-metal-doped zinc oxide nanoparticles as a new lithium-ion anode material. *Chemistry of Materials, 25*(24), 4977–4985.
15. Burstall, F. H., & Nyholm, R. S., (1952). 681. Studies in co-ordination chemistry. Part XIII. Magnetic moments and bond types of transition-metal complexes. *Journal of the Chemical Society (Resumed)*, 3570–3579.
16. Butler, I. S., (1977). Transition-metal thiocarbonyls and selenocarbonyls. *Accounts of Chemical Research, 10*(10), 359–365.
17. Cadar, O., Dippong, T., Senila, M., & Levei, E. A., (2020). Progress, challenges, and opportunities in divalent transition metal-doped cobalt ferrites nanoparticles applications. *Advanced Functional Materials, 7*(4), 1–17.
18. Campen, M. J., Nolan, J. P., Schladweiler, M. C., Kodavanti, U. P., Evansky, P. A., Costa, D. L., & Watkinson, W. P., (2001). Cardiovascular and thermoregulatory effects of inhaled PM-associated transition metals: A potential interaction between nickel and vanadium sulfate. *Toxicological Sciences, 64*(2), 243–252.
19. Campos, J., (2017). Dihydrogen and acetylene activation by a gold (I)/platinum (0) transition metal only frustrated Lewis pair. *Journal of the American Chemical Society, 139*(8), 2944–2947.
20. Carrick, W. L., Kluiber, R. W., Bonner, E. F., Wartman, L. H., Rugg, F. M., & Smith, J. J., (1960). Transition metal catalysts. I. Ethylene polymerization with a soluble catalyst formed from an aluminum halide, tetraphenyltin and a vanadium halide. *Journal of the American Chemical Society, 82*(15), 3883–3887.
21. Chábera, P., Lindh, L., Rosemann, N. W., Prakash, O., Uhlig, J., Yartsev, A., & Persson, P., (2021). Photofunctionality of iron (III)

N-heterocyclic carbenes and related d5 transition metal complexes. *Coordination Chemistry Reviews, 426*, 213517.

22. Chen, Y., Wang, X., Lu, X., Yang, L., Yang, H., Yuan, W., & Chen, D., (2013). Comparison of titanium and polyetheretherketone (PEEK) cages in the surgical treatment of multilevel cervical spondylotic myelopathy: A prospective, randomized, control study with over 7-year follow-up. *European Spine Journal, 22*, 1539–1546.
23. Clancy, S. P., Clarkson, P. M., DeCheke, M. E., Nosaka, K., Freedson, P. S., Cunningham, J. J., & Valentine, B., (1994). Effects of chromium picolinate supplementation on body composition, strength, and urinary chromium loss in football players. *International Journal of Sport Nutrition and Exercise Metabolism, 4*(2), 142–153.
24. Cleare, M. J., & Hoeschele, J. D., (1973). Studies on the antitumor activity of group VIII transition metal complexes. Part I. Platinum (II) complexes. *Bioinorganic Chemistry, 2*(3), 187–210.
25. Clogston, A. M., Matthias, B. T., Peter, M., Williams, H. J., Corenzwit, E., & Sherwood, R. C., (1962). Local magnetic moment associated with an iron atom dissolved in various transition metal alloys. *Physical Review, 125*(2), 541.
26. Colonna, P., Bezzenine, S., Gil, R., & Hannedouche, J., (2020). Alkene hydroamination via earth-abundant transition metal (iron, cobalt, copper, and zinc) catalysis: A mechanistic overview. *Advanced Synthesis & Catalysis, 362*(8), 1550–1563.
27. Corbett, J. D., Daake, R. L., Poeppelmeier, K. R., & Guthrie, D. H., (1978). Metal-metal bonded clusters in transition metal groups 3 and 4. Synthesis and structure of three M6X12-type clusters for scandium and zirconium. *Journal of the American Chemical Society, 100*(2), 652–654.
28. Cotton, F. A., (1955). Alkyls and aryls of transition metals. *Chemical Reviews, 55*(3), 551–594.
29. Daniel, S., Praveena, M. G., & Mohammed, E. M., (2021). Exploration of highly photoluminescent first-row transition metals (manganese, iron, cobalt, nickel, copper, and zinc) co-doped nano carbon dots as energy storage materials. *Materials Science and Engineering: B, 269*, 115145.
30. Dehnicke, K., Krieger, M., & Massa, W., (1999). Phosphoraneiminato complexes of transition metals. *Coordination Chemistry Reviews, 182*(1), 19–65.

31. DeLoach, L. D., Page, R. H., Wilke, G. D., Payne, S. A., & Krupke, W. F., (1996). Transition metal-doped zinc chalcogenides: Spectroscopy and laser demonstration of a new class of gain media. *IEEE Journal of Quantum Electronics, 32*(6), 885–895.

32. Dolg, M., Wedig, U., Stoll, H., & Preuss, H., (1987). A binitio pseudopotential study of the first row transition metal monoxides and iron monohydride. *The Journal of Chemical Physics, 86*(4), 2123–2131.

33. Douka, A. I., Yang, H., Huang, L., Zaman, S., Yue, T., Guo, W., & Xia, B. Y., (2021). Transition metal/carbon hybrids for oxygen electrocatalysis in rechargeable zinc-air batteries. *EcoMat, 3*(1), e12067.

34. Edwards, D. A., & Hayward, R. N., (1968). Transition metal acetates. *Canadian Journal of Chemistry, 46*(22), 3443–3446.

35. Eikey, R. A., & Abu-Omar, M. M., (2003). Nitrido and imido transition metal complexes of groups 6–8. *Coordination Chemistry Reviews, 243*(1, 2), 83–124.

36. Ekkert, O., White, A. J., & Crimmin, M. R., (2016). Trajectory of approach of a zinc–hydrogen bond to transition metals. *Angewandte Chemie International Edition, 55*(52), 16031–16034.

37. Esayed, A. Y., & Northwood, D. O., (1992). Metal hydrides: A review of group V transition metals—Niobium, vanadium, and tantalum. *International Journal of Hydrogen Energy, 17*(1), 41–52.

38. Ferguson, P. P., Martine, M. L., George, A. E., & Dahn, J. R., (2009). Studies of tin–transition metal–carbon and tin–cobalt–transition metal–carbon negative electrode materials prepared by mechanical attrition. *Journal of Power Sources, 194*(2), 794–800.

39. Florio, J. V., Rundle, R. E., & Snow, A. I., (1952). Compounds of thorium with transition metals. I. The thorium–manganese system. *Acta Crystallographica, 5*(4), 449–457.

40. Gagne, R. R., Spiro, C. L., Smith, T. J., Hamann, C. A., Thies, W. R., & Shiemke, A. D., (1981). The synthesis, redox properties, and ligand binding of heterobinuclear transition-metal macrocyclic ligand complexes. Measurement of an apparent delocalization energy in a mixed-valent copper (I) copper (II) complex. *Journal of the American Chemical Society, 103*(14), 4073–4081.

41. Garza-Cervantes, J. A., Chávez-Reyes, A., Castillo, E. C., García-Rivas, G., Ortega-Rivera, O. A., Salinas, E., & Morones-Ramírez, J.

R., (2017). Synergistic antimicrobial effects of silver/transition-metal combinatorial treatments. *Scientific Reports, 7*(1), 903.

42. Geoffrey N áCloke, F., (1993). Zero oxidation state compounds of scandium, yttrium, and the lanthanides. *Chemical Society Reviews, 22*(1), 17–24.
43. Goodfellow, R. J., (1987). Post-transition metals, copper to mercury. In: *Multinuclear NMR* (Vol. 1, pp. 563–589). Boston, MA: Springer US.
44. Gulbiński, W., Suszko, T., Sienicki, W., & Warcholiński, B., (2003). Tribological properties of silver-and copper-doped transition metal oxide coatings. *Wear, 254*(1, 2), 129–135.
45. Han, C. H., Johansson, C. B., Wennerberg, A., & Albrektsson, T., (1998). Quantitative and qualitative investigations of surface enlarged titanium and titanium alloy implants. *Clinical Oral Implants Research, 9*(1), 1–10.
46. Harding, D. J., & Fielicke, A., (2014). Platinum group metal clusters: From gas-phase structures and reactivities towards model catalysts. *Chemistry–A European Journal, 20*(12), 3258–3267.
47. Hathaway, B. J., & Underhill, A. E., (1961). 592. The infrared spectra of some transition-metal perchlorates. *Journal of the Chemical Society (Resumed)*, *10*, 3091–3096.
48. Hillman, T. A., & Shelton, C., (2003). Ossicular chain reconstruction: Titanium versus plastipore. *The Laryngoscope, 113*(10), 1731–1735.
49. Hock, S. J., Schaper, L. A., Herrmann, W. A., & Kühn, F. E., (2013). Group 7 transition metal complexes with N-heterocyclic carbenes. *Chemical Society Reviews, 42*(12), 5073–5089.
50. Holland, P. L., Cundari, T. R., Perez, L. L., Eckert, N. A., & Lachicotte, R. J., (2002). Electronically unsaturated three-coordinate chloride and methyl complexes of iron, cobalt, and nickel. *Journal of the American Chemical Society, 124*(48), 14416–14424.
51. Holleck, H., (1977). Carbon-and boron-stabilized ordered phases of scandium. *J. Less-Common Met. (Switzerland)*, *52*(1), 465–468.
52. Huang, R., Wang, J., Liu, S., Yang, Y., Deng, W., & Su, Y., (2021). Non-precious transition metal based electrocatalysts for vanadium redox flow batteries: Rational design and perspectives. *Journal of Power Sources, 515*, 230640.
53. Inoue, M., & Negishi, H., (1986). Interlayer spacing of 3d transition-

metal intercalates of 1T-cadmium iodide-type titanium disulfide (TiS_2). *The Journal of Physical Chemistry, 90*(2), 235–238.

54. Irving, H. M. N. H., & Williams, R., (1953). 637. The stability of transition-metal complexes. *Journal of the Chemical Society (Resumed)*, 3192–3210.
55. Ishiyama, T., & Miyaura, N., (2000). Chemistry of group 13 element-transition metal linkage—The platinum-and palladium-catalyzed reactions of (alkoxo) diborons. *Journal of Organometallic Chemistry, 611*(1, 2), 392–402.
56. Janssens, E., Neukermans, S., Nguyen, H. M. T., Nguyen, M. T., & Lievens, P., (2005). Quenching of the magnetic moment of a transition metal dopant in silver clusters. *Physical Review Letters, 94*(11), 113401.
57. Jeitschko, W., (1970). Transition metal stannides with MgAgAs and $MnCu_2$ Al type structure. *Metallurgical Transactions, 1*, 3159–3162.
58. Jiang, Y., Zhang, X., Pei, L., Yue, S., Ma, L., Zhou, L., & Gao, J., (2018). Silver nanoparticles modified two-dimensional transition metal carbides as nanocarriers to fabricate acetycholinesterase-based electrochemical biosensor. *Chemical Engineering Journal, 339*, 547–556.
59. Jutzi, P., Neumann, B., Reumann, G., & Stammler, H. G., (1998). Pentamethylcyclopentadienylgallium (Cp*Ga): Alternative synthesis and application as a terminal and bridging ligand in the chemistry of chromium, iron, cobalt, and nickel. *Organometallics, 17*(7), 1305–1314.
60. Kim, J., Byun, S., Smith, A. J., Yu, J., & Huang, J., (2013). Enhanced electrocatalytic properties of transition-metal dichalcogenides sheets by spontaneous gold nanoparticle decoration. *The Journal of Physical Chemistry Letters, 4*(8), 1227–1232.
61. Kohal, R. J., Finke, H. C., & Klaus, G., (2009). Stability of prototype two-piece zirconia and titanium implants after artificial aging: An in vitro pilot study. *Clinical Implant Dentistry and Related Research, 11*(4), 323–329.
62. Krämer, U., Talke, I. N., & Hanikenne, M., (2007). Transition metal transport. *FEBS Letters, 581*(12), 2263–2272.
63. Kuhl, N., Hopkinson, M. N., Wencel-Delord, J., & Glorius, F., (2012). Beyond directing groups: Transition-metal-catalyzed C—H activation

of simple arenes. *Angewandte Chemie International Edition, 51*(41), 10236–10254.

64. Levison, J. J., & Robinson, S. D., (1970). Transition-metal complexes containing phosphorus ligands. Part III. Convenient syntheses of some triphenylphosphine complexes of the platinum metals. *Journal of the Chemical Society A: Inorganic, Physical, Theoretical,* 2947–2954.
65. Li, Y. Y., Yu, S. L., Shen, W. Y., & Gao, J. X., (2015). Iron-, cobalt-, and nickel-catalyzed asymmetric transfer hydrogenation and asymmetric hydrogenation of ketones. *Accounts of Chemical Research, 48*(9), 2587–2598.
66. Lindner, D., Schlothofer-Schumann, K., Kern, B. C., Marx, O., Müns, A., & Meixensberger, J., (2017). Cranioplasty using custom-made hydroxyapatite versus titanium: A randomized clinical trial. *Journal of Neurosurgery, 126*(1), 175–183.
67. Liu, X. C., Zhao, S., Sun, X., Deng, L., Zou, X., Hu, Y., & Ajayan, P. M., (2020). Spontaneous self-intercalation of copper atoms into transition metal dichalcogenides. *Science Advances, 6*(7), 4092.
68. Lomer, W. M., (1962). Electronic structure of chromium group metals. *Proceedings of the Physical Society, 80*(2), 489.
69. Lu, E., Chu, J., & Chen, Y., (2018). Scandium terminal imido chemistry. *Accounts of Chemical Research, 51*(2), 557–566.
70. Ma, M., Cheng, X., Shi, Z., Zhang, C., Li, Y., Yang, Y., & Xie, E., (2022). Role of N in transition-metal-nitrides for anchoring platinum-group metal atoms toward single-atom catalysis. *Small Methods, 6*(7), 2200295.
71. Macdonald, D., & Geerligs, L. J., (2004). Recombination activity of interstitial iron and other transition metal point defects in p-and N-type crystalline silicon. *Applied Physics Letters, 85*(18), 4061–4063.
72. Mertz, W., (1993). Chromium in human nutrition: A review. *The Journal of Nutrition, 123*(4), 626–633.
73. Mi, L., Hou, H., Song, Z., Han, H., Xu, H., Fan, Y., & Ng, S. W., (2007). Rational construction of porous polymeric cadmium ferrocene-1, 1′-disulfonates for transition metal ion exchange and sorption. *Crystal Growth and Design, 7*(12), 2553–2561.
74. Molina, J. A., Jimenez-Jimenez, F. J., Aguilar, M. V., Meseguer, I., Mateos-Vega, C. J., Gonzalez-Munoz, M. J., & Martinez-Para, M. C., (1998). Cerebrospinal fluid levels of transition metals in patients with

Alzheimer's disease. *Journal of Neural Transmission, 105*, 479–488.

75. Moslemzadeh, N., Beamson, G., Tsakiropoulos, P., Watts, J. F., Haines, S. R., & Weightman, P., (2006). The 1s XPS spectra of the 3d transition metals from scandium to cobalt. *Journal of Electron Spectroscopy and Related Phenomena, 152*(3), 129–133.
76. Murray, R. B., & Yoffe, A. D., (1972). The band structures of some transition metal dichalcogenides: Band structures of the titanium dichalcogenides. *Journal of Physics C: Solid State Physics, 5*(21), 3038.
77. Narath, A., (1967). Nuclear spin-lattice relaxation in hexagonal transition metals: Titanium. *Physical Review, 162*(2), 320.
78. Nasch, T., Jeitschko, W., & Rodewald, U. C., (1997). Ternary rare earth transition metal zinc compounds RT2Zn20 with T= Fe, Ru, Co, Rh, and Ni. *Zeitschrift Für Naturforschung B, 52*(9), 1023–1030.
79. Natali, M., & Giordani, S., (2012). Interaction studies between photochromic spiropyrans and transition metal cations: The curious case of copper. *Organic & Biomolecular Chemistry, 10*(6), 1162–1171.
80. Nesmeyanov, A. N., Ustynyuk, N. A., Makarova, L. G., Andre, S., Ustynyuk, Y. A., Novikova, L. N., & Luzikov, Y. N., (1978). Indenyl-and flourenyl-transition metal complexesss: I. Synthesis, structures, and reactions of v5-and v6-derivatives of chromium group metals. *Journal of Organometallic Chemistry, 154*(1), 45–63.
81. Nicholls, B., & Whiting, M. C., (1959). 113. The organic chemistry of the transition elements. Part I. Tricarbonylchromium derivatives of aromatic compounds. *Journal of the Chemical Society (Resumed)*, 551–556.
82. Ogawa, A., (2000). Activation and reactivity of group 16 inter-element linkage—Transition-metal-catalyzed reactions of thiols and selenols. *Journal of Organometallic Chemistry, 611*(1, 2), 463–474.
83. Packheiser, R., Jakob, A., Ecorchard, P., Walfort, B., & Lang, H., (2008). Diphenylphosphinoferrocene gold (I) acetylides: Synthesis of heterotri-And heterotetrametallic transition metal complexes. *Organometallics, 27*(6), 1214–1226.
84. Pan, J., Tian, X. L., Zaman, S., Dong, Z., Liu, H., Park, H. S., & Xia, B. Y., (2019). Recent progress on transition metal oxides as bifunctional catalysts for lithium-air and zinc-air batteries. *Batteries & Supercaps, 2*(4), 336–347.

85. Piamonteze, C., Miedema, P., & De Groot, F. M., (2009). Accuracy of the spin sum rule in XMCD for the transition-metal L edges from manganese to copper. *Physical Review B, 80*(18), 184410.

86. Preuss, H. G., & Anderson, R. A., (1998). Chromium update: Examining recent literature 1997–1998. *Current Opinion in Clinical Nutrition & Metabolic Care, 1*(6), 509–512.

87. Rana, S., Biswas, J. P., Paul, S., Paik, A., & Maiti, D., (2021). Organic synthesis with the most abundant transition metal–iron: From rust to multitasking catalysts. *Chemical Society Reviews, 50*(1), 243–472.

88. Rauchfuss, T. B., (1979). Transition metal activation of aldehydes: Platinum metal derivatives of o-diphenylphosphinobenzaldehyde. *Journal of the American Chemical Society, 101*(4), 1045–1047.

89. Roels, H. A., Ghyselen, P., Buchet, J. P., Ceulemans, E. L. R. R., & Lauwerys, R. R., (1992). Assessment of the permissible exposure level to manganese in workers exposed to manganese dioxide dust. *Occupational and Environmental Medicine, 49*(1), 25–34.

90. Ruther, R. E., Callender, A. F., Zhou, H., Martha, S. K., & Nanda, J., (2014). Raman microscopy of lithium-manganese-rich transition metal oxide cathodes. *Journal of The Electrochemical Society, 162*(1), A98.

91. Samuel, E., Maillard, P., & Giannotti, C., (1977). Photolysis of dialkyl metallocenes of the titanium group; an ESR study. *Journal of Organometallic Chemistry, 142*(3), 289–298.

92. Sanpo, N., Tharajak, J., Li, Y., Berndt, C. C., Wen, C., & Wang, J., (2014). Biocompatibility of transition metal-substituted cobalt ferrite nanoparticles. *Journal of Nanoparticle Research, 16*, 1–13.

93. Stadler, N., Lindner, R. A., & Davies, M. J., (2004). Direct detection and quantification of transition metal ions in human atherosclerotic plaques: Evidence for the presence of elevated levels of iron and copper. *Arteriosclerosis, Thrombosis, and Vascular Biology, 24*(5), 949–954.

94. Stimmelmayr, M., Edelhoff, D., Güth, J. F., Erdelt, K., Happe, A., & Beuer, F., (2012). Wear at the titanium–titanium and the titanium–zirconia implant–abutment interface: A comparative in vitro study. *Dental Materials, 28*(12), 1215–1220.

95. Stork, J. R., Thoi, V. S., & Cohen, S. M., (2007). Rare examples of transition-metal– main-group metal heterometallic metal– organic frameworks from gallium and indium dipyrrinato complexes and silver salts: Synthesis and framework variability. *Inorganic Chemistry, 46*(26), 11213–11223.

96. Sun, Z., Zhu, M., Lv, X., Liu, Y., Shi, C., Dai, Y., & Majima, T., (2019). Insight into iron group transition metal phosphides (Fe_2P, Co_2P, Ni_2P) for improving photocatalytic hydrogen generation. *Applied Catalysis B: Environmental, 246*, 330–336.

97. Tate, D., Knipple, W. R., & Augl, J. M., (1962). Nitrile derivatives of chromium group metal Carbonyls. *Inorganic Chemistry, 1*(2), 433, 434.

98. Thiel, W., & Voityuk, A. A., (1996). Extension of MNDO to d orbitals: Parameters and results for the second-row elements and for the zinc group. *The Journal of Physical Chemistry, 100*(2), 616–626.

99. Tolbert, M. A., & Beauchamp, J. L., (1984). Activation of carbon-hydrogen and carbon-carbon bonds by transition-metal ions in the gas phase. Exhibition of unique reactivity by scandium ions. *Journal of the American Chemical Society, 106*(26), 8117–8122.

100. Trent, L. K., & Thieding-Cancel, D., (1995). Effects of chromium picolinate on body composition. *The Journal of Sports Medicine and Physical Fitness, 35*(4), 273–280.

101. Urbankowski, P., Anasori, B., Hantanasirisakul, K., Yang, L., Zhang, L., Haines, B., & Gogotsi, Y., (2017). 2D molybdenum and vanadium nitrides synthesized by ammoniation of 2D transition metal carbides (MXenes). *Nanoscale, 9*(45), 17722–17730.

102. Uyanik, F., (2001). The effects of dietary chromium supplementation on some blood parameters in sheep. *Biological Trace Element Research, 84*, 93–101.

103. Vayssilov, G. N., (1997). Structural and physicochemical features of titanium silicalites. *Catalysis Reviews, 39*(3), 209–251.

104. Velicky, M., Donnelly, G. E., Hendren, W. R., McFarland, S., Scullion, D., DeBenedetti, W. J., & Huang, F., (2018). Mechanism of gold-assisted exfoliation of centimeter-sized transition-metal dichalcogenide monolayers. *ACS Nano, 12*(10), 10463–10472.

105. Vincent, J. B., Chang, H. R., Folting, K., Huffman, J. C., Christou, G., & Hendrickson, D. N., (1987). Preparation and physical properties of trinuclear oxo-centered manganese complexes of general formulation $[Mn_3O(O_2CR)6L_3]_0$, + (R= methyl or phenyl; L= a neutral donor group) and the crystal structures of $[Mn_3O(O_2CMe)6(pyr)_3](pyr)$ and $[Mn_3O(O_2CPh)6(pyr)_2\ (H_2O)]$. $cntdot._{0.5}$ MeCN. *Journal of the American Chemical Society, 109*(19), 5703–5711.

106. Wang, M., & Jiang, X., (2021). The same oxidation-state introduction of hypervalent sulfur via transition-metal catalysis. *The Chemical Record, 21*(12), 3338–3355.

107. Wang, X., Zhang, X., Li, P., Otake, K. I., Cui, Y., Lyu, J., & Farha, O. K., (2019). Vanadium catalyst on isostructural transition metal, lanthanide, and actinide-based metal–organic frameworks for alcohol oxidation. *Journal of the American Chemical Society, 141*(20), 8306–8314.

108. Watanabe, S., & Kleppa, O. J., (1984). Thermochemistry of alloys of transition metals: Part IV. Alloys of copper with scandium, yttrium, lanthanum, and lutetium. *Metallurgical Transactions B, 15*, 357–368.

109. Weber, E. R., (1983). Transition metals in silicon. *Applied Physics A, 30*, 1–22.

110. Wei, T., Wang, S., Lu, X., Tan, Y., Huang, J., Liu, F., & Yang, S., (2016). Entrapping a group-VB transition metal, vanadium, within an endohedral metallofullerene: V_xSc_3–xN@ I_h-C_{80} (x= 1, 2). *Journal of the American Chemical Society, 138*(1), 207–214.

111. West, D. X., Carlson, C. S., Bouck, K. J., & Liberta, A. E., (1991). Transition metal ion complexes of tiosemicarbazones derived from 2-acetylpyridine. Part 10. A comparison of the chemical and antifungal properties of the copper (II) complexes of 2-acetylpyridine 3-pyrrolidinyl-, 3-piperidinyl-, 3-hexamethyleneiminyl-and 3-azabicyclo [3, 2, 2]-nonylthiosemicarbazones. *Transition Metal Chemistry, 16*, 271–275.

112. Whittingham, M. S., Song, Y., Lutta, S., Zavalij, P. Y., & Chernova, N. A., (2005). Some transition metal (oxy) phosphates and vanadium oxides for lithium batteries. *Journal of Materials Chemistry, 15*(33), 3362–3379.

113. Xia, Y., Campbell, C. T., Roldan, C. B., & Mavrikakis, M., (2021). Introduction: Advanced materials and methods for catalysis and electrocatalysis by transition metals. *Chemical Reviews, 121*(2), 563–566.

114. Xiao, J., Chernova, N. A., & Whittingham, M. S., (2008). Layered mixed transition metal oxide cathodes with reduced cobalt content for lithium-ion batteries. *Chemistry of Materials, 20*(24), 7454–7464.

115. Yasumatsu, N., Yoshikawa, Y., Adachi, Y., & Sakurai, H., (2007). Antidiabetic copper (II)-picolinate: Impact of the first transition metal in the metallopicolinate complexes. *Bioorganic & Medicinal*

Chemistry, 15(14), 4917–4922.

116. Yazaki, Y., Faridi, Z., Ma, Y., Ali, A., Northrup, V., Njike, V. Y., & Katz, D. L., (2010). A pilot study of chromium picolinate for weight loss. *The Journal of Alternative and Complementary Medicine, 16*(3), 291–299.
117. Ye, Y., & Sanford, M. S., (2012). Merging visible-light photocatalysis and transition-metal catalysis in the copper-catalyzed trifluoromethylation of boronic acids with CF3I. *Journal of the American Chemical Society, 134*(22), 9034–9037.
118. Zaanen, J., Westra, C., & Sawatzky, G. A., (1986). Determination of the electronic structure of transition-metal compounds: 2p x-ray photoemission spectroscopy of the nickel dihalides. *Physical Review B, 33*(12), 8060.
119. Zeiss, H. H., & Tsutsui, M., (1957). π-complexes of the transition metals. I. Hein's polyaromatic chromium compounds. *Journal of the American Chemical Society, 79*(12), 3062–3066.
120. Zeng, D., Wang, M., Deng, W. P., & Jiang, X., (2020). The same oxygenation-state introduction of hypervalent sulfur under transition-metal-free conditions. *Organic Chemistry Frontiers, 7*(23), 3956–3966.
121. Zhan, Y., Lu, M., Yang, S., Xu, C., Liu, Z., & Lee, J. Y., (2016). Activity of transition-metal (manganese, iron, cobalt, and nickel) phosphates for oxygen electrocatalysis in alkaline solution. *ChemCatChem, 8*(2), 372–379.
122. Zhao, P., Feng, X., Huang, D., Yang, G., & Astruc, D., (2015). Basic concepts and recent advances in nitrophenol reduction by gold-and other transition metal nanoparticles. *Coordination Chemistry Reviews, 287*, 114–136.
123. Zheng, S., Li, X., Yan, B., Hu, Q., Xu, Y., Xiao, X., & Pang, H., (2017). Transition-metal (Fe, Co, Ni) based metal-organic frameworks for electrochemical energy storage. *Advanced Energy Materials, 7*(18), 1602733.

CHAPTER 5

Lanthanides

Contents

5.1. INTRODUCTION

Upon initial observation (Figure 5.1) as mentioned in the periodic table one is typically inclined to focus on the two rows of boxes situated beneath the primary section of the table, which house elements with relatively unfamiliar nomenclature. Those elements encompassing in the first row, specifically elements 57 to 71, are commonly referred to as "lanthanides." This nomenclature is derived from the initial element in the row, lanthanum, which occupies the 57th position in the periodic table. These elements exhibit distinct photogenic, magnetic, mechanical, as well as nuclear characteristics, and have been extensively utilized in various industries such as glass as well as in the production of ceramic, metalworking, electronics, and agriculture (Ungur, 2018). The term "lanthanum" is derived from the Greek word "lanthanō," which translates to "to lie hidden" in English. The International Union of Pure and Applied Chemistry (IUPAC) has officially adopted terms "lanthanides" or "lanthanoids" as collective names for these elements. It is worth noting in describing term "lanthanoids" conveys the similarity to lanthanum, which is a member of this group. However, the term "lanthanides" has gained more widespread usage as the primary designation for this group of elements. One aspect that warrants consideration is the utilization of the suffix "-ide" in chemistry to designate naming of various anions (such as sulfide, bromide, etc.), that has the potential to generate misconceptions regarding the elemental properties of these substances. Nevertheless, the utilization of this terminology has become widespread in the scientific literature pertaining to chemistry (Liu et al., 2022).

The lanthanides are occasionally regarded as exhibiting characteristics similar to the elements of group 3 due to development of various trivalent cations. One intriguing observation is that the abundance of elements is often greater than what their names might suggest. The lanthanide elements are naturally occurring and can be found in various minerals, such as xenotime, monazite, and bastnaesite, among others. Xenotime and monazite are classified as orthophosphate ($LnPO_4$) salts of the elements. Monazite, a mineral of significance, is known to contain yttrium and the radioactive element thorium, thereby presenting challenges and hazards in its handling. It can be generally observed that lanthanides of even atomic numbers exhibit greater prevalence compared to those with odd atomic numbers. Furthermore, the relative abundance of the most commonly occurring lanthanides diminishes in the sequence of Ce, La, Nd, and Pr (Geoffrey N áCloke, 1993).

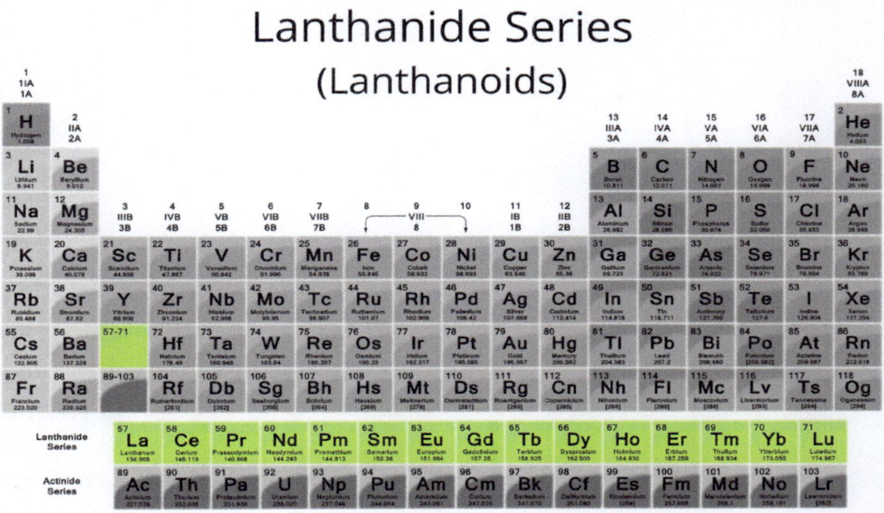

Figure 5.1. Elements of lanthanide series highlighted in the periodic table.

Source: Mohammad Reza Ganjali, creative commons license.

The nomenclature rare earths (REs), rare earth elements (REEs), and rare earth metals (REMs) is employed to encompass the lanthanides, as well as scandium (Sc) and yttrium (Y). These elements possess a diverse array of planned solicitations, covering medical as well as energy knowledge, magnets, laser beams, magnetic resonance imaging (MRI) contrast agents, catalysts, batteries, and alloys. Figure 5.2 presents a visual representation of the estimated global reserves of REEs across various countries (Handl & Gillies, 2005).

According to the US Geological Survey, the global rare earth assets are estimated to be approximately 130 million metric tons (mmt), with a significant portion concentrated in China. Notably, China holds the largest share of rare earth deposits and serves as the primary commercial provider of these valuable resources. Brazil possesses approximately 22 million metric tons (MMT) of these reserves, while other nations with significant stocks of REMs including the United States, Australia, India, and Malaysia. The REEs that are most extensively produced include lanthanides, specifically cerium, lanthanum, and neodymium (Meijerink et al., 2006).

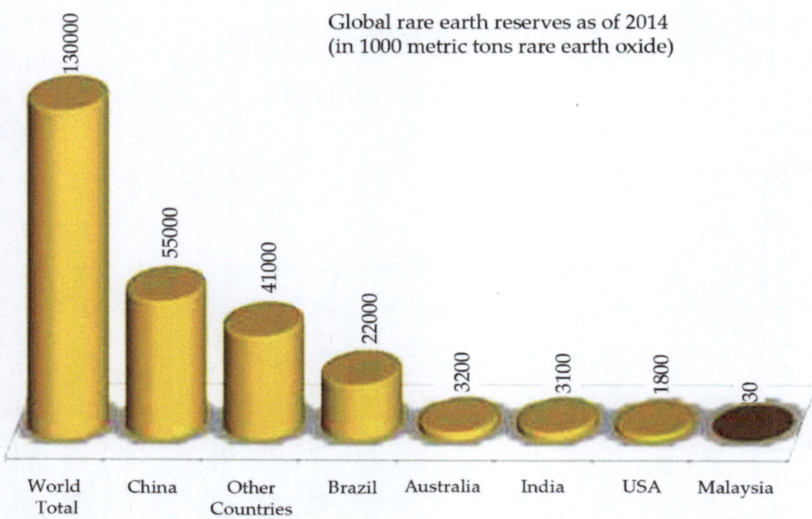

Figure 5.2. The global rare earth metals assets as estimated as in 2014, mention by country.

Source: Vinod Kumar Gupta, creative commons license.

Table 5.1. Electron Configuration and Basic Properties of Lanthanides

Element	Symbol	Atomic Number	Ideal Electronic Configuration	Observed electronic Configuration
Actinium	Ac	89	[Rn] $5f^0\ 6d^1\ 7s^2$	[Rn] $5f^0\ 6d^1\ 7s^2$
Thorium	Th	90	[Rn] $5f^1\ 6d^1\ 7s^2$	[Rn] $5f^0\ 6d^2\ 7s^2$
Protactinium	Pa	91	[Rn] $5f^2\ 6d^1\ 7s^2$	[Rn] $5f^2\ 6d^1\ 7s^2$
Uranium	U	92	[Rn] $5f^3\ 6d^1\ 7s^2$	[Rn] $5f^3\ 6d^1\ 7s^2$
Neptunium	Np	93	[Rn] $5f^4\ 6d^1\ 7s^2$	[Rn] $5f^4\ 6d^1\ 7s^2$
Plutonium	Pu	94	[Rn] $5f^5\ 6d^1\ 7s^2$	[Rn] $5f^6\ 6d^0\ 7s^2$
Americium	Am	95	[Rn] $5f^6\ 6d^1\ 7s^2$	[Rn] $5f^7\ 6d^0\ 7s^2$
Curium	Cm	96	[Rn] $5f^7\ 6d^1\ 7s^2$	[Rn] $5f^7\ 6d^1\ 7s^2$
Berkelium	Bk	97	[Rn] $5f^8\ 6d^1\ 7s^2$	[Rn] $5f^9\ 6d^0\ 7s^2$
Californium	Cf	98	[Rn] $5f^9\ 6d^1\ 7s^2$	[Rn] $5f^{10}\ 6d^0\ 7s^2$
Einsteinium	Es	99	[Rn] $5f^{10}\ 6d^1\ 7s^2$	[Rn] $5f^{11}\ 6d^0\ 7s^2$
Fermium	Fm	100	[Rn] $5f^{11}\ 6d^1\ 7s^2$	[Rn] $5f^{12}\ 6d^0\ 7s^2$
Mendelevium	Md	101	[Rn] $5f^{12}\ 6d^1\ 7s^2$	[Rn] $5f^{13}\ 6d^0\ 7s^2$
Nobelium	No	102	[Rn] $5f^{13}\ 6d^1\ 7s^2$	[Rn] $5f^{14}\ 6d^0\ 7s^2$
Lawrencium	Lw or Lr	103	[Rn] $5f^{14}\ 6d^1\ 7s^2$	[Rn] $5f^{14}\ 6d^1\ 7s^2$

The electron configurations of lanthanides (as shown in Table 5.1) were estimated concluded via experimental investigations as in emission spectra lines. These lines provide information about the energy changes associated with electron transitions between different energy levels. An examination of the electron configuration of those components reveals that the similarity in their properties can be attributed to the consistent presence of an equal number of electrons in both the fifth as well as sixth energy levels for the majority of cases (Banerjee et al., 2005).

5.2. LANTHANUM

The element lanthanum, which occupies the initial position in the series, was initially detected in a compound alongside cerium during the latter part of the 1830s. This discovery was made by the Swedish chemist Carl Gustav Mosander, who employed thermal decomposition techniques on samples of cerium nitrate. However, the isolation of the pure type of the element did not occur until 1923. In relation to the isotopes of the element, it is observed that approximately 99.91% of naturally arising elements called lanthanum consists of the isotope ^{139}La. Furthermore, a total of 38 radioisotopes of lanthanum have been identified, among which the most stable ones are ^{138}La (with a half-life of 1.05×10^{11} years) and ^{137}La (with a half-life of 60,000 years). The remaining isotopes exhibit half-lives that are shorter than hours, and in some cases, even minutes. These isotopes are produced via the process of bombarding atoms with particles (Kulkarni et al., 2006). The element exhibits three distinct allotropic forms, namely α, β, and γ, which are determined by variations in temperature and pressure. The α-phase exhibits a hexagonal structure that is double close-packed, characterized by lattice parameters a = 3.7740 A° and c = 12.171 A°. This phase is dominant at standard room temperature. On the other hand, the β-phase, which emerges at a temperature of 325°C, possesses a face-centered cubic structure with a lattice parameter of a = 5.303 A°. The γ-phase, that is generated at a temperature of 887°C, exhibits a body-centered cubic crystal structure with a lattice parameter of 4.26 A°. The element in question possesses an atomic number of 57, exhibits a silvery white appearance, and possesses a soft consistency that allows it to be readily cut with a knife. Furthermore, it demonstrates a high degree of ductility. Lanthanum (La) is a relatively reactive rare-earth element that undergoes oxidation when exposed to atmospheric oxygen. Similar to the other elements in the series, La exhibits a sluggish reaction rate when exposed to cold water and a relatively rapid reaction rate when subjected to hot water (Morss, 1976). It is known to exhibit oxidation states of 3+ and 2+, with the latter getting less stable, as previously discussed. The element exhibits a high electropositivity and readily undergoes dissolution when exposed to dilute sulfuric acid

(H_2SO_4). This particular element exhibits metal state in direct reactivity with following elements like carbon, silicon, boron, nitrogen, phosphorus, sulfur, selenium, and halogens. Additionally, it serves as the primary constituent of the alloy commonly referred to as Misch metal. Misch metal is a metallic alloy primarily consisting of cerium (Ce) and lanthanum (La), along with minor quantities of neodymium (Nd) and praseodymium (Pr). This alloy finds application in the production of lighter flints. Los Angeles exhibits the highest level of basicity compared to the other elements in its family, thereby enabling Mosander to successfully isolate as well as purify its salt compound (Feng et al., 2006).

The element in question is found persist inside earth's crust at a meditation of 32 parts per million (ppm), indicating that its abundance is comparable to that of copper or zinc, contrary to prevailing beliefs. The element, however, exhibits widespread distribution and is therefore not confined to a single location. The extraction and processing of La are significantly hindered by these factors, leading to a notable escalation in its market value. Monazite and bastnaesite are the predominant ores that contain lanthanum. These ores are specifically composed of the phosphate as well as mixed carbonates-fluoride of cerium, lanthanum, thorium, neodymium, praseodymium, and yttrium, respectively. Loparite is a third mineral that is utilized in the process of obtaining pure lanthanum. The world's largest bastnaesite deposits are primarily located in the US and China, while major monazite deposits can be found in Australia, China, India, Brazil, Malaysia, South Africa, Thailand, Sri Lanka, and the United States. Loparite deposits have also been identified in Russia (Wang et al., 1992).

The initial step of metals extraction involves the mineral ores comminution. One notable improvement of monazite in this context is its magnetic characteristics, that greatly facilitate its separation process. Following the separation of the crushed stage from the residues, the latter are subjected to treatment with hot concentrated sulfuric acid in order to facilitate the dissolution of rare earth content. Subsequently, the resultant solution is subjected to filtration, followed by a partial neutralization of its pH to approximately 3–4 through the addition of sodium hydroxide or alternative basic substances. During this phase, the minerals undergo a process in which the thorium content forms precipitates that can be readily eliminated (Weiss, 1974). Subsequently, the solution is subjected to treatment with an ammonium oxalate solution, resulting in the formation of rare earth oxalates. These oxalates are then transformed into their respective oxides by means of annealing. Subsequently, the oxides are dissolved in nitric acid.

At this particular stage, cerium is isolated due to the insoluble nature of its oxide under the given operating conditions. In this phase, the separation of a mixed salt containing La as well as ammonium nitrate is achieved by means of crystallization, exploiting its comparatively lower solubility in comparison to the corresponding species of other REEs (Kragten & Decnop-Weever, 1987).

An additional effectual method for the extraction of lanthanum (La) is predicated upon utilization of ion exchange mechanism or liquid-liquid extraction techniques. The previous methodology contains the adsorption of target ions onto an appropriate ion exchanger, resulting in their subsequent separation. Subsequently, the desorption process is facilitated by the formation of complexes with ligands like ammonium citrate and nitrilotriacetate. In contrast, the latter method entails utilizing organic solvents such as tributyl phosphate, bis-2-ethylhexyl phosphate, or 2-ethylhexyl ester of 2-ethylhexylphosphonic acid (Kajjumba & Marti, 2022).

The synthesis of metallic La involves the application of thermal energy to its oxide at temperatures ranging from 300 to 400°C, while employing an ammonium halide (either F or Cl) as well as hydrofluoric acid as catalysts.

$La_2 + 3O_3 + 6NH_4Cl \rightarrow 2LaCl_3 + 6NH_3 + 3H_2O$

Additionally, the chloride salt can be further reduced by employing an alkali and alkaline earth metal in a vacuum or inert atmosphere, as depicted in the reaction below, or by subjecting the molten chloride salt to electrolysis (Yakubovich & Alekseev, 2012).

$LaCl_3 + 3Li \rightarrow La + 3LiCl$

5.3. CERIUM

The discovery of the element occurred concurrently with two distinct research teams, one located in Bastnäs, Sweden and the other in Germany, in the year 1803. The Swiss team designated the celestial body as Ceres, drawing inspiration from a planet which had been previously detected in the year 1801. In order to conduct a qualitative assessment of the existence of cerium (Ce) in mixtures of REEs, the sample is treated with ammonia (NH_3) and hydrogen peroxide (H_2O_2). The manifestation of a deep brown hue signifies the existence of the element. According to a reliable source cerium metal exhibits a high degree of pyrophoricity, meaning it has the potential to ignite when subjected to grinding or scratching. Due to its high electropositivity, the elemental form of cerium is not naturally found in abundance (Aiken et al., 1988). Instead, its initial isolated state was in the system of its oxide, known as ceria. Subsequent advancements in electrochemical methods facilitated the production of metallic cerium, along with other elements in the lanthanide series. The element possesses an atomic number of 58 and exhibits the physical characteristics of a silvery metal, resembling iron in both color and luster. However, similar to other elements within its group, it displays properties of softness, malleability, and ductility. The element is characterized by the presence of four isotopes that occur naturally. These isotopes include 140Ce (88.5%), 142Ce (11%), 138Ce (0.3%), and 136Ce (0.2%). Among these isotopes, 140Ce along with others is considered stable, while the have half-lives of last three isotopes is 5×10^{16} years, 1.5×10^{15} years, and 70 trillion years respectively (Krumpel et al., 2008).

In addition to the four isotopes that occur naturally, cerium possesses 35 synthetic isotopes having mass numbers ranging from 119 to 157. These synthetic isotopes exhibit extremely brief half-lives, typically lasting only a few thousandths to one second. Regarding nuclear reactions, it has been observed that heavier isotopes of Ce undergo β-decays, while the lighter isotopes (those with a mass less than 140Ce) exhibit a preference for electron capture and β+ decay reactions (Palizban et al., 2010).

The electronic configuration of cerium exhibits considerable variability due to the nearly equivalent energy levels of the 4f orbital and the valence orbitals. This implies that energy blockade necessary for electrons to transition between levels is significantly reduced, thus enabling the existence of what are commonly referred to as "dual-valency states." Consequently, alterations in the environmental conditions lead to modifications in the physical characteristics of cerium. The metallic cerium exhibits a reduction in volume of approximately 10% under conditions of elevated pressure or reduced temperature. Additionally, there is a potential alteration in its valency from approximately 3 to 4 (Hower et al., 2013).

Cerium exhibits a distinct fundamental characteristic in comparison to other elements within its family, facilitating its straightforward separation. The separation procedure relies on the distinct basicity levels exhibited by the 2 oxidation states. In accordance with this methodology, a solution of nitric acid with low concentration is employed for the purpose of leaching the amalgamated hydroxides. Subsequently, the Ce^{4+} cations are precipitated from a solution containing nitrate or chloride ions of lanthanides through the utilization of potassium permanganate as well as sodium carbonate in a molar ratio of 1:4. Ultimately, the REs nitrate solution is subjected to the process of boiling in the presence of potassium bromate as well as marble chips (Zuo et al., 2008).

Due to its high electro-positivity, cerium exhibits a sluggish reaction when exposed to cold water, as illustrated by the following equation:

$$2Ce\ (s) + 6H_2O\ (l) \rightarrow 2Ce(OH)_3\ (aq) + 3H_2\ (g)$$

The reaction rate exhibits a positive correlation with temperature, whereby an increase in temperature leads to an increase in the rate of the reaction. The metallic form exhibits reactivity with all halogens, as demonstrated by the following comprehensive reaction (Garner & Heppell, 2005):

$$2Ce\ (s) + 3X_2\ (g) \rightarrow 2CeX_3\ (s)$$

The resultant products exhibit a predominantly white appearance, with the exception of the iodide salt, that manifests a slightly yellow hue (Dahle & Arai, 2015).

The metallic state of cerium (Ce) exhibits a high susceptibility to oxidation, resulting in the formation of Ce^{3+} ions. These ions can easily dissolve in dilute sulfuric acid, leading to the creation of $[Ce(OH_2)_9]^{3+}$ complexes. However, the solubility of cerium in methane sulfonic acid is relatively low. Cerium, being the most prevalent element within the lanthanide group, can be found in the earth's crust and seawater at approximate concentrations of 0.0046 weight percent and 1.5 parts per trillion, respectively (Cheisson et al., 2019).

5.4. PRASEODYMIUM

The nomenclature of the chemical element is derived from a combination of two Greek words, namely "prasios" denoting the color "green" and "didymos" signifying "twin." In 1879, Mosander identified an element named Didymia, derived from lanthana. Subsequently, in the same year, Lecoq de Boisbaudran successfully isolated a new element identified as Samaria from Didymia samples derived from samarskite ores. Building upon these findings, von Welsbach, in 1885, was able to discern and separate two distinct elements with different colors from didymium. The individual referred to the green specimen as praseodymia, while designating the other specimen as neodymia. Similar to other rare earths, the elements in question exhibit distinct and narrow spectral absorption bands or lines when present in solution phase. These bands or lines possess a width of only a few Angstroms. Praseodymium, denoted by the symbol Pr and possessing an atomic number of 59, shares similar physical characteristics with the elements Cerium (Ce) and Lanthanum (La). These characteristics include a soft and silvery appearance, as well as malleability and ductility. Additionally, praseodymium exhibits favorable properties in terms of magnetism, electricity, chemistry, and optics. The chemical element exhibits a range of isotopes, characterized by mass numbers ranging from 121 to 159. Among these isotopes, the most prevalent one is denoted as ¹41Pr. Lighter isotopes undergo β+ decay, a process that does not alter the elemental identity of the parent isotope. On the other hand, the β− decay, which is more prevalent in heavier isotopes, results in the formation of neodymium isotopes with 60 protons (Zhang et al., 2016).

The elemental form of this element does not naturally occur due to its high reactivity. Even after being synthetically produced in a pure form, a green oxide layer gradually forms on its surface. Consequently, similar to other elements, it is necessary to store its pure form in light mineral oil. The metallic element Praseodymium (Pr) undergoes a gradual process of tarnishing when exposed to air. Additionally, it exhibits a combustion reaction at a temperature of 150°C, as described by the following equation:

$$12Pr + 11O_2 \rightarrow 2Pr_6O_{11}$$

The chemical element in question exhibits a reaction with cold water, resulting in the formation of a hydroxide. Additionally, it displays similar characteristics to the elements La and Ce when exposed to hot water and halogens, with the exception of the formation of a green product when reacting with any halogen (Ganjali et al., 2007).

$$2Pr\ (s) + 6H_2O\ (l) \rightarrow 2Pr(OH)_3\ (aq) + H_2\ (g)$$

$$2Pr\ (s) + 3X_2\ (g) \rightarrow 2PrX_3\ (s)[green]$$

The reaction of praseodymium with dilute sulfuric acid results in the production of green Pr^{3+} ions, specifically in the way of $[Pr(OH_2)_9]^{3+}$ complexes. This reaction is analogous to the behavior of cerium. Similar to the aforementioned two lanthanides, the oxidation states of praseodymium (Pr) encompass 2+, 3+, or 4+. Pr^{4+} is an exceptionally potent oxidizing agent capable of oxidizing water as well as hydrochloric acid to generate molecular oxygen (O_2) and chlorine gas (Cl_2). Consequently, the tetravalent Praseodymium cation exhibits instability in aqueous environments, leading to its conversion into the stable Praseodymium ion with a distinct yellow-green hue (Bradley et al., 1994). The black praseodymium oxide, denoted as Pr_6O_{11}, is synthesized by igniting the oxalate or carbonate salts of praseodymium. It consists of a combination of Pr^{3+} and Pr^{4+} ions. Additional examples of commonly encountered salts of the element praseodymium (Pr) include praseodymium fluoride (PrF_2, PrF_3, PrF_4), praseodymium chloride ($PrCl_3$), praseodymium bromide ($PrBr_3$, Pr_2Br_5), praseodymium iodide (PrI_2, PrI_3, Pr_2I_5), praseodymium oxide (PrO_2, Pr_2O_3, Pr_6O_{11}), praseodymium sulfide (PrS, Pr_2S_3), praseodymium sulfate ($Pr_2(SO_4)_3$), praseodymium selenide (PrSe), praseodymium telluride (PrTe, Pr_2Te_3), and praseodymium nitride (PrN). Praseodymium found in nature consists solely of a single stable isotope, namely 141Pr. This isotope holds significance in the fields of nuclear magnetic resonance (NMR) as well as electron paramagnetic resonance (EPR) spectroscopies (De Smet et ál., 2001). In total, a total of 38 radioisotopes have been identified and studied. Among these, the radioisotope 143P has been found to be the most stable, with a half-life (t_(1/2)) of 13.57 days. Additionally, the radioisotope 142Pr has a half-life of 19.12 hours. According to the cited source it is observed that the rest of the radioactive isotopes exhibit half-lives that are all below 6 hours. Furthermore, a significant proportion of these isotopes possess half-lives which are less than 10 minutes. While not as plentiful as the two preceding elements (with a concentration of 9.5 parts per million in the Earth's crust), Praseodymium (Pr) is considered a relatively abundant REE. It is found in greater quantities in minerals such as monazite as well as bastnaesite, which are commonly found in China, the US, Brazil, India, Sri Lanka, as well as Australia (Strubelt et al., 1980).

5.5. NEODYMIUM

Neodymium and praseodymium were simultaneously discovered in 1885 by von Welsbach, who successfully isolated these elements from the substance known as didymium. The extraction techniques involved several fractionation steps utilizing ammonium didymium nitrate, as well as nomenclature in element originated from the Greek term "Neos" as well as "didumous," which signifies the novel twin (Fujii et al., 2005).

With an atomic number of 60, the properties of this element exhibit notable similarities to those of praseodymium, resulting in their prolonged inseparability. Neodymium (Nd) is a metallic element that exhibits a silvery-white appearance and possesses a moderate level of reactivity. Upon exposure to air, the metal surface undergoes oxidation, resulting in the formation of a yellowish layer that subsequently flakes off. The elemental manifestation exhibits a hexagonal crystalline structure, possessing a precise density of 7.004 g/cm³. The alternative allotrope of neodymium exhibits a body-centered cubic crystal structure, which undergoes formation at a temperature of approximately 863°C (Cendrowski-Guillaume et al., 2000).

Neodymium samples found in nature consist of isotopes such as 142Nd (constituting 27% of the sample), as well as 143Nd, 145Nd, 146Nd, 148Nd, as well as two radioisotopes, namely 144Nd and 150Nd. In summary, the element possesses a total of 24 additional synthetic radioisotopes, among which the most enduring are 144Nd (characterized by a decay as well as a half-life of 2.29 × 1015 years) as well as 150Nd (exhibiting double b-decay and a half-life of approximately 7 × 1018 years). The residual radioactive isotopes exhibit half-lives that span from several days to under 70 seconds. The reaction between metallic neodymium and atmospheric oxygen occurs at a relatively sluggish pace, however, it undergoes combustion when exposed to temperatures around 150°C (Edelmann et al., 1994).

$$4Nd + 3O_2 \rightarrow 2Nd_2O_3$$

Neodymium, like the aforementioned Lns, exhibits electropositive properties and undergoes varying rates of reaction with water depending on the temperature. This reaction results in the formation of neodymium(III) hydroxide, as depicted by the equation provided:

$$2Nd\ (s) + 6H_2O\ (l) \rightarrow 2Nd(OH)_3\ (aq) + 3H_2\ (g)$$

The element neodymium exhibits similar reactions to halogens, resulting in the formation of reaction products with distinct colors. Specifically, neodymium fluoride (NdF_3) and neodymium bromide ($NdBr_3$) display a violet hue, neodymium chloride ($NdCl_3$) appears mauve, and neodymium iodide (NdI_3) exhibits a green coloration. These color variations arise from the following general reaction (Campazzi et al., 1998):

$$2Nd\ (s) + 3X_2\ (g) \rightarrow 2NdX_3\ (s)$$

Neodymium exhibits solubility in dilute sulfuric acid, resulting in the formation of Nd^{3+} ions that exist as $[Nd(OH_2)_9]^{3+}$ complexes.

$$2Nd\ (s) + 3H_2SO_4\ (aq) \rightarrow 2Nd^{3+}\ (aq) + 3SO_4^{2-}\ (aq) + 3H_2\ (g)$$

Additional neodymium compounds that are frequently encountered in scientific literature comprise NdS, NdN, $Nd_2(SO_4)_3$, $Nd(OH)_3$, Nd_2S_3, NdP, $Nd(NO_3)_3$, as well as NdC_2,. Neodymium exhibits the second highest abundance among all elements, surpassed only by cerium, with a notable presence of nearly 28 parts per million (ppm) inside earth crust. The minerals that encompass neodymium (Nd) are monazite and bastnaesite, primarily found in Brazil, China, the United States, India, Thailand, Sri Lanka, as well as Australia. Loparite, a mineral abundant in Russia, is another example of a mineral that contains neodymium (Nd) (Najafi et al., 2019). The conventional method employed for the synthesis of neodymium metal involves subjecting halide salts containing Nd^{3+} to elevated temperatures in the manifestation of an extra amount of calcium. This reaction results in the formation of calcium fluoride (CaF_2) and the production of neodymium in its pure metallic form, a process commonly referred to as calciothermic reduction. The halide salt that is frequently employed is NdF_3. Its synthesis involves the vaporization of a combination of anhydrous hydrofluoric acid (HF) and 60% acid-free reagent (AR) over Nd^{2+} oxide at a temperature of 700°C for a duration of 16 hours. The fluoride salt, which exhibits a certain level of impurity, with an approximate oxygen content of 300 ppm, undergoes a purification process in the subsequent phase through the application of elevated temperatures (De Almeida et al., 2012).

5.6. PROMETHIUM

The identification of promethium, which was named after Prometheus, the titan from Greek mythology known for stealing fire from the gods and delivering it to humanity, was achieved by a group of American researchers at Oak Ridge National Laboratory in 1945. The element was successfully isolated from the fission products of uranium by Lawrence E. Glendenin, Jacob A. Marinsky, and Charles D. Coryell. However, the announcement of this discovery was delayed for a period of 2 years due to the project's confidentiality concerns. The synthesis of metallic promethium was initially achieved in 1963 by employing the reduction reaction of promethium (III) fluoride with lithium vapor. Presently, the production of the metal involves the utilization of the calcio thermic reduction method, wherein the trihalides of the metal (with anhydrous $PmCl_3$ being the most prevalent) are employed (Pedersen & Albinsson, 1991). In contrast to the stable isotopes observed in other elements within the Lns group, it is notable that Pm (promethium) lacks stable isotopes altogether. Furthermore, the element itself exhibits a relatively low level of stability. This is the reason why the element has been designated as the least stable among the initial 84 elements of the periodic table. The element undergoes a process of decay, resulting in its conversion to neighboring elements, namely neodymium and samarium. Specifically, the decay processes of positron decay as well as electron capture lead to its conversion to neodymium, while beta decay transforms it into samarium. The element under consideration exhibits its highest level of stability in the form of the isotope 145Pr. This particular isotope undergoes electron capture with a half-life of 17.7 years, as documented in reference (Gomes et al., 2014). Due to the presence of 84 neutrons, which exceeds the commonly recognized stable neutron configuration of 82, it is possible for this particular element to undergo alpha decay, resulting in the emission of an alpha particle consisting of 2 neutrons. Consequently, the resulting product of this decay process would be 141Pr, characterized by its 82 neutrons. Therefore, it can be concluded that this particular promethium isotope is the sole one that has been verified to undergo alpha decay. The alpha decay half-life of a 145Pm nucleus is approximately 6.3×10^9 years, with a relative probability of $2.8 \times 10^{-7}\%$ for decay to occur via this pathway. Additionally, there are several other isotopes of Pm, such as 144Pm, 146Pm, 147Pm, among others, that exhibit a positive energy discharge during the process of alpha decay. Although the occurrence of these alpha decays is expected, there is currently no experimental evidence to support their observation. All isotopes of promethium, characterized by an atomic number of 61, exhibit radioactive properties, much like technetium. One notable differentiation between Pm and the aforementioned Lns lies in its inability to generate tetravalent cations. Instead, Pm exclusively engages in compound formation through the relinquishment of two 6s and 14f electrons. In regards to the atomic radius, Promethium (Pm) is positioned as the third largest element within the lanthanide series. Nevertheless, it exhibits a slightly larger atomic radius compared to its neighboring elements, thus deviating from the prevailing trend of lanthanide contraction (Wendt et al., 2014).

The properties of promethium, such as its melting point, 1st 3 ionization energies, and hydration energies, exhibit an intermediate nature, reflecting an average of the corresponding properties of the elements preceding (neodymium) and succeeding (samarium) it in the periodic table. This trend is also observed in its chemical properties. It is important to acknowledge that due to its inherent instability, the element in question has not been extensively investigated in terms of its chemical properties. The production of soluble yellow $PmCl_3$ and pink $Pm(NO_3)_3$ can be achieved by dissolving the element in HCl or HNO_3. The addition of NH_3 to Pm^{3+} solutions results in an elevation of the pH, which subsequently triggers the creation of $Pm(OH)_3$, a water-insoluble compound exhibiting a light-brown coloration. Similar to $Ce_2(SO_4)_3$, $Pm_2(SO_4)_3$ exhibits a slight degree of solubility in water. Additionally, the oxalate compound of Pm demonstrates the lowest solubility in comparison to the oxalates of other lanthanides, as indicated by a previous study. Upon undergoing heating, the oxalate salt undergoes a transformation into a white or lavender amorphous oxide salt. The resulting salt exhibits properties that are more similar to the oxide of samarium, while displaying notable distinctions from those of neodymium. The cubic lattice of this oxide salt remains stable at temperatures up to 600°C. However, subjecting it to annealing at 800°C and subsequent heating up to 1750°C results in the irreversible development of a structure that exhibits a mixture of monoclinic as well as hexagonal phases. The mining of this element is not economically viable due to its low concentration in the earth's crust. Pitchblende, a mineral primarily composed of uranium, is known to contain the element Pm in trace amounts (4 parts per quintillion) resulting from the fission of uranium-238 (Freire et al., 2007).

5.7. SAMARIUM

In 1879, Paul Émile Lecoq de Boisbaudran successfully isolated the element from the mineral samarskite. The mineral in question was named after Vasili Samarsky-Bykhovets, a notable Russian mining official. Samarium, with an atomic number of 62, exhibits a higher degree of hardness in comparison to the previously discussed lanthanides. However, it undergoes oxidation upon exposure to atmospheric air. The element predominantly exhibits an oxidation state of +3, although there are also known compounds in which it assumes an oxidation state of +2 (such as SmO, SmS, SmSe, SmTe, and SmI_2). The hardness as well as density of samarium exhibit similarities to those of zinc. Samarium exhibits radioactive properties with an activity level of 128 Bq/g. Its naturally occurring samples consist of four stable isotopes, namely 144Sm, 150Sm, 152Sm, and 154Sm, alongside radioisotopes characterized by significantly long half-lives. These radioisotopes include 147Sm (with a half-life of 1.06×10^{11} years), 148Sm (with a half-life of 7×10^{15} years), and 149Sm (with a half-life exceeding 2×10^{15} years). The isotope 152Sm is the most prevalent, constituting 26.75% of the total weight. Certain references classify 149Sm as either stable or radioactive. The remaining isotopes exhibit the following half-lives: The half-life of 151Sm is 90 years, while the half-life of 145Sm is 340 days. The remaining isotopes exhibit progressively shorter half-lives, ranging from minutes to seconds. The isotopes 146Sm, 147Sm, as well as 148Sm, which possess relatively extended half-lives, experience alpha decay and give rise to isotopes of neodymium (Nd). The lighter isotopes, on the other hand, undergo electron capture decays which give rise to isotopes of Pm, while the heavier isotopes undergo beta decays and give rise to isotopes of Eu. The process of alpha-decay, specifically the transformation of 147Sm into ^{1}43Nd, serves as the fundamental principle behind the samarium-neodymium dating technique. This decay process has a half-life of 1.06×10^{11} years (Szostak & Procter, 2012).

The element, along with its 3+ oxide (Sm_2O_3, commonly referred to as sesquioxide), exhibits paramagnetic behavior at normal environmental conditions. However, when the temperature is lowered to 14.8 K, the metal demonstrates antiferromagnetic characteristics. Previous studies have documented the successful isolation for each samarium atom by means of encapsulation within fullerene molecules as well as the introduction of samarium atoms into C60 to induce superconducting properties at temperatures below 8 K. The literature has documented the observation of superconducting characteristics at temperatures reaching 56 K in iron-doped with Sm. Like other lanthanides, samarium exhibits a silvery luster when in its fresh state. However, it undergoes tarnishing upon exposure to air and can even undergo ignition at temperatures exceeding 150°C. The metallic state exhibits high reactivity, leading to tarnishing even when stored in the presence of mineral oil. This results in the formation of a surface layer consisting of a mixture of grayish-yellow oxide and hydroxide compounds. The behavior of the metallic form exhibits similarities to the previously examined lanthanides in various aqueous environments, including water, halogens, as well as dilute sulfuric acid solutions. However, it is important to note that the complex created by sulfuric acid differs in color, appearing as a pale green hue. In the divalent state, the cation induces a transformation in the aqueous solutions, resulting in the manifestation of blood-red hues. Samarium oxide (Sm_2O_3) is considered the most thermodynamically stable oxide of samarium, exhibiting various crystalline phases. Among these phases is the trigonal structure, which is attained when the molten salt is cooled. The monoclinic structure is an alternative crystalline arrangement of Sm_2O_3. This particular structure of Sm_2O_3 is synthesized using the flame fusion method, also known as the Verneuil process. This method produces transparent cylindrical crystals that can reach lengths of several centimeters (Molander & Hahn, 1986).

An alternative method involves subjecting the trigonal structure to thermal treatment at a temperature of 1900°C. Additionally, there have been documented reports regarding the cubic Sm_2O_3. The trivalent state of samarium (Sm(III)) is capable of forming salts with sulfide, selenide, and telluride. Samarium is one of the notable lanthanides that have the ability to form monoxides. Samarium monoxide (SmO), possessing a cubic crystal structure, exhibits a lustrous and radiant golden-yellow appearance. This compound can be synthesized by subjecting samarium(III) oxide (Sm_2O_3)

to reduction under the influence of metallic samarium at temperatures exceeding 1000°C and pressures above 50 kilobars. Additional divalent samarium compounds that exhibit a cubic rock-salt crystal structure include samarium sulfide (SmS), samarium selenide (SmSe), and samarium telluride (SmTe). The reduction of samarium trihalides (SmX_3) to dihalides (SmX_2) can be achieved by employing elemental samarium, lithium, and sodium at temperatures ranging from 700 to 900°C. The production of the diiodide salt can be accomplished through the reaction of metallic samarium with 1,2-diiodoethane in anhydrous tetrahydrofuran at standard temperature and pressure (Watt & Gillow, 1969).

$$Sm(s) + ICH_2\text{-} CH_2I \rightarrow SmI_2 + CH_2 = CH_2$$

The reduction of SmX_3 can lead to the generation of various nonstoichiometric crystalline halides, such as Sm_3F_7, $Sm_{14}F_{33}$, and $Sm_{27}F_{64}$, as well as $Sm_{11}Br_{24}$, Sm_5Br_{11}, and Sm_6Br_{13}. An intriguing characteristic observed in the halides of samarium is their proclivity to undergo crystal structure modifications in response to alterations in the halide component. This phenomenon has been elucidated through the identification of multiple halides exhibiting two distinct crystal phases characterized by varying degrees of stability. The conversion from the less stable form to the more stable form can be achieved through the application of pressure and thermal treatment. One example involves the transformation of monoclinic SmI_2 as well as SmI_3 compounds into orthorhombic structures resembling the $PbCl_2$ crystal lattice. This conversion occurs under pressure and results in densities of 5.90 g/cm^3 and 5.97 g/cm^3 for SmI_2 and SmI_3, respectively (Hou et al., 1988).

5.8. EUROPIUM

In 1890, P. É. Lecoq de Boisbaudran postulated the presence of europium, as he observed unidentified spectral lines in samarium-gadolinium mixtures that could not be ascribed to either samarium or gadolinium. The commencement of experimentation by Eugène-Anatole Demarçay, aimed at investigating potential contamination of recently found elemental samarium samples with an unidentified element, occurred in 1896. In the year 1901, the researcher successfully achieved the isolation of a novel element, which was subsequently designated as europium. Similar to samarium, this particular element exhibits characteristics of being a relatively rigid, lustrous, and malleable metal that undergoes oxidation upon exposure to both air and water. The European Union (EU) possesses a crystal lattice structure known as body-centered cubic (BCC). The half-filled electron shell of Eu is responsible for a significant number of its properties. Europium exhibits superconducting properties when subjected to compression exceeding 80 GPa and cooled to a temperature of 1.8 K. The conversion of the divalent atoms to its metallic state to its trivalent state occurs beneath the quantified conditions. The presence of a robust local magnetic moment (J = 7/2) in the divalent state of the element serves as additional evidence, as it exerts a suppressive influence on superconductivity. However, this effect disappears in the trivalent state (J = 0 in Eu^{3+}) (Antonio et al., 2015).

Similar to other lanthanides (Lns), europium (Eu) exhibits oxidation states of 3+ and 2+. The compounds associated with the 2+ oxidation state of europium are known to possess mild reducing properties. The element exhibits negligible biological effects and may be regarded as non-toxic in comparison to supplementary heavy metals. Europium (Eu) is considered to be the most reactive of all REE, exhibiting a reactivity level comparable to that of calcium. Consequently, when exposed to air, Eu undergoes oxidation at a significant rate, leading to complete oxidation of a cubic centimeter of the element within a few days. The elemental behavior inside the manifestation of water, dilute sulfuric acid, as well as air exhibits similarities to that of other Lanthanides, with the exception of its ignition occurring at a relatively higher temperature range of 150 to 180°C. As previously stated, in addition to its trivalent oxidation state, europium exhibits the ability to form divalent compounds, a characteristic that is relatively atypical among the majority of lanthanides. The divalent state exhibits similarities to the barium ion in terms of size as well as coordination number. The Eu^{2+} ions exhibit a notable degree of instability when exposed to atmospheric conditions, readily undergoing oxidation to Eu^{3+}. However, in an oxygen-deprived environment, these ions demonstrate a higher level of stability and consequently engage in exchanges with calcium minerals and other alkaline earth elements. This phenomenon is commonly known as the "negative europium anomaly," which pertains to the occurrence of lower-than-anticipated europium content in various lanthanide minerals. The examination of trace amounts of europium is frequently employed in geochemistry as well as petrology studies to gain insights into the processes involved in the formation of igneous rocks. This is done by examining the characteristics of the europium anomaly (Shapiro et al., 1971).

Europium samples found in nature consist of two isotopes, namely 151Eu and 153Eu. Among these isotopes, 153Eu is more stable and also the most abundant, accounting for approximately 52.2% of the total weight. The isotope 151Eu has been observed to exhibit instability and undergo radioactive decay, with a half-life of approximately 5×10^{18} years. The chemical element possesses an additional 35 synthetic radioisotopes, among which the most stable ones are 150Eu (with a half-life of 36.9 years), 152Eu (with a half-life of 13.516 years), as well as 154Eu (with a half-life of 8.593 years). The remaining synthetic radioisotopes exhibit half-lives ranging from a few years to a few seconds. The isotopes with atomic mass less than 153Eu predominantly undergo electron capture decay, resulting in the production of Sm. Conversely, the isotopes with higher atomic mass generally exhibit beta-minus decay, leading to the production of Gd (Donohue, 1977).

Europium (III) compounds typically exhibit coordination with six to nine ligands, predominantly containing oxygen atoms, such as water. Europium exhibits reactivity with various halogens, resulting in the formation of several compounds. These compounds include trivalent europium fluoride (EuF_3) with a white appearance, europium chloride ($EuCl_3$) with a yellow coloration, europium bromide ($EuBr_3$) with a gray hue, and europium iodide (EuI_3) which is colorless. Additionally, divalent europium fluoride (EuF_2) is produced, displaying a yellow-green shade, along with europium chloride ($EuCl_2$), europium bromide ($EuBr_2$) which is colorless, and europium iodide (EuI_2) which exhibits a green color. All compounds containing Eu^{3+} and the elements of the oxygen group exhibit stability, whereas Eu^{2+} exclusively forms stable compounds with S, Se, as well as Te. The European Union (EU) is known to possess three distinct oxides, namely EuO, Eu_2O_3, and Eu_3O_4, in which the oxidation states of Eu ions exist in both +2 and +3 states. Previous studies have documented the presence of Eu^{2+} in small quantities, which has been observed to enhance the fluorescence properties of certain mineral fluorite (CaF_2) specimens.

The commercial production of rare earths was significantly impacted by the ion-exchange technology pioneered by Frank Spedding during the mid-1950s. In order to obtain a concentrated form of Lns, the ores undergo a roasting process, which is subsequently followed by acidic and basic leaching. If the Ce content is present, it is then precipitated. Subsequent to the initial stages of solvent removals and ion-exchange elution method, further procedures are employed to enhance the concentration of europium. These procedures involve the reduction of the ionic species through the utilization of zinc, zinc/amalgam, electrolysis, or alternative reduction methods capable of converting Eu^{3+} into Eu^{2+} europium(II). Following reduction, the resulting europium(II) may be lead to a kind of carbonate or co-precipitated with barium sulfate. The production of Europium metal can also be achieved by means of electrolysis of molten $EuCl_3$ as well as NaCl (or $CaCl_2$) in a graphite cell, which serves as the cathode as well (Husband et al., 2012).

5.9. GADOLINIUM

Gadolinium (Gd) samples found in nature consist of six stable isotopes: 154Gd, 155Gd, 156Gd, 157Gd, 158Gd, and 160Gd. Additionally, there is a radioisotope, 152Gd, present. Among these isotopes, 158Gd has the highest abundance, accounting for 24.84 weight percent. In addition to the aforementioned isotopes, there have been reports of 28 additional synthetic radioisotopes for the element gadolinium. These include 152Gd with a half-life of 1.08 × 1014 years, and 150Gd with a half-life of 1.79 × 106 years. The isotopes that remain exhibit half-lives ranging from several decades to a few seconds. In general, isotopes with a mass lower than 158Gd tend to undergo electron capture reactions, resulting in the formation of isotopes of Eu. Conversely, isotopes with a higher mass decay through β decay, leading to the production of terbium isotopes (Good et al., 2022).

Gd, as the 64th element in the periodic table, shares similar visual characteristics, such as color and shine, with other lanthanides previously discussed. However, it exhibits distinctive properties when combined with other metals or alloys, leading to significant enhancements in temperature resistance for iron, chromium, and their respective alloys. The aforementioned effects can be observed even in cases where minute quantities of the element (1 wt%) are incorporated into the composition of the alloys. Moreover, due to its significant neutron absorption properties, elemental Gd or its salts are employed in shielding devices for neutron radiography and various nuclear applications.

The Gd^{3+} cation, which possesses a valence of three, exhibits fluorescence properties, making its salts suitable for utilization as green phosphors across various applications. One notable deviation in the conduct of this particular element, in comparison to its counterparts, is its ability to maintain stability within the atmosphere even in a lack of humidity. The mention of the influence of moist air on the development of Gd_2O_3 scales on the surface is unnecessary (Gordon et al., 2004).

$$4Gd + 3O_2 \rightarrow 2Gd_2O_3$$

While the behavior of gadolinium in the atmosphere may exhibit variations compared to the aforementioned elements, it demonstrates similar reactions in both cold and hot water, as well as when diluted in sulfuric acid. In the latter scenario, gadolinium forms $[Gd(H_2O)_9]^{3+}$. The aforementioned statement holds true in regards to the chemical interaction between the substance in question and halides. With the exception of the yellow triiodide, all three halides exhibit a white coloration. The trivalent state of lanthanides exhibits a characteristic inclination to form complexes with high coordination numbers. An illustrative instance of this behavior is its interaction with 1,4,7,10-tetraazacyclododecane-1,4,7,10-tetraacetic acid (DOTA), whose salts find utility in magnetic resonance imaging (MRI) applications. Gadolinium in its elemental state exhibits pronounced reducible characteristics, effectively facilitating the reduction of various metal oxides to their elemental counterparts. The divalent state may be achieved by subjecting a combination of the trivalent cation as well as the metallic form to thermal treatment in a tantalum crucible under an inert atmosphere. One example of an unconventional gadolinium compound is the sesquichloride (Gd_2Cl_3), that has the ability to produce layers resembling graphite-like structures of GdCl when subjected to annealing at a temperature of 800°C (Andruh et al., 1993).

5.10. TERBIUM

Terbium, similar to lanthanum and praseodymium, was discovered in 1843 by C. G. Mosander, a Swedish scientist. The initial observation of the element's existence occurred when it was detected as an impurity within Y_2O_3. However, the isolation of the element was not achievable until the advancement of ion exchange techniques. During his exploratory endeavors, Mosander successfully fractionated yttria into three distinct compounds. Notably, one of these fractions, later identified as terbia, exhibited a distinctive pink hue. Subsequently, it was discovered that the color of this compound is attributed to the presence of erbium, an element known as such in contemporary scientific nomenclature (Mukherjee et al., 2011).

Terbium in its natural state consists solely of its stable isotope, 159Tb, thus earning the designation of mononuclidic or monoisotopic. A total of thirty-six artificial radioisotopes have been thoroughly examined for the element in question, with the most massive and least massive isotopes being 171Tb as well as 135Tb, respectively. The synthetic radioisotopes of terbium that exhibit the highest level of stability are 158Tb, with a half-life of 180 years, and 157Tb, with a half-life of 71 years. The remainder of radioactive isotopes exhibit half-lives that span from less than three months to less than thirty seconds. The predominant mode of decay preceding the most prevalent stable isotope, namely 159Tb, is electron capture. This process leads to the formation of gadolinium isotopes, that subsequently undergo beta decay, resulting in the production of dysprosium isotopes.

The characteristics of the element exhibit a notable resemblance to its adjacent element, gadolinium, and therefore display significant distinctions from the lighter lanthanides. Terbium, the element with atomic number 65 in the periodic table, exhibits two primary allotropes that undergo transformation between each other at temperatures either below or above 1289°C. The emission spectrum of the Tb^{3+} cation exhibits a vibrant lemon-yellow hue. In addition to its prevalent 3+ oxidation state, as observed in compounds such as Tb_2O_3, the element terbium also exhibits a 4+ oxidation state, exemplified by compounds like TbO_2 and TbF_4. The combustion of terbium has been documented to result in the formation of mixed oxides containing Tb^{3+} and Tb^{4+} ions, as indicated by the following reaction (Matijević & Hsu, 1987):

$$8Tb + 7O_2 + \rightarrow 2Tb_4O_7$$

The reactions of the element with water, halogens, and sulfuric acid exhibit similarities to those of the previously discussed lanthanides. These reactions can be succinctly summarized as follows:

$$2Tb + 6H_2O \rightarrow 2Tb(OH)_3 + 3H_2\uparrow$$

$$2Tb + 3X_2 \rightarrow 2TbX_3\ (X = F, Cl, Br, I)$$

$$2Tb(s) + 3H_2SO_4 \rightarrow 2Tb_3+ + 3SO_4^{2-} + 3H_2\uparrow$$

In addition, it has been documented that this particular element can generate $Tb^{2+}$and Tb^{3+} compounds, such as TbH_2, TbH_3, TbB_2, Tb_2S_3, TbSe, TbTe, and TbN, when subjected to high temperatures and reacted with carbon, nitrogen, sulfur, boron, selenium, phosphorus, silicon, as well as arsenic. The possibility of forming Tb^{2+} compounds arises from the co-annealing process involving elemental terbium and 3+ terbium compounds, which occurs under conditions that are comparable to those observed in the case of gadolinium. A further resemblance to gadolinium is the ability to form sesquichloride (Tb_2Cl_3), which can be reduced to TbCl under the same conditions as observed in the case of Gd (Akman et al., 2020).

5.11. DYSPROSIUM

Dysprosium, an element, was first isolated by Paul Émile Lecoq de

Boisbaudran in 1886. The isolation process involved the utilization of holmium oxide, wherein dysprosium oxide was dissolved in an acidic solution. Subsequently, the addition of NH_3 was employed to elevate the pH level, thereby facilitating the precipitation of dysprosium hydroxide. The researchers made over 30 attempts to successfully isolate the element, leading to the adoption of a name derived from the Greek term "dysprositos," denoting the challenging nature of its acquisition.

The acquisition of the element in its pure form, however, had to be postponed until the 1950s when ion-exchange techniques were introduced. Dysprosium found in nature consists of several stable isotopes, namely 156Dy, 158Dy, 160Dy, 161Dy, 162Dy, 163Dy, and 164Dy (Benelli et al., 1990). It is worth noting that only the isotope 156Dy exhibits radioactive decay, with a half-life of 1×10^{18} years. The isotopes 164Dy and 162Dy are the predominant ones, constituting approximately 54% of the overall weight of the samples. A total of 29 synthetic radioisotopes, spanning mass numbers from 138 to 173, have been documented for the aforementioned element. The two most stable isotopes are dysprosium-154 (with a half-life of 3×10^6 years) and dysprosium-159 (with a half-life of 144.4 days). In general, it can be observed that heavier isotopes tend to undergo b+ decays, while lighter isotopes tend to undergo b– decays. However, it is important to note that there are exceptions to this trend. The elemental characteristics are indistinguishable from those of its adjacent counterparts. The lanthanides, specifically dysprosium and holmium, exhibit the most pronounced magnetic properties, particularly when subjected to low temperatures (Ishikawa et al., 2005).

Dysprosium metal exhibits similar reactivity patterns to other elements within the lanthanide family when exposed to O_2, water, and halogens. The trifluoride, triiodide, and tribromide compounds exhibit a green coloration, while the trichloride compound displays a white color. The paramagnetic nature of the sulfate salt is observed when it is dissolved in a solution of dilute sulfuric acid. The main providers of the element are situated in China (106) and Australia (107). According to reports, ores possessing elevated yttrium levels have been found to predominantly consist of dysprosium, which accounts for approximately 7% to 8% of the heavy lanthanide element. The procedure for isolating the element entails the process of purifying of yttrium and commences with the utilization of monazite, a composite of diverse phosphates. Unwanted constituents are eliminated via the implementation of magnetic separation or flotation techniques. Other methods for obtaining pure dysprosium involve ion-exchange procedures or thermal reactions of

DyF_3 or $DyCl_3$ with calcium or lithium metals in the presence of an inert atmosphere, as depicted below:

$$3Ca + 2DyF_3 \rightarrow 2Dy + 3\ CaF_2$$

$$3Li + DyCl_3 \rightarrow Dy + 3LiCl$$

Throughout the procedure, the by-products as well as molten dysprosium are distinguished from one another by taking into account their disparities in density (Biju et al., 2003).

5.12. HOLMIUM

Similar to numerous other lanthanides (Lns), the identification of holmium was predicated upon the observation of distinctive spectral lines within specimens, which had not been previously observed. The aforementioned task was carried out by Marc Delafontaine as well as Jacques-Louis Soret in the year 1878. It was also pursued independently by T. Cleve during his research on erbium oxide. The individual employed C. G. Mosander's technique to eliminate the recognized impurities from erbia, resulting in the production of two novel substances with brown and green hues (Yan & Chen, 2001). These substances were subsequently designated as holmia (in reference to the Latin appellation for Stockholm) and thulia, correspondingly. Subsequent studies revealed that the substance referred to as "holmia" was, in fact, holmium oxide, while the other substance was identified as thulium oxide. Holmium found in nature consists solely of the stable isotope 165Ho. However, there have been reports of synthetic radioisotopes, such as 163Ho (with a half-life of 4570 years). The remaining radioisotopes exhibit significantly brief half-lives, typically ranging from several days to a few hours (Babailov et al., 2014).

At room temperature, holmium exhibits paramagnetic behavior, while at temperatures below 19 K, it displays ferromagnetic properties. Additionally, holmium possesses a significantly high magnetic moment of 10.6 μB and exhibits various unconventional magnetic characteristics. Furthermore, it has the ability to form highly magnetic compounds when combined with yttrium. Similar to its neighboring elements, Ho (with an atomic number of 67) exhibits a high level of resistance to corrosion when exposed to dry air or maintained at ambient temperature. However, when subjected to moist air or elevated temperatures, Ho undergoes oxidation, resulting in the formation of a yellowish compound that causes the metallic, bright silvery appearance of the element to tarnish. The color of this oxide is subject to variations in lighting conditions, exhibiting an orange-red hue when illuminated with trichromatic light (Kay et al., 1972).

Holmium exhibits similar reactions to water, halogens, as well as dilute sulfuric acid solutions. Its longest-lasting oxidation state is 3+. Ho, like other REEs, has a relatively low abundance in the earth's crust, with concentrations of approximately 1.3 parts per million (ppm). In soil, the concentration is around 1 ppm, while in seawater, it is even lower at 400 parts per quadrillion (ppq). This element is typically found in conjunction with other lanthanide compounds in minerals such as gadolinite, monazite, and various lanthanide minerals. These minerals are widely distributed in countries such as China, the United States, Brazil, Sri Lanka, India, as well as Australia. Commercial acquisition of this element is achieved through ion-exchange procedures or the thermal reduction of its trichloride and trifluoride using metallic calcium, while maintaining an inert atmosphere (Blagg et al., 2011).

5.13. ERBIUM

The discovery of the element was credited to C. G. Mosander in 1843. Mosander successfully fractionated "yttria," derived from gadolinite, into three distinct components. The three fractions were denoted as yttria, erbia, as well as terbia by the individual in question. There was initially confusion regarding the last two elements. In the year 1905, Georges Urbain as well as Charles James successfully synthesized Er_2O_3 in its pure form. However, it was not until 1934 that the pure metallic form of erbium was prepared. This achievement was accomplished by Klemm and Bommer, who utilized potassium vapor to reduce the chloride salt of Er(III). The natural erbium samples contain stable isotopes, namely 162Er, 164Er, 166Er, 167Er, 168Er, as well as 170Er. Among these isotopes, 166Er exhibits the highest concentration, accounting for 33.50 weight percent (Ramakrishnan & Prasada Rao, 2006). Approximately 29 synthetic isotopes were documented for erbium, with 169Er (with a half-life of 9.4 days) being the most stable among them. Other notable isotopes include 172Er (with a half-life of 49.3 hours), 160Er (with a half-life of 28.58 hours), 165Er (with a half-life of 10.36 hours), as well as 171Er (with a half-life of 7.516 hours). Additionally, there exist several other isotopes of erbium with significantly shorter half-lives. Electron capture is the predominant decay mode observed in isotopes with a mass lighter than 166Er, resulting in the production of holmium isotopes. Conversely, beta decay is the prevalent decay mechanism for heavier isotopes, leading to the emergence of thulium isotopes. Erbium, with an atomic number of 68, exhibits similar physical properties to other lanthanides. However, unlike its lighter lanthanide counterparts, erbium demonstrates stability in dry air but reacts with moist water and air, resulting in the formation of compounds with a rose coloration. The efficacy of the element as a biological agent has not been substantiated, although it has been observed to exert metabolic effects (Schumann et al., 2002).

Er exhibits distinct magnetic properties depending on temperature. Below 19 K, it demonstrates ferromagnetic characteristics, while in the temperature range of 19 to 80 K, it displays antiferromagnetic behavior. At temperatures exceeding 80 K, Er exhibits paramagnetic properties. With the exception of the pink hue exhibited by its trifluoride compound, the violet color displayed by the remaining trihalides, and the red color observed in the solutions of Er in dilute sulfuric acid, the chemical reactions of erbium with water, oxygen, halogens, as well as acid solutions are consistent with the previously described characteristics. Erbium, a REE, is found in the earth's crust at approximately 2.8 parts per million (ppm) (Paschalidis et al., 2000). However, similar to other REEs, erbium is not naturally occurring in its elemental state. The minerals xenotime and euxenite are known to contain erbium, and the extraction of erbium from these ores follows a procedure similar to that used for other lanthanides. This procedure involves the separation of insoluble cerium oxide via HNO_3, which effectively dissolves the remaining elements. The subsequent step involves filtration of the solution containing cerium oxide, followed by treatment with magnesium nitrate. This process results in the formation of a crystallized mixture of double salts of rare-earth metals. These double salts are then utilized in ion exchange procedures, which are subsequently followed by elution using ligands. Another conventional technique entails the reduction of the oxide or halides by subjecting them to high temperatures in the presence of a reducing agent, such as calcium or lithium vapor, at a temperature of 1450°C, within an argon atmosphere (Faridbod et al., 2007).

5.14. THULIUM

Thulium was discovered by P. T. Cleve in 1879 during his investigation of rare-earth oxides for the presence of impurities. After the successful removal of Er_2O_3, the researcher encountered two oxides with unfamiliar brown and green appearances. These oxides were subsequently named holmia as well as thulia, respectively. The element in question was originally represented by the symbol "Tu," but it is currently denoted by the symbol "Tm." The attainment of pure thulium was significantly challenging due to the limited availability of this element. As a result, the research conducted on the element primarily focused on spectroscopic analysis and centered around the observed amplification in the intensity of the two distinct absorption bands. This amplification was attributed to the progressive elimination of erbium from the system. The initial documentation regarding the acquisition of pure thulium was presented by Charles James in 1911. The researcher employed a method known as bromate fractional crystallization following a series of 15,000 purification procedures. In contrast to the prevailing trend among Lns, the commercial availability of this element was not realized until the latter part of the 1950s and was facilitated by the renowned ion-exchange technique (Moutet et al., 2021).

The isotopes of thulium span the entire range from 146Tm to 177Tm. The isotope 169Tm is considered to be the most stable as well as abundant among its counterparts. Isotopes with a mass less than 169Tm have been observed to go through electron capture reactions, resulting in the conversion of thulium (Tm) to erbium. Conversely, isotopes with a greater mass yield ytterbium through beta emissions. The isotope 169Tm is regarded as stable, despite theoretical predictions suggesting it should undergo decay with an extremely lengthy half-life. Conversely, the remaining isotopes of thulium are classified as unstable. Among these isotopes, the most stable ones are 171Tm (with a half-life of 1.92 years) as well as 170Tm (with a half-life of 128.6 days). The remaining isotopes exhibit half-life values that range from a few minutes to a few seconds. Thulium, the 69th element in the periodic table, exhibits chemical properties that are largely similar to other lanthanides. It is characterized as a silvery, soft, and ductile solid metal, existing in two allotropes: tetragonal α-Tm as well as the more stable hexagonal β-Tm. Thulium gradually undergoes tarnishing when exposed to air, primarily due to its most prevalent oxidation state of 3+. This oxidation state is observed in various thulium compounds, including its oxide, halides, and other chemical compounds. According to the literature, liquid thulium has been observed to exhibit a high degree of volatility (Kubra et al., 2023). The element Tm exhibits ferromagnetic behavior at temperatures below 32 K, antiferromagnetic behavior within the temperature range of 32–56 K, and paramagnetic behavior at temperatures above 56 K. Thulium exhibits reactivity towards various substances, including water, halogens (yielding products of yellow and white appearance), sulfuric acid, as well as moist air. The element exhibits the ability to form various compounds with both metallic and nonmetallic elements. Several examples of thulium compounds are known, such as TmN, TmC_2, Tm_2C_3, TmS, TmH_2, $TmSi_2$, $TmGe_3$, TmH_3, TmB_4, TmB_6, and TmB_{12}. It is worth noting that in certain cases, thulium can adopt the less stable 2+ oxidation states alongside the more common 3+ state. Tm^{2+} compounds exhibit a reddish-purple coloration and can be synthesized by means of reducing Tm^{3+} ions. However, it should be noted that these compounds lack stability. The compound $TmCl_2$ exhibits highly reactive behavior when exposed to water, resulting in the production of hydrogen gas and the formation of $Tm(OH)_3$, which is characterized by a diminishing red hue. The chemical interactions involving thulium and hydrochloric acid (HCl) or nitric acid (HNO_3) yield the production of thulium chloride ($TmCl_3$) and thulium nitrate ($Tm(NO_3)_3$), along with the liberation of hydrogen gas. Thulium is frequently encountered in ores that contain yttrium and gadolinium, such as gadolinite. Additionally, it can be found in monazite, xenotime, as well as euxenite, which are plentiful in various regions including Brazil, Australia, Greenland, India, Tanzania, as well as the United States. The concentrations of this element in the earth's crust and seawater are 50 parts per billion (ppb) and 250 parts per quadrillion (ppq), respectively. The metallic form of lanthanum can be synthesized through the reduction of lanthanum oxide under the influence of calcium, similar to the process observed for other lanthanides (Justice et al., 1969).

5.15. YTTERBIUM

Ytterbium was discovered by Jean Charles Galissard de Marignac in the year 1878 during his research involving gadolinite. The element was designated with the appellation derived from the Swedish locality known as Ytterby, in close proximity to which the element's discovery occurred. Ytterbium is composed of several isotopes, namely ^{168}Yb, ^{170}Yb, ^{171}Yb, ^{172}Yb, ^{173}Yb, ^{174}Yb, as well as ^{176}Yb. Among these isotopes, ^{174}Yb is the most prevalent, accounting for 31.8% of the total weight. This particular chemical element possesses a total of 27 radioisotopes. Notable among these are the relatively stable isotopes 169Yb (with a half-life of 32.0 days), 175Yb (with a half-life of 4.18 days), and 166Yb (with a half-life of 56.7 hours). Additionally, there exist several less stable isotopes with half-lives ranging from a few hours to a few minutes (Tilley et al., 1980).

The 2+ oxidation state exhibits a distinctive stability due to its electron configuration of $4f^{14}$. The divalent ion exhibits a green-yellow hue, possesses strong reducing properties, and undergoes a reaction with water, resulting in the release of H_2 gas and the formation of a colorless ytterbium(III) ion. However, it is worth noting that most of ytterbium compounds are still classified as being in the 3+ oxidation state. Similar to the situation observed with the elements Eu, Sm, as well as Tm, the ytterbium trihalides have the capability to undergo conversion into dihalides through the utilization of pure ytterbium alongside a reducing agent, such as zinc. The physical properties of elemental ytterbium are indistinguishable from those of other elements. Similarly, its reactivity towards mineral acids, halogens, air, as well as water is comparable to that of other elements. Notably, ytterbium reacts with water, resulting in the production of hydrogen gas. Ytterbium undergoes a gradual process of oxidation when exposed to atmospheric conditions, with the rate of this reaction being more rapid for smaller particles of ytterbium (Heckmann & Niemeyer, 2000).

The utilization of ytterbium (III) chloride as a catalyst has been documented in the Aldol as well as Diels–Alder reactions. Ytterbium iodide (YbI_2) has been identified as a viable reducing agent in coupling reactions, as documented in reference. The inert and nontoxic nature of YbF_3 has been documented in literature as a suitable material for dental fillings. Additionally, it has been observed that YbF_3 exhibits a continuous release of fluoride ions, which are beneficial for dental health. Ytterbium, with an abundance of 3 parts per million in the earth's crust, is a constituent of ores such as monazite (0.03 weight percent), as well as euxenite and xenotime, that are distributed across various countries including China, the US, Brazil, Sri Lanka, India, as well as Australia. The utilization of minute quantities of this particular element as dopants in Yb: YAG lasers has been documented. The process is highly time-consuming due to the inherent challenge of isolating the element from other Lns. The ores are first dissolved via inorganic acids and subsequently undergo separation through ion-exchange procedures. An alternative method for separation entails the utilization of a sodium amalgam reduction process. In this procedure, a buffered acidic solution containing the trivalent cation is exposed to a molten alloy of sodium and mercury. Consequently, the cation of the element is converted into its reduced form (Girard et al., 1980).

5.16. LUTETIUM

Lutetium was independently discovered as an impurity in Ytterbia in 1907 by Georges Urbain, Baron Carl Auer von Welsbach, as well as Charles James. Natural samples of lutetium consist of two isotopes: stable lutetium-175 and radioactive lutetium-176, which has a half-life of 3.78 × 10^10 years and undergoes beta decay. The element exhibits a total of 32 artificially produced radioisotopes ranging from 150Lu to 184Lu. Notably, the most stable isotopes within this range are 174Lu with a half-life of 3.31 years and 173Lu with a half-life of 1.37 years. The remaining isotopes have significantly shorter half-lives, ranging from a few days to a few minutes. The isotopes with masses below 175Lu predominantly undergo electron capture, accompanied by some emissions (Xu et al., 2016). In contrast, the heavier isotopes have been observed to decay through alpha and positron emission, resulting in the production of hafnium isotopes. The element exhibits physical characteristics that are indistinguishable from those of its neighboring elements. Typically, the prevailing oxidation state for the element is the colorless 3+, that is observed in numerous compounds of the element. Upon crystallization, the soluble salts of lutetium, namely $Lu(NO_3)_3$, $Lu_2(SO_4)_3$, and $Lu(CH_3COO)_3$, undergo hydration to form hydrates. Other compounds such as lutetium oxide (Lu_2O_3), lutetium hydroxide ($Lu(OH)_3$), lutetium fluoride (LuF_3), lutetium carbonate ($Lu_2(CO_3)_3$), lutetium phosphate ($LuPO_4$), and lutetium oxalate ($Lu_2(C_2O_4)_3$) exhibit limited solubility in aqueous solutions (Flotow & O'Hare, 1981). Lutetium metal exhibits a notable degree of instability when exposed to atmospheric conditions, and, akin to other elements in the lanthanide series, it undergoes combustion at a temperature as low as 150°C. The oxide of lutetium, which is produced in this manner, is widely recognized as a sorbent for carbon dioxide and water. The chemical interactions involving the element with water, halogens, as well as sulfuric acids exhibit similarities to those observed in other Lanthanides. Lutetium exhibits an abundance of 0.5 parts per million (ppm) in the earth's crust, yet its occurrence in minerals such as monazite is characterized by significantly lower concentrations, specifically 0.0001 weight percent (wt%). The isolation of the element from its minerals poses significant challenges due to the ion-exchange procedures and the reduction of its chlorides in the existence of calcium under high temperatures and inert atmospheres (Sessler et al., 1997).

5.17. SUMMARY

The lanthanides refer to a group of 15 elements located on the Periodic Table, characterized by their atomic numbers falling within the range of 57 to 71. These elements are commonly known as REMs and inner transition metals. The placement of these elements below the primary set of elements on the periodic table has resulted in their segregation from the remaining elements. The primary factor contributing to this division is the presence of electrons in the f-sublevel within the lanthanides.

In the realm of lanthanide chemistry, the commonly employed informal chemical symbol "Ln" is utilized to denote any member of the lanthanide series during general discussions. With the exception of one, the lanthanides are classified as f-block elements, which aligns with the occupation of the 4f electron shell. The classification of lutetium as a d-block element, and consequently as a transition metal, has raised concerns regarding its categorization. Nevertheless, it exhibits similar behavior to the other 14 elements, including scandium as well as yttrium in group 3. The trivalent cations, Ln^{3+}, formed by all lanthanide elements, exhibit chemical properties that are primarily influenced by their ionic radius. This ionic radius demonstrates a gradual decrease as one progresses from lanthanum (La) to lutetium (Lu).

REFERENCES

1. Aiken, B. A. R., Hsu, W. P., & Matijević, E., (1988). Preparation and properties of monodispersed colloidal particles of lanthanide compounds: III, yttrium (III) and mixed yttrium (III)/cerium (III) systems. *Journal of the American Ceramic Society, 71*(10), 845–853.
2. Akman, E., Akin, S., Ozturk, T., Gulveren, B., & Sonmezoglu, S., (2020). Europium and terbium lanthanide ions co-doping in TiO_2 photoanode to synchronously improve light-harvesting and open-circuit voltage for high-efficiency dye-sensitized solar cells. *Solar Energy, 202*, 227–237.
3. Andruh, M., Ramade, I., Codjovi, E., Guillou, O., Kahn, O., & Trombe, J. C., (1993). Crystal structure and magnetic properties of [Ln_2Cu_4] hexanuclear clusters (where Ln= trivalent lanthanide). Mechanism of the gadolinium (III)-copper (II) magnetic interaction. *Journal of the American Chemical Society, 115*(5), 1822–1829.
4. Antonio, M. R., McAlister, D. R., & Horwitz, E. P., (2015). An europium (III) diglycolamide complex: Insights into the coordination chemistry of lanthanides in solvent extraction. *Dalton Transactions, 44*(2), 515–521.
5. Babailov, S. P., Dubovskii, P. V., & Zapolotsky, E. N., (2014). Paramagnetic lanthanides as magnetic resonance thermo-sensors and probes of molecular dynamics: Holmium-DOTA complex. *Polyhedron, 79*, 277–283.
6. Banerjee, S., Huebner, L., Romanelli, M. D., Kumar, G. A., Riman, R. E., Emge, T. J., & Brennan, J. G., (2005). Oxoselenido clusters of the lanthanides: Rational introduction of oxo ligands and near-IR emission from Nd (III). *Journal of the American Chemical Society, 127*(45), 15900–15906.
7. Benelli, C., Caneschi, A., Gatteschi, D., Guillou, O., & Pardi, L., (1990). Synthesis, crystal structure, and magnetic properties of tetranuclear complexes containing exchange-coupled dilanthanide-dicopper (lanthanide= gadolinium, dysprosium) species. *Inorganic Chemistry, 29*(9), 1750–1755.
8. Biju, V. M., Gladis, J. M., & Rao, T. P., (2003). Effect of γ-irradiation of ion imprinted polymer (IIP) particles for the preconcentrative separation of dysprosium from other selected lanthanides. *Talanta, 60*(4), 747–754.

9. Blagg, R. J., Tuna, F., McInnes, E. J., & Winpenny, R. E., (2011). Pentametallic lanthanide-alkoxide square-based pyramids: High energy barrier for thermal relaxation in a holmium single molecule magnet. *Chemical Communications, 47*(38), 10587–10589.
10. Bradley, D. C., Chudzynska, H., Hursthouse, M. B., Motevalli, M., & Wu, R., (1994). Volatile fluorinated tertiary alkoxides of some lanthanides, tris-hexafluoro-tertiary butoxides of lanthanum, praseodymium, and europium. *Polyhedron, 13*(1), 1–6.
11. Campazzi, E., Solari, E., Floriani, C., & Scopelliti, R., (1998). The fixation and reduction of dinitrogen using lanthanides: Praseodymium and neodymium meso-octaethylporphyrinogen–dinitrogen complexes. *Chemical Communications*, (23), 2603–2604.
12. Cendrowski-Guillaume, S. M., Le Gland, G., Nierlich, M., & Ephritikhine, M., (2000). Lanthanide borohydrides as precursors to organometallic compounds. Mono (cyclooctatetraenyl) neodymium complexes. *Organometallics, 19*(26), 5654–5660.
13. Cheisson, T., Kersey, K. D., Mahieu, N., McSkimming, A., Gau, M. R., Carroll, P. J., & Schelter, E. J., (2019). Multiple bonding in lanthanides and actinides: Direct comparison of covalency in thorium (IV)-and cerium (IV)-imido complexes. *Journal of the American Chemical Society, 141*(23), 9185–9190.
14. Dahle, J. T., & Arai, Y., (2015). Environmental geochemistry of cerium: Applications and toxicology of cerium oxide nanoparticles. *International Journal of Environmental Research and Public Health, 12*(2), 1253–1278.
15. De Almeida, L., Grandjean, S., Vigier, N., & Patisson, F., (2012). Insights into the thermal decomposition of lanthanide (III) and actinide (III) oxalates–from neodymium and cerium to plutonium. *European Journal of Inorganic Chemistry, 2012*(31), 4986–4999.
16. De Smet, F., Ruiz, P., Delmon, B., & Devillers, M., (2001). Rationalization of the catalytic behavior of lanthanide oxides and praseodymium molybdates in total and selective oxidation of isobutene. *The Journal of Physical Chemistry B*, *105*(49), 12355–12363.
17. Donohue, T., (1977). Photochemical separation of europium from lanthanide mixtures in aqueous solution. *The Journal of Chemical Physics, 67*(11), 5402–5404.
18. Edelmann, F. T., Steiner, A., Stalke, D., Gilje, J. W., Jagner, S., &

Mikael, H., (1994). Lanthanide alkoxides—III. Four-coordinate anionic neodymium (III) alkoxides and amides. *Polyhedron, 13*(4), 539–546.

19. Faridbod, F., Ganjali, M. R., Larijani, B., Norouzi, P., Riahi, S., & Mirnaghi, F. S., (2007). Lanthanide recognition: An asymetric erbium microsensor based on a hydrazone derivative. *Sensors, 7*(12), 3119–3135.
20. Feng, L., Xiao, H., He, X., Li, Z., Li, F., Liu, N., & Chai, Z., (2006). Neurotoxicological consequence of long-term exposure to lanthanum. *Toxicology Letters, 165*(2), 112–120.
21. Flotow, H. E., & O'Hare, P. A. G., (1981). Thermodynamics of the lanthanide trifluorides. IV. The heat capacities of gadolinium trifluoride GdF_3, lutetium trifluoride LuF_3, and yttrium trifluoride YF_3 from 5 to 350 K. *The Journal of Chemical Physics, 74*(5), 3046–3055.
22. Freire, R. O., Da Costa, N. B., Rocha, G. B., & Simas, A. M., (2007). Sparkle/PM3 parameters for the modeling of neodymium (III), promethium (III), and samarium (III) complexes. *Journal of Chemical Theory and Computation, 3*(4), 1588–1596.
23. Fujii, T., Nagai, T., Sato, N., Shirai, O., & Yamana, H., (2005). Electronic absorption spectra of lanthanides in a molten chloride: II. Absorption characteristics of neodymium (III) in various molten chlorides. *Journal of Alloys and Compounds, 393*(1, 2), L1–L5.
24. Ganjali, M. R., Norouzi, P., Mirnaghi, F. S., Riahi, S., & Faridbod, F., (2007). Lanthanide recognition: Monitoring of praseodymium (III) by a novel praseodymium (III) microsensor based on N $'$-(pyridin-2-ylmethylene) benzohydrazide. *IEEE Sensors Journal, 7*(8), 1138–1144.
25. Garner, J. P., & Heppell, P. S. J., (2005). Cerium nitrate in the management of burns. *Burns, 31*(5), 539–547.
26. Geoffrey N. C. F., (1993). Zero oxidation state compounds of scandium, yttrium, and the lanthanides. *Chemical Society Reviews, 22*(1), 17–24.
27. Girard, P., Namy, J. L., & Kagan, H. B., (1980). Divalent lanthanide derivatives in organic synthesis. 1. Mild preparation of samarium iodide and ytterbium iodide and their use as reducing or coupling agents. *Journal of the American Chemical Society, 102*(8), 2693–2698.
28. Gomes, D. J. C., Caires, F. J., Silva, R. C., Treu-Filho, O., & Ionashiro, M., (2014). Synthesis, characterization, thermal and spectroscopic

studies of solid glycolate of light trivalent lanthanides, except promethium. *Thermochimica Acta, 587*, 33–41.

29. Good, N. M., Lee, H. D., Hawker, E. R., Su, M. Z., Gilad, A. A., & Martinez-Gomez, N. C., (2022). Hyperaccumulation of gadolinium by methylorubrum extorquens AM1 reveals impacts of lanthanides on cellular processes beyond methylotrophy. *Frontiers in Microbiology, 13*, 820327.
30. Gordon, W. O., Carter, J. A., & Tissue, B. M., (2004). Long-lifetime luminescence of lanthanide-doped gadolinium oxide nanoparticles for immunoassays. *Journal of Luminescence, 108*(1–4), 339–342.
31. Handl, H. L., & Gillies, R. J., (2005). Lanthanide-based luminescent assays for ligand-receptor interactions. *Life Sciences, 77*(4), 361–371.
32. Heckmann, G., & Niemeyer, M., (2000). Synthesis and first structural characterization of lanthanide (II) aryls: Observation of a Schlenk equilibrium in europium (II) and ytterbium (II) chemistry. *Journal of the American Chemical Society, 122*(17), 4227–4228.
33. Hou, Z., Fujiwara, Y., & Taniguchi, H., (1988). Lanthanides in organic synthesis. Samarium metal promoted selective formation of azoxy compounds. *The Journal of Organic Chemistry, 53*(13), 3118–3120.
34. Hower, J. C., Groppo, J. G., Joshi, P., Dai, S., Moecher, D. P., & Johnston, M. N., (2013). Location of cerium in coal-combustion fly ashes: Implications for recovery of lanthanides. *Coal Combustion and Gasification Products, 5*(4), 73–78.
35. Husband, R. J., Loa, I., Stinton, G. W., Evans, S. R., Ackland, G. J., & McMahon, M. I., (2012). Europium-IV: An incommensurately modulated crystal structure in the lanthanides. *Physical Review Letters, 109*(9), 095503.
36. Ishikawa, N., Sugita, M., & Wernsdorfer, W., (2005). Quantum tunneling of magnetization in lanthanide single-molecule magnets: Bis (phthalocyaninato) terbium and bis (phthalocyaninato) dysprosium anions. *Angewandte Chemie International Edition, 44*(19), 2931–2935.
37. Justice, B. H., Westrum, Jr. E. F., Chang, E., & Radebaugh, R., (1969). Thermophysical properties of the lanthanide oxides. IV. Heat capacities and thermodynamic properties of thulium (III) and lutetium (III) oxides. Electronic energy levels of several lanthanide (III) ions. *The Journal of Physical Chemistry, 73*(2), 333–340.
38. Kajjumba, G. W., & Marti, E. J., (2022). A review of the application

of cerium and lanthanum in phosphorus removal during wastewater treatment: Characteristics, mechanism, and recovery. *Chemosphere, 410*, 136462.

39. Kay, J., Moore, J. W., & Glick, M. D., (1972). Structural studies of bridged lanthanide (III) complexes. Diaquotri (nicotinic acid) holmium (III) hexa (isothiocyanato) chromate (III) dihydrate and diaquotris (isonicotinato) lanthanum (III). *Inorganic Chemistry, 11*(11), 2818–2827.
40. Kragten, J., & Decnop-Weever, L. G., (1987). Hydroxide complexes of lanthanides—VIII: Lanthanum (III) in perchlorate medium. *Talanta, 34*(10), 861–864.
41. Krumpel, A. H., Van, D. K. E., Zeelenberg, D., Bos, A. J. J., Krämer, K. W., & Dorenbos, P., (2008). Lanthanide 4f-level location in lanthanide doped and cerium-lanthanide codoped $NaLaF_4$ by photo-and thermoluminescence. *Journal of Applied Physics, 104*(7), 687–690.
42. Kubra, K. T., Hasan, M. M., Hasan, M. N., Salman, M. S., Khaleque, M. A., Sheikh, M. C., & Awual, M. R., (2023). The heavy lanthanide of thulium (III) separation and recovery using specific ligand-based facial composite adsorbent. *Colloids and Surfaces A: Physicochemical and Engineering Aspects, 667*, 131415.
43. Kulkarni, P., Chellam, S., & Fraser, M. P., (2006). Lanthanum and lanthanides in atmospheric fine particles and their apportionment to refinery and petrochemical operations in Houston, TX. *Atmospheric Environment, 40*(3), 508–520.
44. Liu, Y., Shi, D., & Xu, F., (2022). Mixed-lanthanide clusters: A bridging approach that permits the design of molecular magnetic materials by introducing heavy lanthanides into the 3d-light lanthanide clusters without breaking structural integrity. *Polyhedron, 226*, 116088.
45. Matijević, E., & Hsu, W. P., (1987). Preparation and properties of monodispersed colloidal particles of lanthanide compounds: I. Gadolinium, europium, terbium, samarium, and cerium (III). *Journal of Colloid and Interface Science, 118*(2), 506–523.
46. Meijerink, A., Wegh, R., Vergeer, P., & Vlugt, T., (2006). Photon management with lanthanides. *Optical Materials, 28*(6–7), 575–581.
47. Molander, G. A., & Hahn, G., (1986). Lanthanides in organic synthesis. 4. Reduction of. alpha.,. beta.-epoxy ketones with samarium diiodide. A route to chiral, nonracemic aldols. *The Journal of Organic Chemistry, 51*(13), 2596–2599.

48. Morss, L. R., (1976). Thermochemical properties of yttrium, lanthanum, and the lanthanide elements and ions. *Chemical Reviews, 76*(6), 827–841.
49. Moutet, J., Schleinitz, J., La Droitte, L., Tricoire, M., Pointillart, F., Gendron, F., & Nocton, G., (2021). Bis-cyclooctatetraenyl thulium (II): Highly reducing lanthanide sandwich single-molecule magnets. *Angewandte Chemie, 133*(11), 6107–6111.
50. Mukherjee, P., Shade, C. M., Yingling, A. M., Lamont, D. N., Waldeck, D. H., & Petoud, S., (2011). Lanthanide sensitization in II− VI semiconductor materials: A case study with terbium (III) and europium (III) in zinc sulfide nanoparticles. *The Journal of Physical Chemistry A, 115*(16), 4031–4041.
51. Najafi, M., Sadeghi, C. A., Srivastava, V., & Sillanpää, M., (2019). Augmentation of neodymium ions removal from water using two lanthanides-based MOF: Ameliorated efficiency by synergistic interaction of two lanthanides. *Journal of Chemical & Engineering Data, 64*(7), 3105–3112.
52. Palizban, A. A., Sadeghi-Aliabadi, H., & Abdollahpour, F., (2010). Effect of cerium lanthanide on Hela and MCF-7 cancer cell growth in the presence of transferring. *Research in Pharmaceutical Sciences, 5*(2), 119.
53. Paschalidis, D. G., Tossidis, I. A., & Gdaniec, M., (2000). Synthesis, characterization, and spectra of lanthanide (III) hydrazone complexes: The x-ray molecular structures of the erbium (III) complex and the ligand. *Polyhedron, 19*(26, 27), 2629–2637.
54. Pedersen, K., & Albinsson, Y., (1991). Effect of cell number, pH and lanthanide concentration on the sorption of promethium by she wane Ha putrefaciens. *Radiochimica Acta, 54*(2), 91–96.
55. Ramakrishnan, K., & Prasada, R. T., (2006). Ion imprinted polymer solid phase extraction (IIP-SPE) for preconcentrative separation of erbium (III) from adjacent lanthanides and yttrium. *Separation Science and Technology, 41*(2), 233–246.
56. Schumann, H., Freckmann, D. M., & Dechert, S., (2002). Organometallic compounds of the lanthanides. 157 [1] the molecular structure of Tris (trimethylsilylmethyl) samarium,-erbium,-ytterbium, and-lutetium. *Zeitschrift für Anorganische und Allgemeine Chemie, 628*(11), 2422–2426.

57. Sessler, J. L., Dow, W. C., O'Connor, D., Harriman, A., Hemmi, G., Mody, T. D., & Young, S. W., (1997). Biomedical applications of lanthanide (III) texaphyrins lutetium (III) texaphyrins as potential photodynamic therapy photosensitizers. *Journal of Alloys and Compounds, 249*(1–2), 146–152.
58. Shapiro, B. L., Hlubucek, J. R., Sullivan, G. R., & Johnson, L. F., (1971). Lanthanide-induced shifts in proton nuclear magnetic resonance spectra. I. Europium-induced shifts to higher fields. *Journal of the American Chemical Society, 93*(13), 3281–3283.
59. Strubelt, O., Siegers, C. P., & Younes, M., (1980). The influence of silybin on the hepatotoxic and hypoglycemic effects of praseodymium and other lanthanides. *Arzneimittel-Forschung, 30*(10), 1690–1694.
60. Szostak, M., & Procter, D. J., (2012). Beyond samarium diiodide: Vistas in reductive chemistry mediated by lanthanides (II). *Angewandte Chemie International Edition, 51*(37), 9238–9256.
61. Tilley, T. D., Andersen, R. A., Spencer, B., Ruben, H., Zalkin, A., & Templeton, D. H., (1980). Divalent lanthanide chemistry. Bis (pentamethylcyclopentadienyl) europium (II) and-ytterbium (II) derivatives: Crystal structure of bis (pentamethylcyclopentadienyl) (tetrahydrofuran ytterbium (II)-hemitoluene at 176 K. *Inorganic Chemistry, 19*(10), 2999–3003.
62. Ungur, L., (2018). Introduction to the electronic structure, luminescence, and magnetism of lanthanides. In: *Lanthanide-Based Multifunctional Materials* (Vol. 2, pp. 1–58). Elsevier.
63. Wang, R. C., Zhang, Y., Hu, H., Frausto, R. R., & Clearfield, A., (1992). Preparation of lanthanide arylphosphonates and crystal structures of lanthanum phenyl-and benzylphosphonates. *Chemistry of Materials, 4*(4), 864–871.
64. Watt, G. W., & Gillow, E. W., (1969). Samarium (II) dicyclopentadienide 1-tetrahydrofuranate. *Journal of the American Chemical Society, 91*(3), 775–776.
65. Weiss, G. B., (1974). Cellular pharmacology of lanthanum. *Annual Review of Pharmacology, 14*(1), 343–354.
66. Wendt, K., Gottwald, T., Mattolat, C., & Raeder, S., (2014). Ionization potentials of the lanthanides and actinides–towards atomic spectroscopy of super-heavy elements. *Hyperfine Interactions, 227*, 55–67.
67. Xu, J., Zhou, S., Tu, D., Zheng, W., Huang, P., Li, R., & Chen, X.,

(2016). Sub-5 nm lanthanide-doped lutetium oxyfluoride nanoprobes for ultrasensitive detection of prostate specific antigen. *Chemical Science, 7*(4), 2572–2578.

68. Yakubovich, Y. Y., & Alekseev, V. G., (2012). Hydrolysis constants of tervalent lanthanum and lanthanide ions in 0.1 M KNO_3 solution. *Russian Journal of Inorganic Chemistry, 57*, 911–915.
69. Yan, B., & Chen, Z., (2001). Cyano-bridged aqua (N, N-Dimethylacetamide) (cyanoiron) lanthanides from samarium, gadolinium, or holmium nitrate and potassium hexacyanoferrate: Crystal structures and magnetochemistry. *Helvetica Chimica Acta, 84*(4), 817–829.
70. Zhang, Q., Hu, S. X., Qu, H., Su, J., Wang, G., Lu, J. B., & Li, J., (2016). Pentavalent lanthanide compounds: Formation and characterization of praseodymium (V) oxides. *Angewandte Chemie International Edition, 55*(24), 6896–6900.
71. Zuo, Y., Liu, Y., Chen, J., & Li, D. Q., (2008). The separation of cerium (IV) from nitric acid solutions containing thorium (IV) and lanthanides (III) using pure [C8mim] PF6 as extracting phase. *Industrial & Engineering Chemistry Research, 47*(7), 2349–2355.

CHAPTER 6

Actinides

Contents

6.1. INTRODUCTION

The actinide elements, which include thorium (element 90), protactinium (91), uranium (92), as well as elements 93–103, are classified as such due to their sequential placement in the periodic table following actinium (element 89). Similar to the relationship between lanthanum and the lanthanides (elements 58–71), the actinides exhibit physical and chemical characteristics that are akin to actinium. The electrons located in the outermost shell of elements 90–103 are specifically found in the "5f" orbitals, similar to how the outer electrons in elements 58–71 are situated in the "4f" orbitals. Among the 3 elements, thorium is found in the highest relative abundance. The primary source of thorium is monazite, a mineral in which approximately 10% of its composition is thorium, while the remaining metallic components primarily include cerium, lanthanum, as well as neodymium. Actinium and protactinium are commonly encountered in uranium ores (Kaltsoyannis, 2003).

6.2. ACTINIUM, THORIUM, AND PROTACTINIUM

The production of elements beyond the lanthanides occurs through the process known as the r-process, wherein heavier nuclei are synthesized in supernova explosions by the rapid and consecutive capture of free neutrons by lighter nuclei. In the context of stellar spectra, it is worth noting that the majority of actinides and trans-actinides exhibit limited observability due to their relatively brief radiative decay half-lives within the astrophysical framework (Wagman, 1981). These elements have undergone transformation since their initial formation (Figures 6.1–6.3).

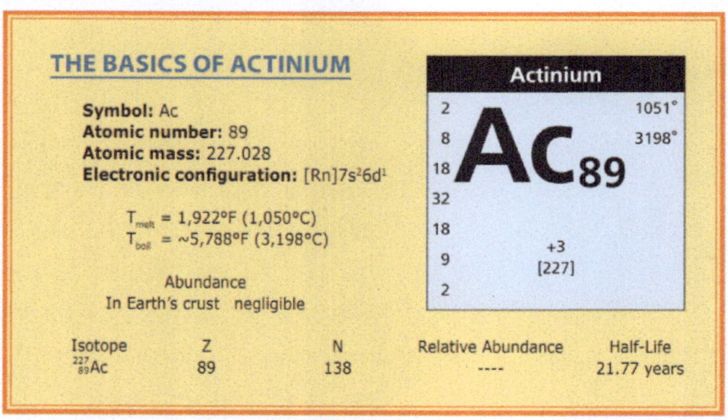

Figure 6.1. Basics of actinium.

Source: Kenneth Street, creative commons license.

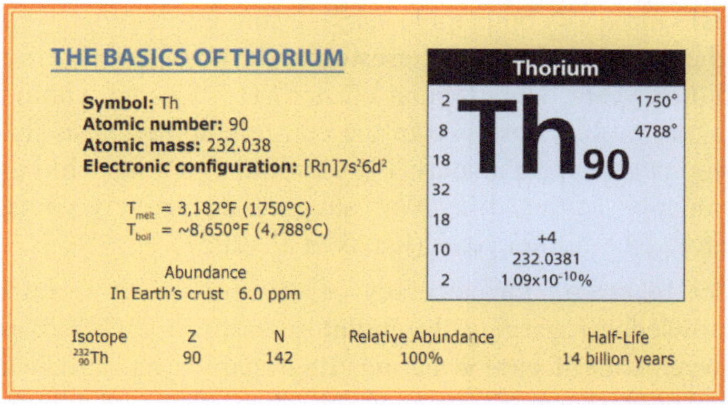

Figure 6.2. Basics of thorium.

Source: Monica Halka, creative commons license.

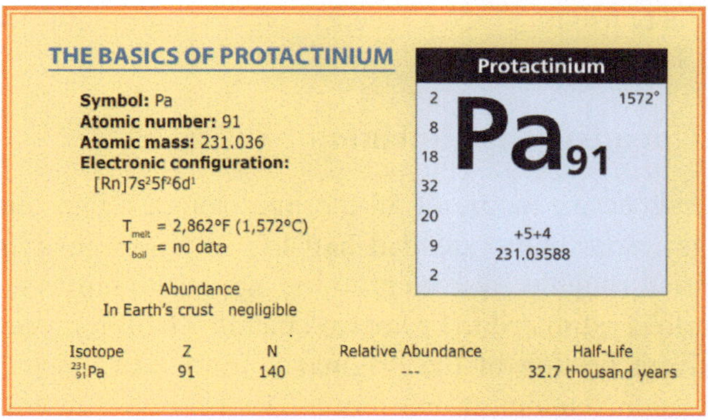

Figure 6.3. Basics of protactinium.

Source: Howard Bryant, creative commons license.

In the context of the most ancient stars, it is primarily the nuclei of thorium-232 and uranium-238, which possess half-lives of 14 billion as well as 4.5 billion years respectively, that could potentially persist in sufficient quantities to allow for the observation of their emission spectra. The ratio between these isotopes in stars is commonly referred to as a Cosmo chronometer due to its ability to provide astrophysicists with valuable insights into the age of stars where this ratio can be observed (Radchenko et al., 2016). Starting from the moment of stellar formation, these elements undergo a process of decay characterized by varying half-lives. The isotopic ratio undergoes corresponding variations, thereby enabling the estimation of the age of a star or even a whole galaxy (by calculating the mean value from a substantial sample of comparable stars). Using this approach, certain stars located in the galactic halo have been estimated to possess an age ranging from 11 to 15 billion years. The globular cluster M15, located within the Milky Way galaxy as well as observed in the constellation Pegasus, is estimated to have originated approximately 14±4 billion years ago. In contrast, the age determination of the Milky Way suggests a relatively younger age of approximately 12.5 billion years (Kmak et al., 2021).

In order to ensure the accuracy of the method, it is imperative to possess knowledge regarding the initial quantity of each isotope, similar to the requirements of carbon dating. Regrettably, the domain of Cosmo chronology predominantly depends on theoretical models instead of empirical astrophysical data for establishing initial values, a circumstance that is suboptimal. Advancements in the field will be facilitated by additional observations and an enhanced comprehension of supernova phenomena, consequently contributing to a deeper understanding of the r-process (Molinari & Snodgrass, 1990).

6.2.1. The Chemistry of Thorium

Despite the radioactive nature of all thorium isotopes, thorium 232 stands out due to its remarkably extended half-life of approximately 14 billion years, a duration roughly equivalent to the age of the universe. Given that the age of Earth is estimated to be approximately 4.6 billion years, it follows that a significant portion of the original thorium that existed during the planet's formation remains in existence. The primary sources of thorium are the minerals thorite as well as monazite (Ortu et al., 2016). Thorium is frequently acquired as a secondary product during the extraction processes of uranium and rare earth minerals. Thorium compounds predominantly exhibit the oxidation state of +4. Illustrative instances include thorium fluoride (ThF_4), thorium dioxide (ThO_2), thorium chloride ($ThCl_4$), as well as thorium nitride (Th_3N_4). The nuclear characteristics of thorium are likely to hold greater significance than its chemical makeup. The fissionability of thorium is inadequate for its utilization as a nuclear fuel. Nevertheless, within the context of a nuclear reactor, it has the potential to undergo a conversion process resulting in the formation of uranium, a substance that possesses the capability to undergo fission. The production of uranium can be elucidated through a series of reactions as depicted below (Day & Stoughton, 1950):

$^{232}_{90}Th + ^{1}_{0}n \rightarrow ^{233}_{90}Th + \gamma$ (where γ is a gamma ray);

$$^{233}_{90}Th \rightarrow \ ^{233}_{91}Pa + \beta;$$

$$^{233}_{91}Pa \ \rightarrow \ ^{233}_{92}U + \beta.$$

Uranium-233 (U-233) exhibits fissionability. The conversion of thorium into uranium for utilization in nuclear reactors serves to enhance the global stockpile of nuclear fuel resources (Langmuir & Herman, 1980) (Figure 6.4).

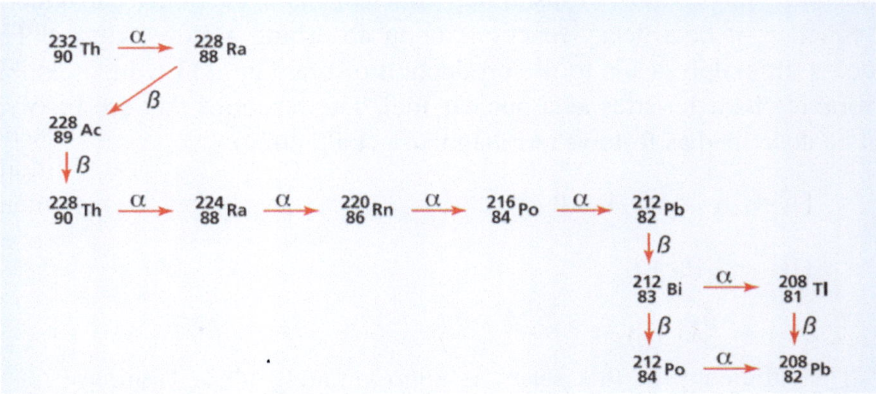

Figure 6.4. Thorium decay chain.

Source: Brian Nordstrom, creative commons license.

Thorium-232 exhibits a radioactive decay series. A series of ten radioactive decays, involving both alpha (α) as well as beta (β) decays, occur, ultimately leading to the production of stable lead 208. Every instance of alpha decay results in the production of a daughter nucleus that occupies a position two elements to the left of the parent nucleus on the periodic table (Adams et al., 1959). Additionally, the mass number of the daughter nucleus is reduced by four units. In the process of beta decay, the resulting daughter nucleus is observed to occupy a position one unit to its right of the nucleus that gave rise to it, while the mass number remains unaffected. When Bismuth-212 is reached, it is possible for both alpha decay and beta decay to occur. Except for thorium-232, the half-lives of the decay processes are generally brief. Polonium 212, an isotope characterized by a half-life of 30 microseconds, exhibits the shortest half-life among all known isotopes. Radium 228 possesses the longest half-life, which spans a duration of 6.7 years (Santschi et al., 2006).

6.2.2. Thorium Fission Reactors

In the realm of sustainable energy solutions, thorium-fueled fission reactors have the potential to meet a significant portion of the global energy demand in a more efficient and less environmentally hazardous manner compared to existing uranium enrichment reactors. Uranium reactors depend on the utilization of low-energy neutrons to initiate the process of nuclear fission. Although natural thorium, primarily in the form of the thorium-232 isotope, is not inherently capable of sustaining a nuclear chain reaction, it can undergo a sequence of beta decay reactions upon absorbing a slow neutron. This process ultimately leads to the production of uranium-233, which exhibits favorable characteristics as a nuclear fuel. The reactions that are involved can be described as follows (Arzhannikov et al., 2020):

$$^{\text{ü}}_{90}\mathrm{Th} + ^{1}_{0}\mathrm{n} \rightarrow ^{\text{ü}}_{90}\mathrm{Th} + \beta^{-}$$

$$^{233}_{90}\mathrm{Th} \rightarrow ^{233}_{91}\mathrm{Pa} + \beta^{-}$$

$$^{233}_{91}\mathrm{Pa} \rightarrow ^{233}_{92}\mathrm{U} + \beta^{-}$$

The efficiency of this series is approximately 90%. Uranium 234 is formed through the absorption of a neutron by protactinium 233 (233Pa) approximately 10% of the time. The initiation of uranium 233 fission can be achieved through the utilization of low-energy neutrons, deliberately generated in reactors to induce the fission of uranium 235. The scientific understanding of this particular reactor type is comprehensive, as evidenced by the successful testing conducted by Oak Ridge National Laboratory on the Molten Salt Experiment during the late 1960s, spanning several years. During the era of the Cold War, there was a preference for utilizing uranium-235 as the primary fuel due to its ability to generate electrical power, while also producing plutonium stockpiles for weapons. The decay chain of uranium-233 does not yield plutonium (Shmelev et al., 2015). The availability of uranium is currently limited, resulting in a significant increase in its market value compared to previous decades. Specifically, the price of one pound of uranium experienced a twenty-fold surge between 2001 and 2006. Moreover, concerns regarding the potential unauthorized acquisition of plutonium have heightened the appeal of the thorium reactor alternative. Thorium is abundantly present in numerous countries, with deposits that necessitate mining techniques that are comparatively less detrimental to the environment when compared to those employed for uranium ore extraction. The waste generated by thorium reactors consists of radioactive by-products with relatively short half-lives, typically spanning a few decades. Consequently, the task of ensuring secure storage for this waste would not be burdensome. Moreover, the utilization of these waste materials for the production of nuclear weapons would pose significantly greater challenges and costs compared to the production of weapons using plutonium. However, the transition to thorium is not a straightforward process. In order to encourage companies to engage in mining activities, it will be necessary to provide them with a compelling incentive (Galahom, 2017). The impetus for this transition is expected to originate primarily from the public sector, as individuals become aware of the availability of this alternative. In April 2008, India commenced the operation of a heavy test for water reactor that integrates passive safety features. The country intends to increasingly depend on these reactors to fulfill a significant portion of its power requirements in the foreseeable future. In March 2009, the United States Congress recommended the initiation of an inquiry into the utilization of thorium-based fuels for nuclear submarines. The Energy Independence and Security Act, a legislative proposal put forth in 2008 by Senators Orrin Hatch as well as Harry Reid, has included provisions advocating for the investigation of thorium reactors as a means to ensure energy independence. However, this proposal has been postponed for further consideration. In the present time, ongoing research in this field is being conducted in multiple nations that possess significant thorium reserves, including India, Norway, Russia, as well as Canada (Kamei, 2012).

6.2.3. Technology and Current uses of Actinium, Thorium, and Protactinium

Actinium holds significant value for researchers due to its utility as a neutron source and its application as a tracer for investigating the dynamics of deep-sea water a widespread movement. The exclusive application of this material lies in the field of satellites, where its radioactive decay chain is harnessed to generate thermoelectric power. Currently, actinium does not possess any prevalent consumer applications (Phuoc et al., 2016).

Thorium presents a contrasting narrative. In its most abundant state, this particular element has the potential to emerge as the predominant global energy source within a span of several decades. In the realm of sustainable energy solutions, thorium-fueled fission reactors have the potential to significantly contribute to meeting the global energy demand in a more efficient and environmentally benign manner compared to conventional uranium enrichment reactors. Ongoing research in this field is being conducted in multiple countries that possess significant thorium reserves, notably India, Norway, Russia, as well as Canada (Vesnovskii, 2003). The United States Congress has recently undertaken certain measures in this regard, including directing the navy to conduct an investigation into the feasibility of utilizing thorium as a power source for submarines. At present, thorium oxide exhibits various consumer applications. These include its utilization in high-refractive-index glass, which ensures optical fidelity in lenses employed in scientific instruments. Additionally, thorium oxide is employed in lamps to enhance brightness and serves as a catalyst in the production of petroleum and sulfuric acid (Radchenko et al., 2021).

Similar to actinium, protactinium holds significant importance in scientific research. It has been extensively utilized as a tracer in oceanic studies, specifically aiding in the comprehension of the dynamics of North Atlantic streams throughout glacial thaw. In light of the prevailing concerns regarding global warming, this particular element has the potential to function as an indicator for subsequent glacial thawing (Jenkins & Sneddon, 1954).

6.3. URANIUM

Uranium, classified as element 92 in the periodic table, is recognized as the most massive naturally occurring element. Indeed, at the time of Dmitri Mendeleev's publication of his periodic table of elements in 1869, the absence of artificially synthesized elements was notable, and uranium was widely regarded as the final element within the table. The pursuit of identifying heavier elements did not commence until the 1930s, when scientists initiated their investigations. This endeavor ultimately led to the detection of the inaugural transuranium element, neptunium, in 1940 (Francis, 1998). Presently, a comprehensive compilation of 25 transuranium elements has been documented, and the pursuit of high atomic number elements remains a vibrant domain of scientific endeavor. The term uranium elicits a diverse range of emotions within the broader populace. It is widely recognized that uranium possesses radioactive properties, and individuals who are consistently exposed to it, such as workers in uranium mines, may experience significant detrimental health consequences. It is widely recognized that uranium serves as the primary fuel source in nuclear power plants, and possesses the potential to be utilized in the production of atomic weapons. Diverse emotional responses are elicited as a result of individuals' perceptions regarding nuclear power. Advocates of nuclear power often perceive uranium as a valuable asset that contributes to the collective welfare (Fattal et al., 2015). Nevertheless, a significant portion of the population harbors apprehensions regarding the potential hazards of nuclear power due to concerns surrounding the radioactivity of uranium and frequently due to misconceptions pertaining to the comprehensive risks related to nuclear power facilities. Furthermore, the prolonged period of 50 years characterized by tensions of the Cold War among the United States and the former Soviet Union, which entailed a genuine potential for a catastrophic nuclear conflict between these two global powers, also played a significant role in instilling a sense of apprehension among individuals towards anything associated with nuclear energy. The apprehension continues to endure in contemporary times, owing to the presence of terrorists and antagonistic foreign governments. The objective of this chapter is to provide an impartial examination of uranium, encompassing its historical background, chemical properties, and its significance in the realms of power generation and weapons development (Irving & Edgington, 1960) (Figure 6.5).

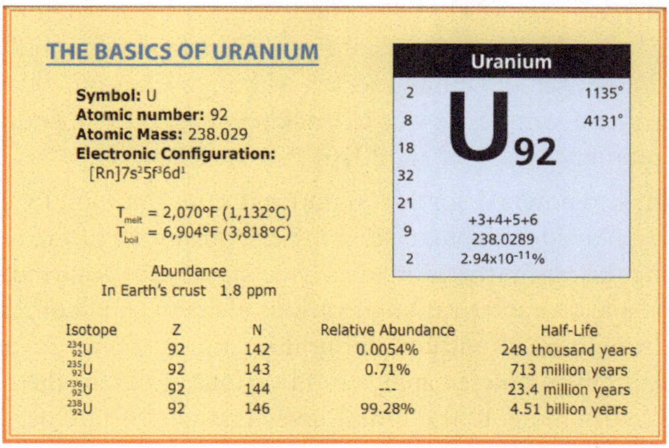

Figure 6.5. Basics of uranium.

Source: Greg Nobles, creative commons license.

6.3.1. Discovery and Naming of Uranium

The utilization of uranium oxide as a means to bestow a green hue upon glass can be traced back to the initial century C.E. The identification of uranium as an element, however, was not made until the latter part of the 18th century (Dolg & Cao, 2009).

Martin Heinrich Klaproth (1743–1817) emerged as a prominent figure in the field of chemistry during the late 18th as well as early 19th centuries in Germany. During that period, pitchblende was a widely recognized mineral that chemists hypothesized to be a composite of zinc and iron. In the year 1789, during the analysis of pitchblende, Klaproth arrived at the conclusion that an unidentified metallic element was present within it. Klaproth subjected a specimen of pitchblende to dissolution in nitric acid. Upon the addition of potassium hydroxide, a solid of yellow color precipitated. Upon subjecting the yellow solid to heat in the presence of charcoal, Klaproth successfully generated a black powder, which he postulated to be an entirely novel element of utmost purity.

The planet Uranus was discovered by the English astronomer William Herschel in 1781 (Durakovic, 2016). This discovery holds significant importance in the field of astronomy as it marked the first planetary discovery since ancient times. Klaproth bestowed the name "uranium" upon his newly discovered element as a tribute to Uranus. The realization that Klaproth had merely isolated an oxide of uranium, rather than the pure metal, was not achieved until the year 1841. The yellow solid is likely sodium diuranate, $Na_2U_2O_7{\cdot}6H_2O$, which is a prevalent form of uranium commonly known as yellowcake in the uranium mining sector. Although Klaproth did not succeed in obtaining a pure sample, he was still acknowledged as the individual who discovered uranium (Wang et al., 2016).

Klaproth is renowned for his significant contributions to the field of chemistry. He played a pivotal role in the establishment of the discipline of analytical chemistry. Furthermore, he made significant contributions to the identification and characterization of various elements, including zirconium, titanium, chromium, strontium, tellurium, and cerium, in addition to uranium. Klaproth conducted analyzes on various minerals, thereby making significant contributions to the initial investigations of rare earth elements (REEs) (Su & Chen, 2015).

Undoubtedly, the characteristics of uranium that have garnered significant recognition among the general populace pertain to its radioactivity and capacity for fission. However, it is worth noting that these particular attributes were not unveiled until a considerable period subsequent to Klaproth's demise. The extent of radiation exposure experienced by scientists such as Klaproth in the period preceding the discovery of radioactivity and the subsequent understanding of its associated health risks remains a subject of speculation. Given that Klaproth's lifespan extended to 73 years, it is plausible that he did not experience adverse consequences resulting from his interaction with uranium, although the precise circumstances surrounding his demise remain undisclosed (Weigel & Hoffmann, 1976).

6.3.2. The Chemistry of Uranium

There exists a variety of minerals that possess uranium. One notable illustration is uraninite, with pitchblende being a specific variant of this mineral. Uranium is the sole significant metal present in uraninite. In addition to uranium, it is common for other minerals to contain multiple other metals. Uraninite is the mineral that is frequently extracted in mining operations within the western region of the United States. Additional examples of uranium-bearing minerals include euxenite, which is extracted from mining operations in Idaho; carnotite and coffinite, which are obtained through mining activities in Colorado; autunite, sourced from mining operations in France; uranophane, extracted from mining activities in Africa; and brannerite, obtained through mining operations in Canada (Katz & Rabinowitch, 1961). The Earth's crust is predominantly composed of rocks that contain trace amounts of uranium. One example of a rock that frequently contains uranium is granite. It is possible for building materials to contain minute quantities of uranium. Hence, individuals occasionally express apprehension regarding potential radiation exposure arising from housing materials, as well as from the soils and rocks upon which their residences are constructed. Another frequently raised concern relates to the presence of radon gas.

Given that radon is a byproduct of the decay series of uranium and is the sole gaseous element within this series, there exists a possibility of inhaling radon, particularly in confined spaces such as basements. Radon is classified as an alpha emitter (Selbin & Ortego, 1969). Alpha particles present a significant hazard when they are located outside the human body. However, the act of inhaling radon gas leads to the emission of alpha particles within the delicate tissues of the lungs, thereby creating a potential for harm. Research on the potential risks posed by radon gas in residential settings has yielded inconclusive findings regarding its impact on human health. The chemistry of uranium in non-nuclear reactions is primarily characterized by compounds and ions whereby uranium exhibits oxidation states of +4 or +6. The isolation of pure uranium metal poses significant challenges, as evidenced by the historical failure of chemists to achieve this feat during Martin Klaproth's era. The conventional method involves the utilization of magnesium metal to carry out the reduction of uranium tetrafluoride to uranium metal under elevated temperatures, as demonstrated by the ensuing chemical reaction (Loiseau et al., 2014):

$$UF_4 \text{ (l)} + 2Mg \text{ (l)} \rightarrow U \text{ (l)} + 2MgF_2 \text{ (l)}.$$

However, the need for metallic uranium is minimal. The utilization of uranium typically entails the utilization of its various compounds. In addition to uranium tetrafluoride, other uranium compounds in the oxidation state of +4 involve uranium tetrachloride (UCl_4) as well as uranium dioxide (UO_2). Examples of compounds in the "+6" oxidation state include uranyl chloride (UO_2Cl_2), uranyl sulfide (UO_2S), uranium trioxide (UO_3), uranium hexachloride (UCl_6), as well as uranium hexafluoride (UF_6). Uranium hexafluoride holds significant importance within the realm of chemical compounds. The substance can be readily heated to transition into a gaseous state, which is then employed in the process of gaseous diffusion for the purpose of separating uranium isotopes. The process of gaseous diffusion capitalizes on the principle that the average velocities of molecules in the gaseous phase are inversely proportional to their respective masses. In essence, it can be observed that lighter molecules generally exhibit higher average velocities compared to their heavier counterparts. The isotopes of uranium exhibit a minor disparity in mass, which is duly manifested in the masses of the various isotopes of uranium present in UF6 molecules. The process of gaseous diffusion separation is characterized by its intricate nature, involving a substantial number of sequential stages. However, the dissimilarity in molecular velocities can be effectively exploited to achieve the separation of distinct isotopes. The field of nuclear chemistry pertaining to uranium primarily revolves around two fundamental processes, namely radioactivity and nuclear fission, which lead to the transformation of uranium into different elements. The topic of discussion will initially revolve around radioactivity and the radioactive decay series of uranium (Parker et al., 2018).

The existence of elements with atomic numbers greater than that of bismuth is solely observed in the form of radioactive isotopes. All of these elements experience a sequence of radioactive decay processes, ultimately resulting in their transformation into isotopes of lead (element 82) and bismuth (element 83). Uranium is characterized by the presence of three primary isotopes in its natural state: U-234, U-235, and U-238. These isotopes exhibit distinct half-lives, namely 248 thousand years, 713 million years, and 4.51 billion years, respectively. The majority of uranium, specifically 99.3%, consists of U-238 (Závodská et al., 2008) (Figure 6.6).

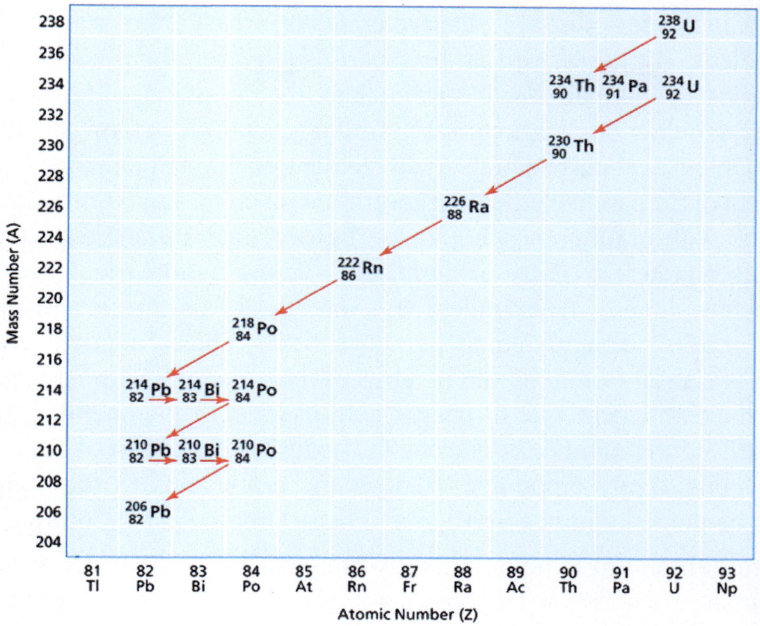

Figure 6.6. Uranium decay chain.

Source: Brian Nordstrom, creative commons license.

Uranium-238 (U-238) and its progeny experience a total of 14 decay events, encompassing both alpha (α) as well as beta (β) decay processes, ultimately resulting in the formation of lead-206 (Pb-206), an isotope that exhibits stability. Once the isotope polonium-218 is attained, there exist isotopes that can undergo decay through multiple modes. However, ultimately, the decay process leads to the formation of lead-206. The durations of decay half-lives can range from milliseconds to billions of years, exemplified by the case of U-238. The decay series that produces additional isotopes of uranium exhibits comparable characteristics. With the exclusion of thorium, the isotopes of all the elements ranging from bismuth to uranium possess significantly shorter half-lives compared to the age of the Earth. Consequently, the original quantities of these substances no longer persist. The presence of trace quantities in the Earth's crust or mantle can be attributed to their continuous replenishment through the decay series of uranium and thorium. During their formation, it is noteworthy that the majority of these entities undergo rapid decay (Berthet & Ephritikhine, 1998).

Radioactive decay is a phenomenon that can manifest across all elements in the periodic table, characterized by an asymmetrical breakdown of a nucleus. Alpha as well as beta particles exhibit significantly smaller dimensions in comparison to the nucleus that is undergoing decay. The resulting daughter nucleus exhibits a similar magnitude in terms of size as its parent nucleus and corresponds to an element positioned one or two atomic spaces out of the parent element. In contrast, nuclear fission is limited to heavy elements, with uranium serving as the lightest exemplification. Moreover, fission is characterized by a significantly more symmetrical breakdown of a nucleus. The nucleus undergoes a process of fission, resulting in the formation of two roughly equal-sized fragments. These fragments exhibit a notable deviation from uranium in terms of their position within the periodic table. Due to the principle of charge conservation, it is imperative that the combined atomic numbers, denoting the quantity of protons, of the resultant entities consistently amount to 92, which corresponds to the number of protons contained within a uranium nucleus (Grenthe et al., 2006).

Natural uranium is predominantly composed of U-238, accounting for approximately 99.3% of its isotopic composition. However, it is important to note that U-238 is not capable of undergoing fission. The isotope U-235, which is capable of undergoing fission naturally, is found in natural uranium at a concentration of approximately 0.7 percent. The process of uranium enrichment is necessary in order to augment the concentration of U-235 within natural uranium. After the production of a significant amount of U-235, various pairs of products can be generated through the fission of its nuclei. The fission process commences with the capture of a neutron by a uranium nucleus. In addition to the generation of two newly formed elements, the nuclear reaction also results in the emission of two to three neutrons. Due to the capacity of these neutrons to be assimilated by adjacent uranium nuclei, a greater degree of fission transpires. A cascade effect ensues when the rate of nuclear fission experiences exponential growth (Mullane et al., 2014).

Several reactions serve as examples of fission pathways, which are as follows:

$$^{235}_{92}U + ^{1}_{0}n \rightarrow ^{236}_{92}U \rightarrow ^{89}_{36}Kr + ^{ü}_{56}Ba + 3\ ^{1}_{0}n;$$

$$^{235}_{92}U + ^{1}_{0}n \rightarrow ^{236}_{92}U \rightarrow ^{92}_{36}Kr + ^{142}_{56}Ba + 2\ ^{1}_{0}n;$$

$$^{235}_{92}U + ^{1}_{0}n \rightarrow ^{236}_{92}U \rightarrow ^{92}_{38}Sr + ^{14}_{54}Xe + 3\ ^{1}_{0}n.$$

One of the challenges associated with nuclear reactors pertains to the radioactive nature of the fission fragments generated during the process. The half-lives of these substances are of sufficient duration to enable their storage for extended periods, potentially spanning thousands of years.

While U-238 does not possess the ability to undergo fission, it does exhibit the capacity to capture neutrons. The aforementioned product, U-239, is classified as a beta emitter, resulting in the production of Np-239. Neptunium-239 is known to exhibit beta decay, resulting in the production of plutonium-239. Plutonium-239 exhibits fissionability. Plutonium is the preferred material for nuclear weapons due to its lower requirement of 11 pounds (5 kilograms) compared to uranium's 33 pounds (15 kilograms) for the construction of a nuclear warhead (Alwan & Williams, 1980).

6.3.3. Uranium Fission Reactors: Natural and Human-Made

Nuclear fission reactors currently contribute to approximately 14% of the global electricity supply, while accounting for around 20% of the electricity generated within the United States. Additionally, these reactors are responsible for the entirety of the plutonium utilized in the production of nuclear weapons. In contrast to weapons, reactors necessitate a controlled environment in which the generation of heat energy occurs without any occurrence of an explosive nature. The application of thermal energy is employed to facilitate the vaporization of water, subsequently generating steam which in turn powers turbines, thereby supplying electricity to a global population exceeding 100 million individuals. The initial step in establishing a functional reactor involves attaining the appropriate concentration of uranium 235. In its natural state, this is found to be merely 0.720 percent, with the exception of fossil reactors, which will be further elaborated upon (Wheatley et al., 2016). The prevailing isotope of uranium, denoted as ^{238}U, lacks the requisite characteristics to sustain a nuclear chain reaction within the typical operational conditions of reactors. The optimal operating conditions encompass the following factors:

- The presence of slow neutrons to initiate the process of fission;
- The capacity of regular water to mild the reaction and provide cooling for the reactor
- The sustenance of a consistent neutron population within the core (Øvergaard, 2008).

In order to meet these requirements, it is necessary to have a minimum concentration of 3 percent uranium 235. This concentration can be achieved through the method of uranium enrichment, in which mass separation plays a crucial role. The diffusion rate of uranium 235 is higher than that of uranium 238 due to its relatively lower atomic mass. The gaseous diffusion procedure utilizes pressure differentials to facilitate the separation of 235U into a distinct compartment. The residual material is characterized by a reduced concentration of U-235 in comparison to natural uranium, thus earning the designation of depleted uranium (Siegrist & Sütterlin, 2014).

A more recent and energy-efficient approach to isotope separation involves utilizing centrifugal force to separate U-235, which is lighter, from U-238, which is heavier. This separation occurs as the isotopes move towards the outer edge of a cylinder that is turning. Heat is occasionally employed to elevate the desired U-235, causing it to separate from the drained gas. The centrifuge method presently constitutes a majority of the global U-235 enrichment process (Cuttler, 2001).

After obtaining the appropriate fraction of U-235, the fuel elements are put together and subsequently inserted into a chamber containing water or another suitable moderator. Both the Uranium-235 and Uranium-238 isotopes undergo spontaneous fission, resulting in the fragmentation of their nuclei into smaller fragments with lower atomic masses. Additionally, this process involves the emission of fast neutrons, heat, as well as gamma rays. The primary benefit of U-235 relative to U-238 lies in its susceptibility to induced fission through collisions with neutrons. In contrast, U-238 exhibits limited efficiency in undergoing fission with naturally generated fast neutrons due to their high mobility and propensity to escape. The deceleration of these particles can be effectively achieved by their interaction with water or molten salt, which are commonly referred to as moderators. These moderators play a vital role in ensuring the optimal functioning of nuclear power systems. A potential issue may arise if an excessive number of neutrons were present at a particular moment. This scenario has the potential to result in a significantly elevated quantity of nuclear fissions within the reactor core, leading to a self-sustaining chain reaction or a state of supercriticality. The issue at hand can be resolved through the utilization of control rods that have the capability to be inserted or removed. Nuclear reactor control rods are commonly composed of materials like silver, indium, cadmium and or boron, which possess a high affinity for neutrons, effectively sequestering them and preventing their participation in fission reactions. The provided diagram depicts the primary constituents of a nuclear reactor (Manion & Evan, 2001). Remarkably, in ancient times, the Earth possessed all the essential elements required for the operation of a self-sustaining nuclear power reactor over a period spanning tens of thousands of years. In the present-day African nation of Gabon, there exists a sandstone deposit containing sedimentary rock veins that seemingly underwent a criticality event approximately 1.7 billion years ago. This event occurred in a minimum of 15 separate locations within three distinct ore deposits. This phenomenon was feasible due to the relatively elevated U-235 abundance during that period, approximately reaching the threshold of 3 percent required to initiate a reaction (Uppal, 2016) (Figure 6.7).

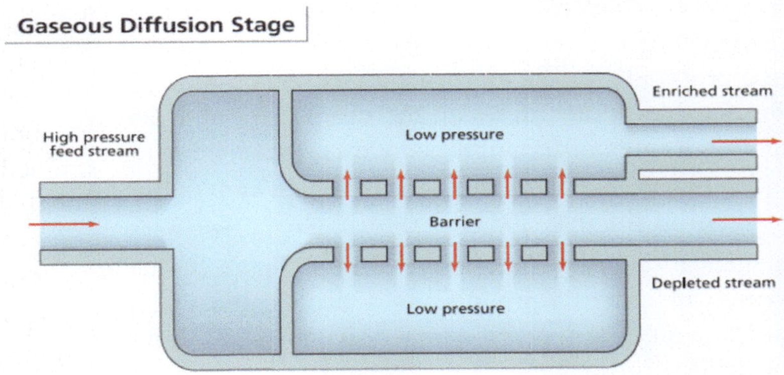

Figure 6.7. Uranium diffusion stage.

Source: Emilio Segrè, creative sources license.

Under the appropriate circumstances, the presence of water flowing via the rock acted as a moderator, enabling the reactor to achieve criticality in a cyclic pattern characterized by remarkable consistency, with a duration of 30 minutes of operation followed by a resting period of two and a half hours. The Oklo Fossil Reactors, formerly known as the reactors, underwent an automatic shutdown mechanism upon the conversion of water into steam. This transformation rendered the steam incapable of moderating the fast neutrons generated through the process of spontaneous fission of uranium. When the temperature decreased and water persisted in traversing the rock, responses could recommence. The production of a substantial quantity of plutonium-239 (239Pu) would have occurred through the reaction described below (Davidson & Baum, 1990):

$$n + {}^{238}_{92}U \rightarrow {}^{239}_{92}U + \gamma \xrightarrow{23.5} {}^{239}_{93}Np + e^{-} \xrightarrow{2.35d} {}^{239}_{94}Pu + e^{-}.$$

The half-lives of U-239 as well as Pu-239 are indicated above the arrows denoting the radioactive decay events. Antineutrinos are also generated during the process of decay. Plutonium-240 would have been generated through the process of neutron absorption by plutonium-239. This implies that plutonium, rather than uranium, is the most massive naturally occurring element on Earth. However, the scarcity of 235U in ore was insufficient to initiate such occurrences for a minimum of 1.5 billion years. The United States possesses the highest number of operational nuclear reactors (104) compared to any other nation, with Japan and France closely following with 59 and 54 reactors, respectively (Giel, 1998). The imperative to decrease carbon emissions and enhance energy autonomy has, nevertheless, stimulated proposals for the construction of numerous additional facilities in the foreseeable future. Between the years 2007 and January 2010, the U.S. Nuclear Regulatory Commission was presented with a total of 21 fresh applications for construction, resulting in a cumulative estimate of 31 anticipated units (Shibuya, 2021).

The nuclear industry faces several significant challenges that must be addressed in order to progress as a dependable and widely adopted electricity generation source. It is imperative to establish public confidence by ensuring that the potential for a self-sustaining reaction leading to a nuclear incident is effectively mitigated. The industry is required to identify suitable locations for the disposal of radioactive waste or explore alternative methods to mitigate its hazardous properties. Potential solutions may be found in the utilization of next-generation reactors that integrate inherently safer characteristics. The Very-High Temperature Reactor and the Sodium-Cooled Fast Reactor are currently subjects of ongoing investigation. Both of these proposed models exhibit significantly higher levels of efficiency compared to existing models. However, it is unlikely that either of these models will be readily available above the prototype stage prior to the year 2025. Thorium reactors have the potential to emerge as the most optimal option for future advancements (Romanchuk et al., 2013).

6.4. THE TRANSURANIUM ELEMENTS

Throughout the course of history, a primary objective pursued by alchemists, and subsequently by contemporary chemists and physicists, has been the exploration and identification of novel elements. Ninety out of the initial 92 elements are naturally present on Earth. The elements technetium and promethium deviate from the norm as they can only be generated through artificial means. This is due to the fact that both elements solely comprise radioactive isotopes in relatively brief times of radioactive decay. Following the discovery of element 92, uranium, all subsequent elements have been exclusively synthesized through artificial means. Negligible quantities, typically in concentrations of parts per trillion or lower, of certain elements such as plutonium are generated through subterranean nuclear processes (Seaborg, 1946). However, these minuscule amounts will be disregarded due to their inherent irretrievability. At the time of authoring this publication, a comprehensive compilation of chemical and physical investigations has yielded a collective count of 118 distinct elements. Since the year 1940, laboratories in the US, Germany, and Russia have successfully synthesized a total of 25 elements beyond uranium, known as the transuranium elements. This remarkable achievement commenced with the creation of element 93, neptunium. The transuranium elements comprise approximately 30% of the elements that are currently known, thereby constituting a substantial expansion of the periodic table beyond the naturally occurring elements. The pursuit of elements beyond uranium in contemporary times has been predominantly motivated by the scientific community's desire to uncover novel entities, thereby expanding the periodic table. Nevertheless, the identification and characterization of novel elements hold greater importance beyond their mere inclusion in the periodic table (Munzenberg, 1988). The periodic table's utility to scientists stems from its systematic organization of elements into vertical groups, known as families, which exhibit analogous chemical and physical characteristics. The observed trends in established substances can be utilized to make predictions regarding the existence and characteristics of yet-to-be-discovered elements. However, it is worth considering the possibility that the perception of a structured pattern may only hold true for elements of minimal atomic weight. What would be the implications if there were notable deviations in the patterns of physical and chemical characteristics as increasingly heavier elements are uncovered? Scientists have found great satisfaction in the discovery that the characteristics of transuranium elements provide substantial confirmation for scientific theories concerning the fundamental essence of matter (Huill, 1938) (Figures 6.8–6.17).

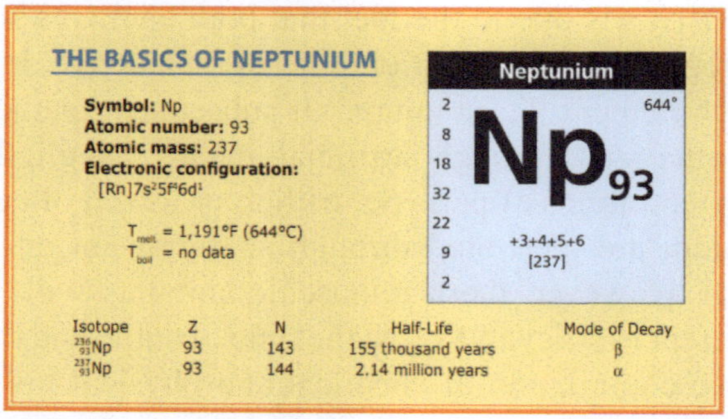

Figure 6.8. Basics of neptunium.

Source: Jodie Rhodes, creative commons license.

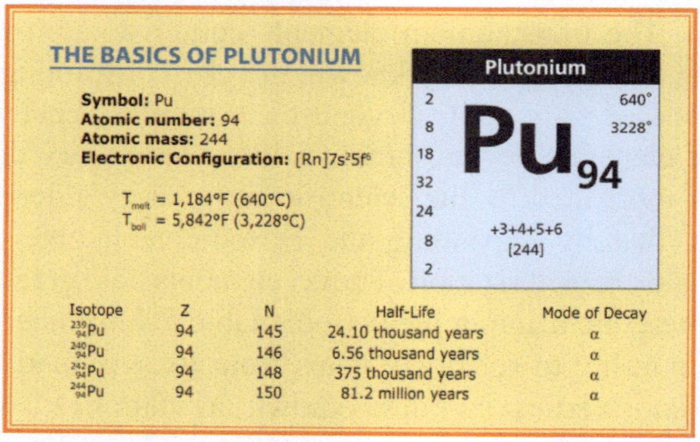

Figure 6.9. Basics of plutonium.

Source: Jodie Rhodes, creative commons license.

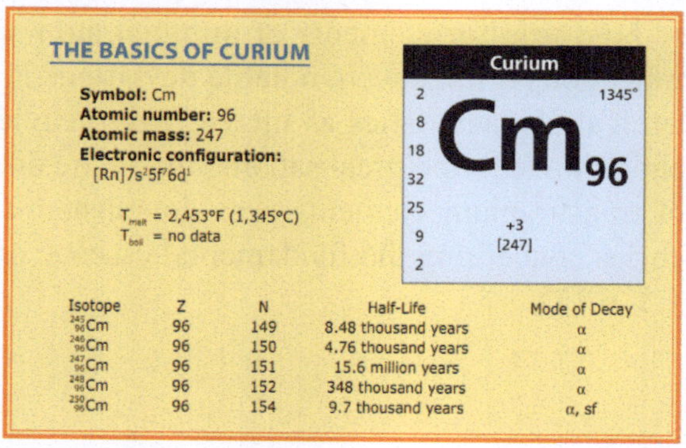

Figure 6.10. Basics of curium.

Source: Frank Darmstadt, creative commons license.

THE BASICS OF BERKELIUM

Symbol: Bk
Atomic number: 97
Atomic mass: 247
Electronic configuration: [Xe]$7s^2 5f^9$

T_{melt} = 2,349°F (1,287°C)
T_{boil} = 4,480°F (2,471°C)

Berkelium
Bk 97
1050°
2 8 18 32 27 8 2
+3+4
[247]

Isotope	Z	N	Half-Life	Mode of Decay
$^{247}_{97}Bk$	97	150	1.4 thousand years	α
$^{249}_{97}Bk$	97	152	320 days	α

Figure 6.11. Basics of berkelium.

Source: Frank Darmstadt, creative commons license.

THE BASICS OF CALIFORNIUM

Symbol: Cf
Atomic number: 98
Atomic mass: 251
Electronic configuration: [Rn]$7s^2 5f^{10}$

T_{melt} = 1,650°F (900°C)
T_{boil} = no data

Californium
Cf 98
900°
2 8 18 32 28 8 2
+3
[251]

Isotope	Z	N	Half-Life	Mode of Decay
$^{249}_{98}Cf$	98	151	351 years	α
$^{250}_{98}Cf$	98	152	13.1 years	α
$^{251}_{98}Cf$	98	153	~890 years	α
$^{252}_{98}Cf$	98	154	2.65 years	α

Figure 6.12. Basics of californium.

Source: Nicole Leonard, creative commons license.

THE BASICS OF EINSTEINIUM

Symbol: Es
Atomic number: 99
Atomic mass: 252
Electronic configuration: [Rn]$7s^2 5f^{11}$

T_{melt} = ~1,580°F (~860°C)
T_{boil} = no data

Einsteinium
Es 99
860°
2 8 18 32 29 8 2
+3
[252]

Isotope	Z	N	Half-Life	Mode of Decay
$^{252}_{99}Es$	99	153	1.29 years	α
$^{254}_{99}Es$	99	155	276 days	α
$^{255}_{99}Es$	99	156	40 days	β

Figure 6.13. Basics of einsteinium.

Source: Nicole Leonard, creative commons license.

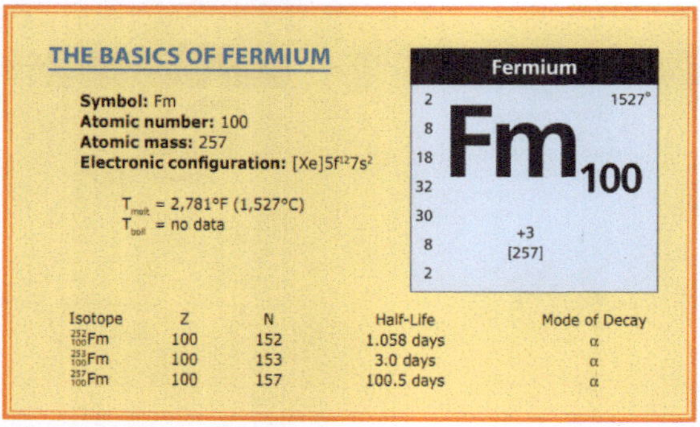

Figure 6.14. Basics of fermium.

Source: Greg Nobles, creative commons license.

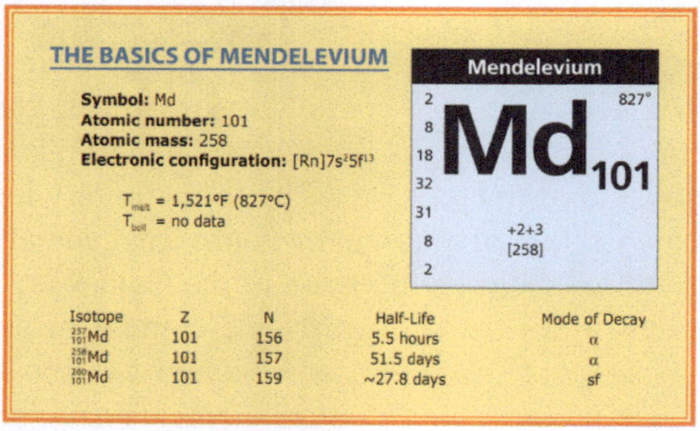

Figure 6.15. Basics of mendelevium.

Source: Nicole Leonard, creative commons license.

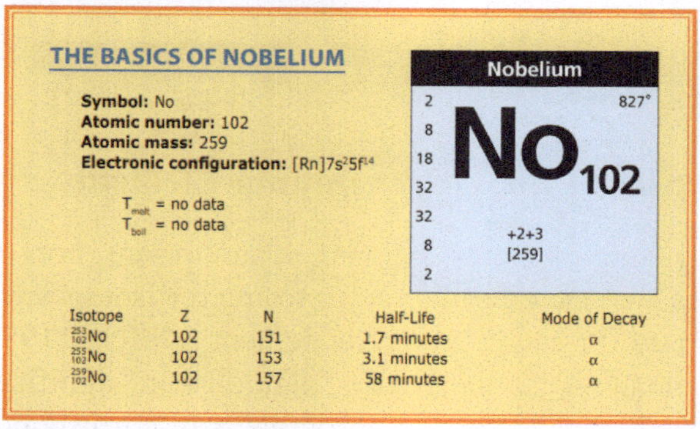

Figure 6.16. Basics of nobelium.

Source: Joe Palca, creative commons license.

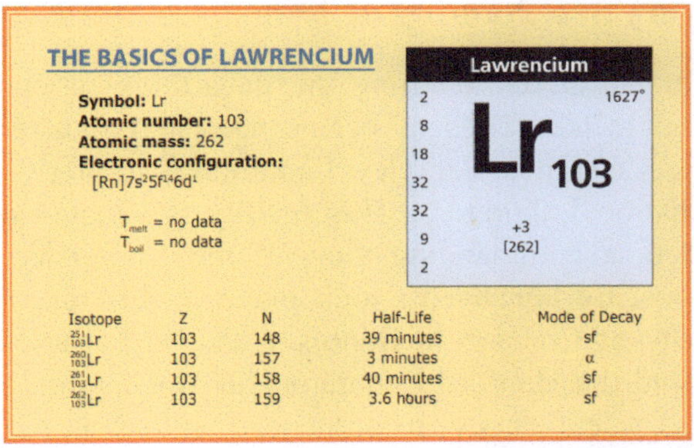

Figure 6.17. Basics of lawrencium.

Source: Joe Palca, creative commons license.

One further incentive for the pursuit of novel elements lies in the potential discovery of hitherto unknown substances that could yield practical implications in the realms of science and technology (Chapman, 2020). The fissionable characteristics of plutonium are widely recognized, and a significant quantity of plutonium has been generated on a global scale, particularly throughout the duration of the Cold War. The subsequent element, americium, has been utilized in the context of smoke detectors. Regrettably, the elements succeeding americium exhibit such brief half-lives that only negligible amounts of these elements have been synthesized (Mayer, 1941). According to theoretical considerations, it is postulated that the existence of elements with greater atomic masses than those currently synthesized could potentially exhibit substantially prolonged half-lives, spanning several centuries or even millennia. If elements of such a nature were to be discovered, it is conceivable that they may exhibit unforeseen and advantageous characteristics (Hyde & Seaborg, 1956).

6.4.1. The Basics of the Transuranium Elements

Naturally occurring elements beyond uranium are not observed. The presented tables provide information regarding the isotopes with the longest observed lifespans for each respective element (Nitsche, 1990). Given that transuranium elements do not occur naturally in significant quantities, the atomic mass indicated for a specific element corresponds to the mass number of the isotope with the longest half-life within that element (Griveau & Colineau, 2014).

6.4.2. Discovery and Naming of Transuranium Elements

The subsequent narratives detailing the unearthing of elements 93–103 primarily chronicle the series of occurrences that transpired within the confines of the Lawrence Berkeley Laboratory, formerly known as the Lawrence Radiation Laboratory. This facility is situated on an elevated terrain, providing a commanding view of the University of California, Berkeley campus. Certain events took place in different locations, such as the University of Chicago in Illinois, Oak Ridge National Laboratory in Tennessee, and the Hanford Laboratory in Washington State (Clark & Hobart, 2019). These events were influenced, at least in part, by activities that transpired during the wartime period. Glenn Seaborg, a prominent figure in the field of chemistry, made significant contributions to the scientific community during his tenure at the University of California, Berkeley. Notably, he is credited with the discovery of plutonium, as well as his involvement in the exploration and identification of numerous other transuranium elements. One of the prominent scientists who has made significant contributions to the identification of transuranium elements is Albert Ghiorso, a longstanding collaborator of Seaborg. As of the time of writing, Ghiorso, born in 1915, continues to actively engage in research in this field. Indeed, it is worth noting that Albert Ghiorso currently holds the esteemed distinction of holding the Guinness World Record for his remarkable discovery of a total of twelve elements (Bagnall, 1964).

During the medieval period, alchemists harbored the aspiration to transmute base metals, such as lead, into the noble metal of gold. The failure of all their endeavors can be attributed to the fact that the conversion or transmutation of an element into another necessitates a nuclear reaction. The alchemists possessed no understanding of atomic structure, nor did they possess any knowledge regarding the constituents comprising atoms. The execution of nuclear transmutations was beyond their technical capacity (Sime, 2000). The aforementioned scenario underwent a transformation following the advancements in scientific knowledge pertaining to elementary particles, specifically protons and electrons, as well as the understanding of radioactivity. Ernest Rutherford (1871–1937) achieved the notable distinction of being the first individual in England to successfully effectuate the transformation of one chemical element into another. Rutherford had already gained prominence due to his contributions to the field of radioactivity and his pivotal discovering of atomic nuclei. Rutherford was awarded the Nobel Prize in chemistry in 1908 in recognition of his contributions to the field of radioactivity. During that period, neutrons had not yet been discovered, thus Rutherford was not aware of their existence (Ewing, 2011). However, he postulated a nuclear atomic model wherein the protons (along with a portion of the electrons) were confined within a minuscule central region referred to as the nucleus. The remaining electrons were positioned at a significant distance from the nucleus. Rutherford's observation pertained to the correlation between an atom's identity and the quantity of protons present within its nucleus. Hence, altering the quantity of protons would result in a modification of the atom's fundamental characteristics (Nagame & Hirata, 2011).

The primary challenge that needed to be addressed pertains to the physical nature of protons, which possess positive charges, resulting in mutual repulsion among them. The magnitude of repulsive force experiences a significant escalation as the protons move closer to each other. Nevertheless, in the event that these entities are able to approach each other within a certain proximity, they possess the capability to merge and form a novel central core (Nash et al., 2006). This phenomenon occurs due to the dominance of the stronger nuclear force of attraction at sufficiently short distances, although Rutherford's comprehension of nuclear forces was limited during that period. In the year 1919, Rutherford successfully accomplished the initial transmutation of an element. Through the process of hitting nitrogen nuclei with helium nuclei, commonly referred to as alpha particles, the researcher successfully achieved the conversion of nitrogen atoms into oxygen atoms. This transformation was accomplished by means of the following reaction (Hoffman, 1998):

$$^{14}_{7}N + ^{4}_{2}He \rightarrow ^{17}_{8}O + ^{1}_{1}H,$$

The utilization of subscripts to denote charges on species, as well as superscripts to indicate mass numbers is a common practice. The principle of charge conservation necessitates that the total charge remains constant. Consequently, the aggregate charge on the left-hand side is equivalent to 9, which is also the total of the charges on the side to the right. The subscript in the context of an element serves the additional purpose of indicating the atomic number of said element, which is directly associated with the quantity of protons present within the atoms of said element. In cases where a species is not a component, the subscript represents the charge on that particular species. The mass number of a subatomic particle is determined by the combined total of its protons and neutrons (Eisenbud et al., 1984). The conservation of the sum is also a requirement, thus the sum of the superscripts on the left-hand side is equal to 18, just as the sum of the superscripts on the opposite side is eighteen. The most prevalent technique employed during subsequent years involved the bombardment of atomic nuclei with alpha particles as a means of inducing transmutation in elements. The aforementioned circumstance underwent a significant transformation in the year 1932 when the neutron was discovered by James Chadwick, an English physicist who was affiliated with the Cavendish Laboratory and conducted his research under the guidance of Ernest Rutherford. Chadwick was awarded the Nobel Prize in physics in 1935 in recognition of his discovery (Sterne et al., 1998). Neutrons, being devoid of electric charge, do not experience repulsion from atomic nuclei. Researchers promptly recognized the inherent impartiality of neutrons, rendering them highly suitable as projectiles for nuclear collisions. In a broad sense, the production of lighter transuranium elements has been achieved through the process of subjecting uranium and other elements to bombardment by particles of relatively low mass, such as neutrons, deuterons, and alpha particles (Choppin & Nash, 1995). The production of more substantial isotopes of the parent element occurs, subsequently leading to beta decay and the resulting creation of the next element to be placed in the periodic table. This technique exhibits efficacy within a limited range. In due course, the scarcity of the most weighty target material necessitates the utilization of lighter targets as well as heavier projectiles, such as carbon, oxygen, or even more massive nuclei. The selection of targets and projectiles is dictated by the condition that the charges (specifically, atomic numbers in the context of elements) of both species must combine to equal the charge (atomic number) on the nucleus of the intended product (Carrott et al., 2016).

In the production of element 93, specifically neptunium, the reaction employed was as follows:

$$^{238}_{92}\mathrm{U} + {}^{1}_{0}\mathrm{n} \rightarrow {}^{239}_{92}\mathrm{U} \rightarrow {}^{239}_{93}\mathrm{Np} + {}^{0}_{-1}\hat{\mathrm{a}}\ .$$

This equation shows that nuclei of U-238 were bombarded with neutrons (${}^{1}_{0}n$ is the symbol for a neutron) to produce unstable U-239 nuclei. The U-239 nuclei underwent beta decay (is the symbol for a beta particle) resulting in Np-239 (Collins et al., 2007).

The equation demonstrates the utilization of neutrons ${}^{1}_{0}n$ in the bombardment of U-238 nuclei, resulting in the formation of U-239 nuclei, which are inherently unstable. The U-239 nuclei underwent beta decay, denoted by the symbol ${}_{-1}^{0}b$, leading to the formation of Np-239 (Wolf, 2006).

6.4.3. The Chemistry of Transuranium Elements

The transuranium elements are not naturally occurring and exhibit radioactive properties in all of their various forms. As one proceeds horizontally in the row in the table, the half-lives exhibit a progressive decrease, ultimately reaching such a diminutive duration that the production of the elements has been limited to microgram quantities or less (Kim, 1992). For instance, the initial compound of berkelium, denoted as $BkCl_3$, was obtained in a minuscule quantity of merely three nanograms, equivalent to three billionths of a gram. Despite their extremely low abundance, there is a substantial amount of scientific literature available on the chemistry of the transuranium elements (Grödler et al., 2023).

Several overarching observations can be discerned. An observation of note is that the initial elements within the row display chemical properties reminiscent of transition metals, as they have the ability to form compounds and ions with varying oxidation states. The subsequent elements within the row exhibit a greater resemblance to their respective lanthanide counterparts, as they predominantly exist in the "+3" oxidation state (Singh et al., 1983). The transuranium elements exhibit characteristics of active metals, rendering them prone to oxidation. The oxidation states of the elements demonstrate a discernible pattern. The oxidation states of elements span a range from "+2" to "+7," with the "+3" state being the most prevalent, akin to the lanthanides. The elements located in the center of the row demonstrate a wider spectrum of oxidation states compared to the elements situated towards the start or the end of the row. The disparity in oxidation states between actinides and lanthanides can be attributed to the contrasting nature of the "4f" electrons in lanthanides and the "5f" electrons in actinides. The electrons in the "4f" orbital of the lanthanides are located in inner energy levels of the atom, resulting in reduced accessibility for chemical bonding. The 5f electrons exhibit a higher degree of exposure, thereby enabling a larger quantity to engage in chemical bonding (Nelson et al., 2008).

Compounds involving the elements americium, californium, as well as berkelium in their +2 oxidation state have been observed to form when combined with chlorine, bromine, as well as iodine. One illustrative instance is the compound known as americium chloride ($AmCl_2$). Einsteinium is also capable of forming halide compounds with a +2 oxidation state, specifically $EsBr_2$ and EsI_2. In contrast to the typical behavior observed in the greater actinides, nobelium exhibits the formation of a fluoride with a charge of +2, denoted as NoF_2, as well as an oxide with a charge of +2, denoted as NoO (White & Marsh, 2019).

The oxidation state of "+3" is frequently observed and can be considered as the most prevalent state. The majority of actinides in the trivalent state have the ability to react with fluorine, bromine, chlorine, as well as iodine to form compounds. Plutonium has the capability to undergo a chemical reaction resulting in the formation of a compound known as plutonium nitride (PuN) (Spitsyn & Katz, 2018). Trioxides are formed by the elements americium, curium, einsteinium, fermium, mendelevium, and nobelium, denoted as Am_2O_3, Cu_2O_3, Es_2O_3, Fm_2O_3, Md_2O_3, and No_2O_3, respectively. Oxychlorides can be formed by the elements berkelium, californium, and einsteinium, resulting in the compounds BkOCl, CfOCl, and EsOCl. Complex ions can be formed by elements, wherein the element exists in a state of positive three charge. One example that can be considered is the AmO^{2+} ion (Noh et al., 2014).

In the state denoted as "+4," neptunium and plutonium exhibit the formation of tetrafluorides, specifically NpF_4 and PuF_4. A number of elements, namely plutonium, americium, curium, neptunium, and berkelium, are capable of forming dioxides, specifically PuO_2, NpO_2, AmO_2, CmO_2, and BkO_2. Elements can exist in the form of a "+4" ion, such as americium (Am^{4+}), or they may form fluoride complexes, as is also the case with americium (AmF_8^{4-}) (Hudry et al., 2012).

Neptunium is the primary transuranium element that demonstrates an oxidation state of "+5." The state of neptunium is exemplified by the compound neptunium pentoxide (Np_2O_5) and the neptunium oxyfluoride ($NpOF_3$) (Kumar & Sood, 1980).

A number of transuranium elements demonstrate the oxidation state of "+6." As an illustration, it is observed that neptunium exhibits the ability to generate neptunium hexafluoride (NpF_6), along with two oxyfluorides denoted as NpO_2F_2 and $NpOF_4$, both of which possess a +6 oxidation state. Berkelium and californium are capable of forming trioxides, specifically BkO_3 and CfO_3. Americium exhibits a valence state of +6, resulting in the formation of a cation with a positive charge. Specifically, Americium forms the ion (AmF_8^{4-}) (Kanellakopulos, 1990).

The elements neptunium and plutonium are the only known elements that display the oxidation state of +7. The oxyanions known as NpO_5^{3-} and $PuO5_5^{3-}$ have been observed, although their stability is relatively low (O'Donnell, 1990).

6.4.4. Technology and Current Uses of Transuranium Elements

Although the practical applications of the majority of transuranium elements are limited, they do find some utilization in research laboratories. One example of a reliable neutron source is the combination of americium and beryllium. During the process of decay, americium emits alpha particles which subsequently engage in a reaction with beryllium, as described below (Fernandez et al., 2009):

$$^{4}_{2}a + ^{9}_{4}Be \rightarrow ^{1}_{0}n + ^{12}_{6}C.$$

Neutrons can also serve as a valuable tool in nondestructive testing of machinery and medical diagnostics due to their distinct reflection and absorption characteristics in various materials. Plutonium is generated in substantially greater quantities compared to the remaining transuranium elements. During the Cold War, a significant quantity of plutonium, amounting to hundreds of tons, was produced, and a substantial portion of this stockpile remains stored globally. Plutonium serves as a fissile substance within nuclear reactors and plays a crucial role in the initial phase of contemporary thermonuclear weaponry. It is noteworthy that a significant number of these weapons are presently being dismantled in accordance with a nonproliferation agreement established by the United Nations. The issue of effectively isolating plutonium from such weapons will continue to be a significant field of study and advancement for several decades to come. Due to the possibility of diversion for terrorist purposes, the military enforces stringent security measures to safeguard plutonium supplies, thereby restricting its civilian applications. One example of a practical application of plutonium-238 is its utilization in the production of compact power generators designed for use in satellites (Berlizov et al., 2008).

One prominent application of americium is its utilization as the chemical compound responsible for detection in smoke detectors. The majority of smoke detectors depend on the utilization of alpha particles generated through the decay of americium 241, which is securely enclosed within americium oxide (AmO_2). This mechanism serves as a means to activate an alarm if smoke particles are detected. Alpha particles, which exhibit a high absorption rate within the plastic container of smoke detectors, effectively cease their movement upon contact. Consequently, these particles typically pose no threat to the well-being of humans or animals (Uozumi et al., 2011). In the event of their potential escape, it is noteworthy that these entities lack the requisite energy to successfully breach the epidermal layer of the human integumentary system. In the event that an individual inhales smoke emitted by a burning detector, there may be potential risks associated with this action. Another risk that warrants consideration is the final disposition of detectors for smoke in landfills. Americium 241 has the potential to serve as a tool for diagnosis in the examination of bone disorders. Moreover, americium is utilized as a primary supplier of neutrons as well as gamma rays. Amidst its role as a neutron the source, americium finds utility in the realm of oil well technological advances and in the assessment of soil density and moisture content. The significance of this particular source lies in its utilization as a portable equipment for gamma ray emission. One potential application that is currently being researched involves the utilization of the metastable state of americium 242 as a means of energy generation for interplanetary travel within the solar system. Neptunium is generated as a secondary product during the production of plutonium and lacks significant applications in the commercial sector. Curium finds applications in the field of medical devices, specifically pacemakers, as well as in power supply systems for satellites. The transcurium elements, due to their extremely limited production quantities, possess no practical applications beyond their utilization in scientific research (Uhlir et al., 2002).

6.5. THE TRANSACTINIDE ELEMENTS

The transactinide elements refer to the elements in the periodic table that possess atomic numbers exceeding 103. At the atomic number of 103, lawrencium (Lr), the electronic configuration of the element reaches a state where the "5f" subshell is fully occupied. Commencing from element 104, the subshell denoted as "6d" initiates the process of electron filling. The inclusion of a "d" subshell in the electron configuration is a defining characteristic of transition metals. Consequently, it is appropriate to classify the initial 10 elements starting from atomic number 104 as transition metals. Nevertheless, these elements have been incorporated into this publication due to their shared characteristics with the actinides, namely their artificial nature and radioactivity. Furthermore, the techniques employed in their discovery originated from the methods utilized in synthesizing the actinides

(Schädel, 2002).

The subshell denoted as "6d" attains complete occupancy upon the attainment of element 112. Starting from element 113, the subshell that undergoes progressive electron occupation is denoted as "7p," thereby classifying these elements as either post-transition metals or nonmetals (Edelstein et al., 2006).

As of the composition of this literary work, the conclusive documentation regarding the discovery of element 117 has not been disclosed. In addition, it is worth noting that there have been a total of 15 transactinides that have been identified thus far. The initial nine elements have been designated as follows: dubnium (Db, 105), seaborgium (Sg, 106), rutherfordium (Rf, element 104), bohrium (Bh, 107), meitnerium (Mt, 109), darmstadtium (Ds, 110), roentgenium (Rg, 111), hassium (Hs, 108), and copernicium (Cn, 112). The reports regarding the finding of elements 113–118 are currently documented; however, no designations in terms of names or symbols have been allocated thus far. The majority of these elements necessitate additional verification to ascertain the accuracy of the discoverers' assertions (Omtvedt et al., 2007).

Due to their significantly brief half-lives, the properties of elements with respect to boiling as well as melting point temperatures, as well as electronic configuration, remain unknown. Consequently, tables containing fundamental properties are not provided for these elements, unlike their counterparts (Katz, 2007).

6.5.1. Rutherfordium

Following the discovery of the final element in the actinide group, lawrencium (atomic number 103), researchers initiated a quest to identify element 104 (also known as eka-hafnium). It was anticipated that this element would exhibit characteristics akin to zirconium and hafnium, thus classifying it as a transition metal. In the year 1964, Georgi Nikolaevich Flerov and his colleagues conducted an experiment at Dubna in which they subjected a plutonium target to neon ions. They subsequently asserted the discovery of element 104, supported by the nuclear reaction provided below (Schumann & Dressler, 2016):

$$^{242}_{94}Pu + ^{22}_{10}Ne \rightarrow {}^{263}104 + ^{1}_{0}n.$$

After the detection of lawrencium (atomic number 103), the last element in the actinide series, scientists embarked on a pursuit to ascertain the characteristics of element 104, commonly referred to as eka-hafnium. The expectation was that this particular element would demonstrate properties similar to those of zirconium and hafnium, thereby categorizing it into the category of a transition metal. In 1964, an experiment was conducted by Georgi Nikolaevich Flerov as well as his colleagues at Dubna. The objective of the experiment was to subject a plutonium target to neon ions. The researchers subsequently claimed the identification of element 104, substantiated by a nuclear reaction presented hereafter (Kasamatsu et al., 2021):

$$^{\text{ü}}_{98}\mathrm{Cf} + {}^{12}_{6}\mathrm{C} \rightarrow {}^{257}104 + 4\,{}^{1}_{0}\mathrm{n};$$

$$^{249}_{98}\mathrm{Cf} + {}^{13}_{6}\mathrm{C} \rightarrow {}^{259}104 + 3\,{}^{1}_{0}\mathrm{n};$$

$$^{248}_{96}\mathrm{Cm} + {}^{16}_{8}\mathrm{O} \rightarrow {}^{258}104 + 6\,{}^{1}_{0}\mathrm{n}.$$

The team led by Ghiorso proposed the designation of element 104 as rutherfordium (Rf) as a tribute to the renowned physicist Ernest Rutherford (1871–1937), who was originally from New Zealand and later became a British citizen. The International Union of Pure and Applied Chemistry (IUPAC) acknowledged the precedence of discovery by the scientists from Berkeley and subsequently endorsed their proposal to name the element as rutherfordium. Element 104 was frequently denoted as "kurchatovium" in periodic tables published between the years 1964 and 1969.

Rutherfordium is classified as a member of the titanium group of elements as well as is positioned below hafnium in the periodic table. Extensive investigations have been conducted on rutherfordium, establishing its tetravalent nature, characterized by its propensity to engage in chemical reactions in the "+4" oxidation state. Consequently, rutherfordium exhibits greater similarities to zirconium and hafnium rather than the actinide elements. A total of 14 isotopes have been artificially produced, all of which exhibit relatively short half-lives. The observed decay rate of rutherfordium-254 exhibits a half-life of a mere 23 microseconds. Rutherfordium-265 exhibits the longest observed half-life, lasting for a duration of approximately 13 hours (Czerwinski, 1992).

6.5.2. Dubnium

Similar to the case of element 104, the discovery and naming of element 105 were also accompanied by controversy. In the year 1968, Georgi Nikolaevich Flerov and his colleagues at Dubna made a scientific announcement regarding the creation of two distinct isotopes of element 105, also known as eka-tantalum. This achievement was accomplished through a series of reactions involving americium and neon ions.

$$ {}^{243}_{95}\text{Am} + {}^{22}_{10}\text{Ne} \rightarrow {}^{260}105 + 5\ {}^{1}_{0}\text{n}; $$

$$ {}^{\text{ü}}_{95}\text{Am} + {}^{22}_{10}\text{Ne} \rightarrow {}^{261}105 + 4\ {}^{1}_{0}\text{n}. $$

Nevertheless, upon further examination by different research teams, it was determined that the Soviet group had made a mistake in their data analysis.

The synthesis of element 105 was reported in 1970 by Albert Ghiorso and his colleagues Matti Nurmia, Kari Eskola, Jim Harris, as well as Pirkko Eskola at Berkeley. This synthesis involved bombarding a californium 249 target with nitrogen 15 ions, as represented by the following equation:

$$ {}^{249}_{98}\text{Cf} + {}^{15}_{7}\text{N} \rightarrow {}^{260}105 + 4\ {}^{1}_{0}\text{n}; $$

In the subsequent year, the group successfully synthesized an additional pair of isotopes for element 105. The findings of the scientists from Berkeley were deemed acceptable and they were acknowledged for their priority in making the discovery.

The Soviet group proposed the designation "nielsbohrium" as a tribute to Niels Bohr, the Danish physicist who lived from 1885 to 1962. The nomenclature committee of the United States of America proposed the designation "hahnium" (Ha) as a tribute to Otto Hahn, a renowned chemist from Germany. Initially, the appellation "hahnium" and its corresponding symbol (Ha) were deemed acceptable, and these designations were employed in scientific literature for a duration of approximately 25 years. During the mid-1990s, the IUPAC made the decision to designate element 105 as dubnium, in reference to the geographical location of the Joint Institute for Nuclear Research (JINR) in Russia. The symbol adopted by the (IUPAC was Db. Simultaneously, the IUPAC opted to abbreviate the name "nielsbohrium" to "bohrium" and designated this nomenclature for element 107, denoted by the symbol Bh. The pentavalent nature of dubnium was demonstrated through experiments conducted at the University of California, Berkeley during the 1980s by Kenneth E. Gregorich, Darleane Hoffman, as well as their colleagues. To clarify, it can be categorized as a member of the vanadium group of elements and exhibits comparable chemical behavior to niobium as well as tantalum when it comes to the formation of compounds in the oxidation state of +5. Surprisingly, despite the limited half-life of only

34 seconds for dubnium's longest-lived isotope, Db-262, studies in aqueous solution can still be conducted (Despotopulos et al., 2015).

6.5.3. Seaborgium

Element 106, also referred to as eka-tungsten, represents the most substantial identified constituent within the chromium group. The initial documentation of the production of element 106 was published in 1974 by a group of researchers led by Albert Ghiorso, Kenneth Hulet, as well as Glenn Seaborg at the Lawrence Berkeley Laboratory. The target material, specifically californium-249, underwent bombardment with oxygen-18, as evidenced by the ensuing reaction:

$$^{249}_{98}Cf + ^{18}_{8}O \rightarrow {}^{263}106 + 4\,^{1}_{0}n\,.$$

An inquiry of merit pertains to the methodology employed by scientists in discerning elements composed of individual atoms and characterized by their ephemeral existence prior to undergoing radioactive decay. The observed phenomenon is the occurrence of a radioactive decay series. Consider a scenario where the initial substance is denoted as Sg-263. The element in question undergoes an alpha decay process, resulting in the formation of Rf-259. Rutherfordium undergoes radioactive decay, resulting in the formation of No-255. The decays are succinctly outlined in the subsequent series.

$$^{263}_{106}Sg \xrightarrow{\text{á}} {}^{259}_{104}Rf \xrightarrow{\text{á}} 25\,^{\text{ü}}_{\text{ü}}\ No\,.$$

Every instance of alpha decay is associated with a pre-established half-life. Furthermore, it is worth noting that every alpha particle possesses a distinct and identifiable energy level. Once scientists commence the detection of alpha decays that can be attributed to established isotopes, they are able to retroactively ascertain the initial isotope by tracing back through the decay series. The confirmation of the synthesis of element 106 is established by nuclear scientists through the positive evidence provided by the genetic relationship among isotope 263 of this element, its daughter (_104^259)Rf, as well as its granddaughter (_102^255)No (Nash & Bursten, 1999).

The acceptance of the discovery took a considerable amount of time, spanning several years. However, in 1993, a team from Berkeley, under the leadership of Ken Gregorich and Darleane Hoffman, provided confirmation of the findings initially reported by the original team. The IUPAC granted permission to the scientists from Berkeley to propose a name for the newly discovered element, coinciding with its 20th anniversary. Element

106, known as seaborgium (symbol Sg), was named in honor of Glenn T. Seaborg, a distinguished chemist and Nobel laureate from the University of California, Berkeley. Seaborg had played a significant role in the discovery or co-discovery of numerous transuranium elements, starting with plutonium in 1940. In a gesture of tribute, the discoverers of element 106 convened a meeting without Seaborg's awareness and unanimously agreed to bestow the name seaborgium upon this newly identified element. During the national meeting of the American Chemical Society in March 1994, Ken Hulet made an announcement in front of the Division of Nuclear Chemistry regarding the proposal of the name "seaborgium" for element 106 to the IUPAC.

While it was not entirely unusual for an element to be designated with the name of a living individual (as exemplified by the cases of Einstein and Fermi, who were both alive when the suggestion was made to name elements 99 and 100 after them), the initial rejection of the proposal by the IUPAC was based on the fact that Seaborg was still alive. The name garnered significant support, prompting the IUPAC to ultimately reach a consensus in favor of its adoption. Dr. Glenn T. Seaborg had the opportunity to witness the formal recognition of the element named after him, seaborgium, prior to his demise in 1999 as a result of stroke-related complications.

Isotopes of seaborgium with longer half-lives compared to the isotope 263-Sg have been artificially produced. Seaborgium 265 exhibits a half-life of approximately 15 seconds, while seaborgium 271 demonstrates a half-life of roughly two minutes. The duration of half-lives associated with seaborgium is adequate for conducting studies on its chemical properties. Seaborgium demonstrates chemical behavior akin to the oxidation state chemistry of its homologous elements, namely chromium, molybdenum, as well as tungsten, with a +6 oxidation state. This is evident in the formation of compounds such as SgO_2Cl_2 in the gaseous phase, which bears resemblance to CrO_2Cl_2, MoO_2Cl_2, as well as WO_2Cl_2 (Hübener et al., 2001).

6.5.4. Bohrium

In 1981, a group of scientists, headed by Peter Armbruster (1931) as well as Gottfried Münzenberg (1940), based at GSI in Darmstadt, Germany, published their findings on the identification of element 107, also known as eka-rhenium. An alternative methodology was employed, deviating from the conventional practice of subjecting transfermium elements to a continuous barrage of lightweight projectiles. Subsequently, researchers commenced employing lead as well as bismuth targets as a means to subject them to bombardment by projectiles of intermediate weight. In this particular instance, the element of interest was bismuth-209, while the element utilized as the projectile was chromium-54, as evidenced by the subsequent reaction:

$$^{209}_{83}Bi + ^{54}_{24}Cr \rightarrow ^{262}107 + ^{1}_{0}n.$$

A total of five atoms of the isotope $^{262}107$ were produced, all of which went through alpha decay resulting in the formation of lighter isotopes. This phenomenon is exemplified by the following instance:

$$^{262}107 \rightarrow ^{258}_{105}Db + ^{4}_{2}He \rightarrow ^{254}_{103}Lr + ^{4}_{2}He.$$

The genetic correlation among isotope 262 and its subsequent daughter and granddaughter isotopes has once again substantiated the synthesis of element 107.

Element 107 was initially designated as nielsbohrium, paying tribute to the esteemed Danish physicist Niels Bohr. However, in 1997, the IUPAC decided to abbreviate the name to bohrium, denoted by the chemical symbol Bh. Element 107 is identified as the most massive constituent within Group VIIA. To date, the synthesis of bohrium has yielded a limited number of atoms, characterized by notably brief half-lives. Among these isotopes, Bh-272 exhibits the longest half-life, approximately one second. Consequently, our current understanding of bohrium's chemical properties remains largely unknown (Gibson, 2002).

6.5.5. Hassium, Meitnerium, and Darmstadtium

Each one of these components were synthetically generated at the GSI facility located in Darmstadt, Germany. Hassium, denoted by the chemical symbol Hs, corresponds to element 108 in the periodic table, commonly referred to as eka-osmium. The initial production of the element in question occurred in 1984 through the collaborative efforts of Peter Armbruster, Gottfried Münzenber, as well as their colleagues. This was achieved by subjecting bismuth 209 to bombardment with manganese 55, and lead 207 as well as lead 208 to bombardment with iron 58, as visually depicted (Armbruster & Münzenberg, 2012).

$$^{209}_{83}Bi + ^{55}_{25}Mn \rightarrow ^{263}108 + ^{1}_{0}n$$

$$^{207}_{82}Pb + ^{58}_{26}Fe \rightarrow ^{264}108 + ^{1}_{0}n$$

$$^{208}_{82}Pb + ^{58}_{26}Fe \rightarrow ^{265}108 + ^{1}_{0}n$$

The production of these isotopes in detectable amounts is not feasible due to their emission of alpha particles and their short half-lives, which last for two milliseconds or less. Following that, several additional isotopes have been generated. Hassium-277 possesses the most protracted half-life documented to date, spanning a duration of 11 minutes.

The element known as hassium was designated with its name taken from the Latin term "Hassia," denoting the German region of Hesse.

Meitnerium, denoted by the chemical symbol Mt, is classified as element 109 in the periodic table, following the eka-iridium series. The discovery of element 109 took place in 1982, conducted by the team at GSI. This achievement involved the production of a single atom of element 109 through the bombardment of a bismuth target with a beam of iron nuclei. The resulting nuclear reaction is as follows:

$$^{209}_{83}\mathrm{Bi} + ^{58}_{26}\mathrm{Fe} \rightarrow {}^{266}109 + ^{1}_{0}\mathrm{n}\,.$$

Several additional isotopes of meitnerium were artificially produced, albeit with limited stability. The current record for the longest observed half-life stands at a mere 0.72 seconds, as exemplified by the isotope Mt-276 (Audi et al., 2012).

The element with atomic number 109 was designated as meitnerium in recognition of the notable achievements of Lise Meitner (1878–1968), an Austrian nuclear physicist who played a pivotal role in the advancement of knowledge surrounding nuclear fission.

The discovery of Element 110, also known as eka-platinum, took place in 1994 at the GSI facility by a team of researchers led by Armbruster, Münzenber, Sigurd Hofmann, as well as their colleagues. Element 110 was synthesized through the process of bombarding a lead target with beams consisting of nickel nuclei, encompassing both nickel 62 and nickel 64 isotopes. The production of darmstadtium atoms was achieved through the implementation of the subsequent nuclear reactions:

$$^{ü}_{82}\mathrm{Pb} + ^{62}_{28}\mathrm{Ni} \rightarrow {}^{269}110 + ^{1}_{0}\mathrm{n}$$

$$^{208}_{82}\mathrm{Pb} + ^{64}_{28}\mathrm{Ni} \rightarrow {}^{271}110 + ^{1}_{0}\mathrm{n}$$

A number of supplementary isotopes of darmstadtium were identified. The majority of half-lives are shorter than one millisecond. The isotope Ds-281 possesses the lengthiest half-life, which spans a duration of 11 seconds. The element with the symbol Ds, known as darmstadtium, derives its name from the city where its discovery took place. The isotope Darmstadtium-269 undergoes alpha decay, resulting in the formation of Hassium-265. Hassium, subsequently, undergoes an additional alpha decay process resulting in the formation of Sg-261, which subsequently undergoes decay to yield Rf-247. Rutherfordium undergoes radioactive decay, resulting in the formation of No-253. The decays are succinctly outlined in the subsequent series.

$$^{269}_{110}\mathrm{Ds} \xrightarrow{\text{á}} {}^{265}_{108}\mathrm{Hs} \xrightarrow{\text{á}} {}^{261}_{106}\mathrm{Sg} \xrightarrow{\text{á}} {}^{257}_{104}\mathrm{Rf} \xrightarrow{\text{á}} {}^{253}_{102}\mathrm{No}\ .$$

Every instance of alpha decay is accompanied by a pre-established half-life. Furthermore, it is worth noting that every alpha particle possesses a distinct and identifiable energy. Once scientists commence the detection of alpha decays that can be ascribed to recognized isotopes, they possess the ability to retroactively determine the initial isotope by tracing back along the decay series (Wieser, 2006).

6.5.6. Roentgenium

Element 111, commonly referred to as eka-gold, was discovered at the GSI (Gesellschaft für Schwerionenforschung) facility in 1994. The group of scientists responsible for this discovery had previously identified darmstadtium through the process of bombarding a bismuth 209 target in nickel 64 ions (Hancock et al., 2006).

$$^{209}_{83}\mathrm{Bi} + {}^{64}_{28}\mathrm{Ni} \rightarrow 272111 + {}^{1}_{0}\mathrm{n}\ .$$

The element with the chemical symbol Rg, known as roentgenium, was designated in honor of Wilhelm Conrad Roentgen (1845–1923), a prominent German physicist credited with the discovery of X-rays in the year 1895. The element Roentgenium-272 exhibits a half-life of approximately two milliseconds. The isotope with the greatest longevity is Rg-280, which possesses a half-life of approximately 3.6 seconds (Griffith, 2008).

6.5.7. Elements 112–118

The investigation into the existence of element 112, also known as eka-mercury, was carried out during a span of 24 days in January and February of 1996. The initial formation of the initial atom of element 112 took place on February 9th. A target composed of lead-208 was subjected to bombardment by zinc-70 nuclei, leading to the formation of an item with a mass number of 277, as illustrated in the subsequent equation:

$$^{208}_{82}Pb + ^{70}_{30}Zn \rightarrow ^{277}112 + ^{1}_{0}n .$$

The atoms of element 277 exhibit alpha-emission characteristics, possessing a notably brief half-life of merely 0.7 milliseconds. Consequently, the initial atom experienced an exceedingly transient existence (Pitzer, 1975).

In the year 2000, the research team at GSI conducted a replication of the experiment, resulting in the synthesis of an additional atom of Uub-277. The experiment was replicated at RIKEN, a Japanese laboratory known as the Institute of Physical as well as Chemical Research (Rikagaku Kenkyūsho). The IUPAC determined in both 2001 and 2003 that the evidence presented by GSI was inadequate to substantiate their assertion of discovery. In May 2009, researchers from GSI and RIKEN provided additional evidence to support their claim, which was subsequently acknowledged by the IUPAC. As a result, GSI was officially recognized as the first institution to have made the discovery.

As of the year 2009, a total for five isotopes of element 112 had been synthesized. The isotope with the greatest longevity possesses a mass number of 285 and exhibits a half-life of 29 seconds. In the year 2010, the IUPAC made an official determination to designate the element with atomic number 112 with the name "Copernicium" in honor of Nicolaus Copernicus. Copernicus, a renowned astronomer from Poland in the 16th century, is widely recognized for his groundbreaking theory proposing a heliocentric model of the universe, which brought about a significant paradigm shift in the field of science. The element with atomic number 112 has been officially named copernicium, denoted by the chemical symbol Cn.

In February 2004, a group of scientists hailing from LLNL in California as well as JINR in Dubna, Russia, made public their findings regarding the identification of two previously unknown elements, specifically element 113 (also known as eka-thallium) and element 115 (also known as eka-bismuth). The reaction involving the bombardment of a target composed of americium 243 with calcium 48 nuclei led to the synthesis of 4 atoms of element 115.

$$^{243}_{95}Am + ^{48}_{20}Ca \rightarrow ^{288}115 + 3 \quad .$$

The atoms synthesized possessed a mass number of 288 as well as exhibited a decay process resulting in the formation of element 113, with a relatively short half-life of approximately 0.1 second.

$$^{288}115 \rightarrow ^{284}113 + ^{4}_{2}He .$$

The atomic nuclei of element 284113 underwent decay with a half-life of approximately 0.5 seconds.

There is knowledge of the existence of isotopes of element 113. Each of the elements mentioned exhibit alpha decay and have half-lives that are shorter than one second. While the experimental data may seem persuasive, it is crucial to await independent verification of the data. During the interim period, scientists will utilize temporary names, specifically ununtrium (Uut) for element 113 and ununpentium (Uup) for element 115, as permanent names have not yet been assigned. The initial assertion regarding the identification of element 114 was put forth in the early stages of 1999 by researchers hailing from Dubna, Russia. These scientists communicated the successful creation of a solitary atom through synthesis. Following that, additional atomic syntheses were documented. Each experimental trial had a duration of slightly over one month. The synthesis of Element 114 involves the utilization of calcium 48 nuclei to bombard a plutonium 244 target, resulting in the following series of reactions ((Zagrebaev et al., 2012):

$$^{244}_{94}\mathrm{Pu} + ^{48}_{20}\mathrm{Ca} \rightarrow {}^{288}114 + 4\ ^{1}_{0}\mathrm{n}$$

$$^{244}_{94}\mathrm{Pu} + ^{48}_{20}\mathrm{Ca} \rightarrow {}^{289}114 + 3\ ^{1}_{0}\mathrm{n}.$$

Both isotopes exhibit radioactive decay with half-lives of less than one minute, specifically five seconds for mass number 288 as well as 30 seconds for mass number 289. While the half-lives of these elements may be considered short in a certain context, they are relatively lengthy when compared to the typically brief half-lives observed in transactinide elements, which frequently exhibit half-lives measured in milliseconds or even shorter durations. For a number of years, nuclear chemists had made predictions regarding the potential discovery of heavy elements exhibiting half-lives lasting seconds or longer. Consequently, the discovery of these two specific half-lives has elicited significant enthusiasm. Subsequently, the discovery of element 114 ensued subsequent to the synthesis of element 118, that subsequently undergoes a rapid succession of alpha decays, as depicted in the subsequent sequence.

$$^{293}118 \rightarrow {}^{289}116 \rightarrow {}^{285}114 \rightarrow {}^{281}112.$$

The duration of each decay's half-life is shorter than one millisecond. Due to the current limitation of producing only one atom of element 118 at a time, it is improbable, at least in the foreseeable future, to generate appreciable quantities of these elements.

The initial report provided by the scientists from Dubna in 1999 was subsequently determined to be inaccurate. Nevertheless, additional observations conducted between 2000 and 2004 were found to be consistent and reliable upon careful examination. In 2009, researchers at the Lawrence Berkeley National Laboratory in Berkeley, California, substantiated the subsequent reports originating from Dubna. The researchers from the University of California, Berkeley successfully synthesized a pair of atoms belonging to element 114 by employing the subsequent techniques developed by the scientists from Dubna. This achievement effectively corroborated the precedence of discovery attributed to Dubna. The provisional designation assigned to element 114 is ununquadium (Uuq). The chemical elements with atomic numbers 116 and 118, provisionally designated as ununhexium (Uuh) as well as unuoctium (Uuo) accordingly, have a complex and eventful past. The discoveries of elements 116 and 118 were reported by scientists at the Lawrence Berkeley National Laboratory in 1999. Nevertheless, during the years 2000 and 2001, when the aforementioned experiments were replicated at the University of California, Berkeley, as well as concurrently at the Institute for Heavy Ion Research in Darmstadt, Germany, as well as the Atomic Energy Research Institute in Japan, researchers encountered difficulties in substantiating identical outcomes. Perplexed by the findings, the scientists from Berkeley conducted a thorough reassessment of the data collected in 1999. Their subsequent report indicated that one of the scientists involved in the initial research team purportedly manipulated the data, which had initially indicated the successful synthesis of elements 116 and 118. As a result, the scientists from Berkeley retracted their assertion.

Although the occurrence of this incident is unfortunate, it does effectively exemplify the functioning of scientific research. The process of peer review as well as replication by other researchers is integral to the validation of any scientific discovery. The validity of a discovery is contingent upon its replication and acceptance by the scientific community. When it comes to the discovery of novel elements, it is imperative that multiple teams of scientists synthesize the element in question before it is formally recognized and incorporated into the periodic table by the international scientific community (Hamilton et al., 2013).

Although the occurrence of this incident is unfortunate, it effectively exemplifies the functioning of scientific research. The process of peer review as well as replication by other scientists is an essential requirement for validating any scientific discovery. The validation of a discovery is contingent upon the replication of an experiment and the acceptance of its results by the scientific community. In the context of elemental discovery, it is imperative that a given element undergoes synthesis by multiple teams of scientists prior to its official inclusion in the periodic table by the international scientific community.

Following this, in the years 2002 and 2005, the production of elements 116 and 118 was achieved through a collaborative endeavor involving scientists from the JINR at Dubna as well as the Lawrence Livermore Laboratory. The experimental methodology employed involved subjecting a californium 249 target to a bombardment of calcium 48 nuclei, as represented by the subsequent equation:

$$ {}^{\text{ü}}_{98}Cf + {}^{48}_{20}Ca \rightarrow {}^{294}118 + 3\,{}^{1}_{0}n. $$

The Uuo-294 isotope undergoes rapid alpha decay, resulting in the formation of Uuh-116, as depicted in the following chemical equation:

$$ {}^{294}118 \rightarrow {}^{290}116 + {}^{4}_{2}He. $$

The Uuo-294 isotope exhibits a rapid process of alpha decay, leading to the creation of Uuh-116, as illustrated by the subsequent chemical equation:

$$ {}^{248}_{96}\mathrm{Cm} + {}^{48}_{20}\mathrm{Ca} \rightarrow {}^{292}116 + 4\,{}^{1}_{0}\mathrm{n}. $$

The observed half-life of Uuh-292 is a mere 47 milliseconds. In April 2010, a scientific collaboration between researchers from Russia as well as the United States announced their successful production of six atoms of element 117 (provisionally named ununseptium, Uus) at the JINR in Dubna, Russia. The production of this elusive element was successfully achieved within a controlled laboratory setting through the process of bombarding a target composed of berkelium 249 atoms, which were obtained from Oak Ridge National Laboratory, with calcium 48 ions. As a consequence of this interaction, a series of reactions occurred, leading to the desired outcome.

$$ {}^{48}_{20}\mathrm{Ca} + {}^{\text{ü}}_{97}\mathrm{Bk} \xrightarrow{78\mathrm{ms}} {}^{294}117 + 3\,{}^{1}_{0}\mathrm{n} $$

$$ {}^{48}_{20}\mathrm{Ca} + {}^{249}_{97}\mathrm{Bk} \xrightarrow{14\mathrm{ms}} {}^{293}117 + 4\,{}^{1}_{0}\mathrm{n} $$

The decay times are observed and recorded above the response arrows. A total of five Uus-293 atoms were detected, along with one Uus-294 atom. As is customary, the process of official recognizing will not commence until the discovery has been formally replicated. It is highly probable that the scientists at JINR will persist in their efforts to achieve the synthesis of elements 119 as well as 120 in forthcoming endeavors. In the event that element 120 is successfully identified, it is expected to exhibit characteristics consistent with those of alkaline earth elements, specifically barium and radium (Oganessian & Utyonkov, 2015).

6.6. SUMMARY

The actinides are a collection of chemical elements that are situated in the number seven period of the periodic table, occupying the region among actinium (element 89) as well as lawrencium (element 103). These elements exhibit analogous electron configurations as well as chemical properties, so they possess the attribute of radioactivity. The actinides are classified as members of the "f-block" or "inner transition metals" group within the periodic table of elements.

The actinide series encompasses a total of fifteen elements, commencing with actinium (Ac) and concluding with lawrencium (Lr). The composition encompasses both elements that occur naturally and those that are artificially produced. All elements in the actinide series exhibit radioactive properties, indicating that their atomic nuclei are inherently unstable and undergo decay processes, resulting in the emission of diverse forms of radiation. Certain actinides exhibit extended half-lives, whereas others manifest pronounced instability and undergo rapid decay. Numerous actinides possess significant utility in the realms of nuclear industry and technology. Uranium as well as plutonium are integral constituents of both nuclear power reactors as well as nuclear weapons, exemplifying their critical significance. Certain elements belonging to the actinide series, such as uranium and thorium, exhibit a comparatively higher prevalence within the Earth's crust. Conversely, elements like neptunium as well as americium are predominantly generated via nuclear reactions. The actinides are widely acknowledged to possess significant toxicity, primarily attributable to their radioactive properties and the potential for integration into biological tissues. It is imperative to implement stringent safety protocols when dealing with these elements. The complex chemistry observed in actinides can be attributed to the inclusion of f-orbitals in their configurations of electrons. The existence of 5f electrons in their outside electronic shells is considered to be one of the contributing factors to the parallels in chemical behavior between these elements. Transuranic elements, which are elements beyond uranium, are typically generated through artificial means via nuclear reactions and have been successfully synthesized within laboratory settings.

The actinides are a group of elements that possess significant scientific interest and practical relevance, particularly in the field of nuclear technology. These elements exhibit distinctive and complex chemical properties, making them a subject of great fascination. Given their inherent radioactivity and associated risks, these materials necessitate meticulous management and appropriate disposal measures to safeguard the well-being of individuals and the environment.

REFERENCES

1. Adams, J. A., Osmond, J. K., & Rogers, J. J., (1959). The geochemistry of thorium and uranium. *Physics and Chemistry of the Earth, 3*, 298–348.
2. Alwan, A. K., & Williams, P. A., (1980). The aqueous chemistry of uranium minerals. Part 2. Minerals of the liebigite group. *Mineralogical Magazine, 43*(329), 665–667.
3. Armbruster, P., & Münzenberg, G., (2012). An experimental paradigm opening the world of superheavy elements. *The European Physical Journal H, 37*(2), 237–309.
4. Arzhannikov, A. V., Shmakov, V. M., Modestov, D. G., Bedenko, S. V., Prikhodko, V. V., Lutsik, I. O., & Shamanin, I. V., (2020). Facility to study neutronic properties of a hybrid thorium reactor with a source of thermonuclear neutrons based on a magnetic trap. *Nuclear Engineering and Technology, 52*(11), 2460–2470.
5. Audi, G., Kondev, F. G., Wang, M., Pfeiffer, B., Sun, X., Blachot, J., & MacCormick, M., (2012). The NUBASE2012 evaluation of nuclear properties. *Chinese Physics C, 36*(12), 1157.
6. Bagnall, K. W., (1964). The transuranium elements. *Science Progress (1933-)*, 66–83.
7. Berlizov, A. N., Malyuk, I. A., Sajeniouk, A. D., Tryshyn, V. V., Petrov, V. V., Savin, A. I., & Tashimova, F. A., (2008). Transuranium elements and fission products in technological channels of unit No. 2 of Chernobyl nuclear power plant. *Journal of Radioanalytical and Nuclear Chemistry, 277*, 49–57.
8. Berthet, J. C., & Ephritikhine, M., (1998). New advances in the chemistry of uranium amide compounds. *Coordination Chemistry Reviews, 178*, 83–116.
9. Carrott, M., Maher, C., Mason, C., Sarsfield, M., & Taylor, R., (2016). "TRU-SANEX": A variation on the EURO-GANEX and I-SANEX processes for heterogeneous recycling of actinides Np-Cm. *Separation Science and Technology, 51*(13), 2198–2213.
10. Chapman, K., (2020). The transuranic elements and the island of stability. *Philosophical Transactions of the Royal Society A, 378*(2180), 20190535.
11. Choppin, G. R., & Nash, K. L., (1995). Actinide separation science. *Radiochimca Acta, 70*(7), 225–236.

12. Clark, D. L., & Hobart, D. E., (2019). Discovery of the transuranium elements inspired rearrangement of the periodic table and the approach for finding new elements. *The Periodic Table I: Historical Development and Essential Features, 104*, 225–259.

13. Collins, E. D., DelCul, G. D., Renier, J. P., & Spencer, B. B., (2007). Preliminary multicycle transuranic actinide partitioning-transmutation studies. *ORNL/TM-2007/24, 12*(35), 4987–4989.

14. Cuttler, J. M., (2001). Why are we so afraid of nuclear radiation? *Canadian Nuclear Society Bulletin, 22*, 123–125.

15. Czerwinski, K. R., (1992). *Studies of Fundamental Properties of Rutherfordium (Element 104) Using Organic Complexing Agents* (Vol. 87, pp. 887–889). University of California, Berkeley.

16. Davidson, L. M., & Baum, A., (1990). Posttraumatic stress in children following natural and human-made trauma. *Handbook of Developmental Psychopathology, 10*(670), 251–259.

17. Day, Jr. R. A., & Stoughton, R. W., (1950). Chemistry of thorium in aqueous solutions. I. Some organic and inorganic complexes. *Journal of the American Chemical Society, 72*(12), 5662–5666.

18. Despotopulos, J. D., Gostic, J. M., Bennett, M. E., Gharibyan, N., Henderson, R. A., Moody, K. J., & Shaughnessy, D. A., (2015). Characterization of group 5 dubnium homologs on diglycolamide extraction chromatography resins from nitric and hydrofluoric acid matrices. *Journal of Radioanalytical and Nuclear Chemistry, 303*, 485–494.

19. Dolg, M., & Cao, X., (2009). Accurate relativistic small-core pseudopotentials for actinides. Energy adjustment for uranium and first applications to uranium hydride. *The Journal of Physical Chemistry A, 113*(45), 12573–12581.

20. Durakovic, A., (2016). Medical effects of internal contamination with actinides: Further controversy on depleted uranium and radioactive warfare. *Environmental Health and Preventive Medicine, 21*(3), 111–117.

21. Edelstein, N. M., Fuger, J., Katz, J. J., & Morss, L. R., (2006). Summary and comparison of properties of the actinide and transactinide elements. *The Chemistry of the Actinide and Transactinide Elements, 658*, 1753–1835.

22. Eisenbud, M., Krauskopf, K., Franca, E. P., Lei, W., Ballad, R.,

Linsalata, P., & Fujimori, K., (1984). Natural analogues for the transuranic actinide elements: An investigation in Minas Gerais, Brazil. *Environmental Geology and Water Sciences, 6*, 1–9.

23. Ewing, R. C., (2011). Safe management of actinides in the nuclear fuel cycle: Role of mineralogy. *Comptes Rendus Geoscience, 343*(2, 3), 219–229.
24. Fattal, E., Tsapis, N., & Phan, G., (2015). Novel drug delivery systems for actinides (uranium and plutonium) decontamination agents. *Advanced Drug Delivery Reviews, 90*, 40–54.
25. Fernandez, A., McGinley, J., Somers, J., & Walter, M., (2009). Overview of past and current activities on fuels for fast reactors at the institute for transuranium elements. *Journal of Nuclear Materials, 392*(2), 133–138.
26. Francis, A. J., (1998). Biotransformation of uranium and other actinides in radioactive wastes. *Journal of Alloys and Compounds, 271*, 78–84.
27. Galahom, A. A., (2017). Minimization of the fission product waste by using thorium-based fuel instead of uranium dioxide. *Nuclear Engineering and Design, 314*, 165–172.
28. Gibson, J. K., (2002). Gas-phase chemistry of actinide ions: Probing the distinctive character of the 5f elements. *International Journal of Mass Spectrometry, 214*(1), 1–21.
29. Giel, R., (1998). Natural and human-made disasters. *Adversity, Stress, and Psychopathology, 6*, 66–76.
30. Grenthe, I., Drożdżynński, J., Fujino, T., Buck, E. C., Albrecht-Schmitt, T. E., & Wolf, S. F., (2006). Uranium. *The Chemistry of the Actinide and Transactinide Elements*, 253–698.
31. Griffith, B. W., (2008). The periodic table and the platinum group metals. *Platinum Metals Review, 52*(2), 114–119.
32. Griveau, J. C., & Colineau, E., (2014). Superconductivity in transuranium elements and compounds. *Comptes Rendus Physique, 15*(7), 599–615.
33. Grödler, D., Sperling, J. M., Rotermund, B. M., Scheibe, B., Beck, N. B., Mathur, S., & Albrecht-Schönzart, T. E., (2023). Neptunium alkoxide chemistry: Expanding alkoxides to the transuranium elements. *Inorganic Chemistry, 62*(6), 2513–2517.
34. Hamilton, J. H., Hofmann, S., & Oganessian, Y. T., (2013). Search for superheavy nuclei. *Annual Review of Nuclear and Particle Science,*

63, 383–405.

35. Hancock, R. D., Bartolotti, L. J., & Kaltsoyannis, N., (2006). Density functional theory-based prediction of some aqueous-phase chemistry of superheavy element 111. Roentgenium (I) is the 'softest'metal ion. *Inorganic Chemistry, 45*(26), 10780–10785.
36. Hoffman, D. C., (1998). the transuranium elements: From neptunium and plutonium to element 112. *Actinides and the Environment, 41*, 3.
37. Hübener, S., Taut, S., Vahle, A., Dressler, R., Eichler, B., Gäggeler, H. W., & Brüchle, W., (2001). Physico-chemical characterization of seaborgium as oxide hydroxide. *Radiochimica Acta, 89*(11, 12), 737–742.
38. Hudry, D., Apostolidis, C., Walter, O., Gouder, T., Courtois, E., Kübel, C., & Meyer, D., (2012). Non-aqueous synthesis of isotropic and anisotropic actinide oxide nanocrystals. *Chemistry–A European Journal, 18*(27), 8283–8287.
39. Huill, L. L., (1938). The transuranium elements. *Chemical Reviews, 23*(1), 87–155.
40. Hyde, E. K., & Seaborg, G. T., (1956). *The Transuranium Elements* (No. UCRL-3312) (Vol. 123, No. 2, pp. 789–793). Lawrence Berkeley National Lab. (LBNL), Berkeley, CA (United States).
41. Irving, H. M. N. H., & Edgington, D. N., (1960). Synergic effects in the solvent extraction of the actinides—I uranium (VI). *Journal of Inorganic and Nuclear Chemistry, 15*(1, 2), 158–170.
42. Jenkins, E. N., & Sneddon, G. W., (1954). *The Analysis of Urine for Traces of Americium and Other Alpha Emitters (No. AERE-C/R-1399)* (Vol. 87, No. 56, pp. 987–990). United Kingdom Atomic Energy Authority. Research Group. Atomic Energy Research Establishment, Harwell, Berks, England.
43. Kaltsoyannis, N., (2003). Recent developments in computational actinide chemistry. *Chemical Society Reviews, 32*(1), 9–16.
44. Kamei, T., (2012). Recent research of thorium molten-salt reactor from a sustainability viewpoint. *Sustainability, 4*(10), 2399–2418.
45. Kanellakopulos, B., (1990). Fifty years of neptunium and plutonium chemistry, fifty years effort to understand it. In: *American Chemical Society National Meeting, Division of Nuclear Chemistry and Technology*.
46. Kasamatsu, Y., Toyomura, K., Haba, H., Yokokita, T., Shigekawa, Y.,

Kino, A., & Shinohara, A., (2021). Co-precipitation behavior of single atoms of rutherfordium in basic solutions. *Nature Chemistry, 13*(3), 226–230.

47. Katz, J. J., & Rabinowitch, E., (1961). *The Chemistry of Uranium* (Vol. 5, pp. 34–40). Рипол Классик.
48. Katz, J. J., (2007). *The Chemistry of the Actinide and Transactinide Elements* (Vol. 1, pp. 216–220). Springer Science & Business Media.
49. Kim, J. I., (1992). The chemical behavior of transuranium elements and barrier functions in natural aquifer systems. *MRS Online Proceedings Library (OPL)*, *294*, 3.
50. Kmak, K. N., Shaughnessy, D. A., & Vujic, J., (2021). Batch and column studies of radium, actinium, thorium, and protactinium on CL resin in nitric acid, hydrochloric acid, and hydrofluoric acid. *Journal of Radioanalytical and Nuclear Chemistry, 328*, 225–233.
51. Kumar, S. V., & Sood, D. D., (1980). Non-aqueous chemistry in nuclear fuel reprocessing. In: *Proceedings of the Symposium on Chemical Reactions in Non-Aqueous Media and Molten Salts [Held at] Hyderabad, March 6–8, 1978*.
52. Langmuir, D., & Herman, J. S., (1980). The mobility of thorium in natural waters at low temperatures. *Geochimica et Cosmochimica Acta, 44*(11), 1753–1766.
53. Loiseau, T., Mihalcea, I., Henry, N., & Volkringer, C., (2014). The crystal chemistry of uranium carboxylates. *Coordination Chemistry Reviews, 266*, 69–109.
54. Manion, M., & Evan, W. M., (2001). Chapter 1-natural and human-made disasters. *International Journal of Risk Assessment and Management, 2*(1, 2), 6–35.
55. Mayer, M. G., (1941). Rare-earth and transuranic elements. *Physical Review, 60*(3), 184.
56. Molinari, J., & Snodgrass, W. J., (1990). The chemistry and radiochemistry of radium and the other elements of the uranium and thorium natural decay series. *The Environmental Behavior of Radium, 1*, 11–56.
57. Mullane, K. C., Lewis, A. J., Yin, H., Carroll, P. J., & Schelter, E. J., (2014). Anomalous one-electron processes in the chemistry of uranium nitrogen multiple bonds. *Inorganic Chemistry, 53*(17), 9129–9139.
58. Munzenberg, G., (1988). Recent advances in the discovery of

transuranium elements. *Reports on Progress in Physics, 51*(1), 57.

59. Nagame, Y., & Hirata, M., (2011). Production and properties of transuranium elements. *Radiochimica Acta, 99*(7, 8), 377–393.
60. Nash, C. S., & Bursten, B. E., (1999). Prediction of the bond lengths, vibrational frequencies, and bond dissociation energy of octahedral seaborgium hexacarbonyl, Sg (CO){sub 6}. *Journal of the American Chemical Society, 121*(46), 6740–6745.
61. Nash, K. L., Madic, C., Mathur, J. N., & Lacquement, J., (2006). Actinide separation science and technology. In: *The Chemistry of the Actinide and Transactinide Elements* (Vol. 2, pp. 2622–2798). Dordrecht: Springer Netherlands.
62. Nelson, A. G. D., Bray, T. H., Zhan, W., Haire, R. G., Sayler, T. S., & Albrecht-Schmitt, T. E., (2008). Further examples of the failure of surrogates to properly model the structural and hydrothermal chemistry of transuranium elements: Insights provided by uranium and neptunium diphosphonates. *Inorganic Chemistry*, *47*(11), 4945–4951.
63. Nitsche, H., (1990). Basic research for assessment of geologic nuclear waste repositories: What solubility and speciation studies of transuranium elements can tell us. *MRS Online Proceedings Library (OPL)*, *212*, 368–370.
64. Noh, S., Kwak, D., Lee, J., Kang, J., & Han, B., (2014). First-principles calculations of the thermodynamic properties of transuranium elements in a molten salt medium. *Journal of the Korean Physical Society, 64*, 806–812.
65. O'Donnell, T. A., (1990). Cationic speciation in non-aqueous media of uranium, neptunium, and plutonium in oxidation states III and IV. In: *American Chemical Society National Meeting, Division of Nuclear Chemistry and Technology*.
66. Oganessian, Y. T., & Utyonkov, V. K., (2015). Super-heavy element research. *Reports on Progress in Physics, 78*(3), 036301.
67. Omtvedt, J. P., Alstad, J., Bjørnstad, T., Düllmann, C. E., Gregorich, K. E., Hoffman, D. C., & Zheng, L., (2007). Chemical properties of the transactinide elements studied in liquid phase with SISAK. *The European Physical Journal D*, *45*, 91–97.
68. Ortu, F., Formanuik, A., Innes, J. R., & Mills, D. P., (2016). New vistas in the molecular chemistry of thorium: Low oxidation state complexes. *Dalton Transactions, 45*(18), 7537–7549.

69. Øvergaard, S., (2008). *Issue Paper: Definition of Primary and Secondary Energy* (Vol. 7, pp. 456–459). Statistics Norway, Oslo.
70. Parker, B. F., Zhang, Z., Rao, L., & Arnold, J., (2018). An overview and recent progress in the chemistry of uranium extraction from seawater. *Dalton Transactions, 47*(3), 639–644.
71. Phuoc, T. X., Wang, P., & McIntyre, D., (2016). Detection of rare earth elements in powder river basin sub-bituminous coal ash using laser-induced breakdown spectroscopy (LIBS). *Fuel, 163*, 129–132.
72. Pitzer, K. S., (1975). Are elements 112, 114, and 118 relatively inert gases? *The Journal of Chemical Physics, 63*(2), 1032–1033.
73. Radchenko, V., Engle, J. W., Wilson, J. J., Maassen, J. R., Nortier, M. F., Birnbaum, E. R., & Fassbender, M. E., (2016). Formation cross-sections and chromatographic separation of protactinium isotopes formed in proton-irradiated thorium metal. *Radiochimica Acta, 104*(5), 291–304.
74. Radchenko, V., Morgenstern, A., Jalilian, A. R., Ramogida, C. F., Cutler, C., Duchemin, C., & Li, Y., (2021). Production and supply of α-particle–emitting radionuclides for targeted α-therapy. *Journal of Nuclear Medicine, 62*(11), 1495–1503.
75. Romanchuk, A. Y., Slesarev, A. S., Kalmykov, S. N., Kosynkin, D. V., & Tour, J. M., (2013). Graphene oxide for effective radionuclide removal. *Physical Chemistry Chemical Physics, 15*(7), 2321–2327.
76. Santschi, P. H., Murray, J. W., Baskaran, M., Benitez-Nelson, C. R., Guo, L. D., Hung, C. C., & Roy-Barman, M., (2006). Thorium speciation in seawater. *Marine Chemistry, 100*(3), 250–268.
77. Schädel, M., (2002). The chemistry of transactinide elements-experimental achievements and perspectives. *Journal of Nuclear and Radiochemical Sciences, 3*(1), 113–120.
78. Schumann, D., & Dressler, R., (2016). Development of an extraction system for the separation of dubnium from rutherfordium using MIBK and HCl/HF solutions. *Radiochimica Acta, 104*(1), 41–49.
79. Seaborg, G. T., (1946). The transuranium elements. *Science, 104*(2704), 379–386.
80. Selbin, J., & Ortego, J. D., (1969). Chemistry of uranium (V). *Chemical Reviews, 69*(5), 657–671.
81. Shibuya, K., (2021). A risk management on demographic mobility of evacuees in disasters. In: *Encyclopedia of Information Science and*

Technology (5th edn., Vol. 1, pp. 1612–1622). IGI Global.

82. Shmelev, A. N., Kulikov, G. G., Kurnaev, V. A., Salahutdinov, G. H., Kulikov, E. G., & Apse, V. A., (2015). Hybrid fusion–fission reactor with a thorium blanket: Its potential in the fuel cycle of nuclear reactors. *Physics of Atomic Nuclei, 78*, 1100–1111.
83. Siegrist, M., & Sütterlin, B., (2014). Human and nature-caused hazards: The affect heuristic causes biased decisions. *Risk Analysis, 34*(8), 1482–1494.
84. Sime, R. L., (2000). The search for transuranium elements and the discovery of nuclear fission. *Physics in Perspective, 2*, 48–62.
85. Singh, R. N., Saraswat, I. P., & Suri, S. K., (1983). Synthesis of $CsPuF_5$ and Cs_2PuF_6 from non-aqueous media. *Synthesis and Reactivity in Inorganic and Metal-Organic Chemistry, 13*(1), 21–28.
86. Spit͡syn, V. I., & Katz, J. J., (2018). *Proceedings of the Moscow Symposium on the Chemistry of Transuranium Elements*. Elsevier.
87. Sterne, P. A., Gonis, A., & Borovoi, A. A., (1998). *Actinides and the Environment* (Vol. 41, pp. 387–390). Springer Science & Business Media.
88. Su, J., & Chen, J., (2015). MOFs of uranium and the actinides. *Lanthanide Metal-Organic Frameworks, 10*, 265–295.
89. Uhlir, J., Hron, M., Priman, V., & Frejtich, Z., (2002). Current status of Czech R&D program in partitioning and transmutation. In: *Proc. of OECD/NEA 7th Information Exchange Meeting on Actinide and Fission Product Partitioning and Transmutation, Jeju, Korea* (Vol. 456, No. 30, pp. 845–849).
90. Uozumi, K., Iizuka, M., Kurata, M., Inoue, T., Koyama, T., Ougier, M., & Glatz, J. P., (2011). Recovery of transuranium elements from real high-level liquid waste by pyropartitioning process. *Journal of Nuclear Science and Technology, 48*(2), 303–314.
91. Uppal, S., (2016). Nuclear technology and diseases. *Academicia: An International Multidisciplinary Research Journal, 6*(3), 10–14.
92. Vesnovskii, S. P., (2003). RFNC-VNIIEF capabilities to production high pure isotopes for scientific and medical applications. *Journal of Radioanalytical and Nuclear Chemistry, 257*, 27–31.
93. Wagman, D. D., (1981). *Selected Values of Chemical Thermodynamic Properties: Compounds of Uranium, Protactinium, Thorium, Actinium, and the Alkali Metals* (Vol. 270, pp. 258–260). US Department of

Commerce, National Bureau of Standards.

94. Wang, L., Yuan, L., Chen, K., Zhang, Y., Deng, Q., Du, S., & Shi, W., (2016). Loading actinides in multilayered structures for nuclear waste treatment: The first case study of uranium capture with vanadium carbide MXene. *ACS Applied Materials & Interfaces, 8*(25), 16396–16403.
95. Weigel, F., & Hoffmann, G., (1976). The phosphates and arsenates of hexavalent actinides. Part I. Uranium. *Journal of the Less Common Metals, 44*, 99–123.
96. Wheatley, S., Sovacool, B. K., & Sornette, D., (2016). Reassessing the safety of nuclear power. *Energy Research & Social Science, 15*, 96–100.
97. White, F. D., & Marsh, M. L., (2019). Recent advances in chemistry of transuranium elements in non-aqueous media. *Handbook on the Physics and Chemistry of Rare Earths, 55*, 123–158.
98. Wieser, M. E., (2006). Atomic weights of the elements 2005 (IUPAC technical report). *Pure and Applied Chemistry, 78*(11), 2051–2066.
99. Wolf, S. F., (2006). Trace analysis of actinides in geological, environmental, and biological matrices. In: *The Chemistry of the Actinide and Transactinide Elements* (Vol. 2, pp. 3273–3338). Dordrecht: Springer Netherlands.
100. Zagrebaev, V. I., Karpov, A. V., & Greiner, W., (2012). Possibilities for synthesis of new isotopes of superheavy elements in fusion reactions. *Physical Review C, 85*(1), 014608.
101. Závodská, L., Kosorinova, E., Scerbakova, L., & Lesny, J., (2008). Environmental chemistry of uranium. *HV ISSN, 87*(3), 1418–7108.

CHAPTER 7

Boron, Carbon, Nitrogen, and Oxygen Groups

Contents

7.1. INTRODUCTION

The Boron group comprises five naturally occurring elements and one synthetic element known as nihonium (Nh). Despite their low reactivity, none of these elements occur naturally in a pure state. Boron (B), the initial constituent, is classified as a metalloid (an element that exhibits characteristics of both metals and non-metals), but the remaining elements in the series are categorized as metals. Aluminum (Al), the second element in the periodic table, is very prevalent in the composition of Earth's rocks (Wang et al., 2021).

The Carbon group is comprised of a single non-metal, two semi-metals, and three metals. Carbon (C), a non-metallic element, serves as the primary constituent in all biological entities. The semi-metals, namely silicon (Si) and germanium (Ge), exhibit characteristics that are characteristic of both metals and non-metals. These elements play a crucial role in the field of electronics. For centuries, humanity have utilized two metallic elements, namely tin (Sn) and lead (Pb). Flerovium (Fl), an artificially synthesized element, currently lacks any established practical use (Bowen, 1927).

The Nitrogen group encompasses a variety of natural elements, including non-metals, semi-metals, and dense metals, alongside moscovium (Mc), an artificially synthesized element. The collective is alternatively referred to as "pnictogens" in academic literature. The term "pnígein" originates from the Greek language, specifically meaning "to choke." It is used to describe the possible toxicity associated with nitrogen (N) in specific forms (Yuan et al., 2012).

The Oxygen group is devoid of any naturally occurring metallic elements. Oxygen (O) and sulfur (S) are two non-metallic elements that are commonly found in nature. The three remaining elements in the natural world can be classified as semi-metals. The sole manufactured element, livermorium (Lv), is now hypothesized to possess metallic properties; nonetheless, its definitive classification as a metal remains uncertain within the field of chemistry (Clark & Hoard, 1943).

7.2. THE BORON GROUP

All constituents of this assemblage possess a total of three electrons in the valence shell of each individual atom. Certain elements possess isotopes that exhibit instability. With the exception of boron, all elements exhibit the characteristic of being solid and possessing a shining appearance. With the exception of boron (B), which is recognized as one of the most rigid elements, all constituents within this assemblage exhibit a relatively low level of hardness. The majority of chemical elements do not exhibit a reactive behavior when exposed to water. Aluminum (Al) undergoes the formation of an oxide layer when exposed to water, and subsequently reacts with the water upon corrosion of this layer. Compounds are generated through the process of electron loss to other elements. Each of these substances undergoes a reaction with oxygen (O) through the formation of chemical bonds with three oxygen atoms (Haussermann et al., 2003) (Figure 7.1).

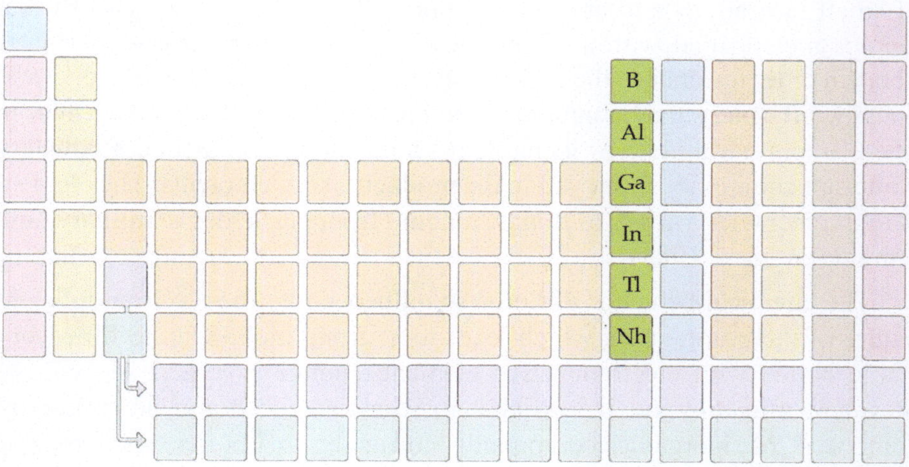

Figure 7.1. The boron group.

Source: Tom Jackson, creative commons license.

7.2.1. Boron

Certain boron compounds exhibit exceptional hardness, comparable to that of diamonds, rendering them among the most resilient synthetic materials known to humankind. This particular element exhibits exceptional hardness, which is further augmented through its reaction with carbon or nitrogen. Various minerals, such as ulexite and kernite, can serve as sources for the extraction of pure boron. The great demand for this particular element was of such magnitude that individuals relocated to reside in the exceedingly hot region of Death Valley, USA, with the intention of engaging in boron mining activities. The presence of boron compounds in soil is crucial for promoting optimal growth and development of plants. Boron is a commonly utilized element in domestic settings on a daily basis (Bolanos et al., 2004).

Sodium borate, commonly known as borax, had previously been utilized a millennium before. In the year 1808, Joseph Louis Gay-Lussac and Louis Jacques Thénard, both of French origin, successfully obtained pure boron through the process of boiling borax in the presence of potassium (Keren, 1996).

Boron is utilized to enhance the durability of heat-resistant glassware, such as measuring cups, by increasing their toughness. Boric acid possesses inherent antimicrobial properties and can be employed for the treatment of small lacerations and abrasions. Thin LCD screens for televisions and laptops are fortified with a pliable coating comprising boron-based glass fibers in order to enhance their durability. Certain types of modeling clay and bouncy silly putty may contain compounds containing the element boron. The element boron derives its name from the compound borax, a powdery white salt commonly employed in the production of detergents. The element is found in a wide variety of things, ranging from pesticides to military tank armor (Bingham, 1983).

The chemical behavior of boron is influenced by the electron-deficient nature of the boron atom, which leads to a strong inclination to form very stable bonds with electronegative elements, notably oxygen. Hence, the prevalent minerals and industrial items that include boron are primarily composed of inorganic compounds consisting of boron and oxygen, commonly referred to as borates. The aforementioned characteristic is also observed in the proclivity of boron to undergo expansion from trigonal covalent bonding through the occupation of its fourth valence orbital, resulting in the formation of tetrahedral structures (Parks & Edwards, 2005).

The presence of both trigonal and tetrahedral boron atoms is observed in all naturally occurring borate minerals. The chemical formula of borax, $Na_2[B_4O_5(OH)_4]8H_2O$, provides a more accurate representation of the distinct hydroxyl groups and water molecules that are present in the crystal structure. Additionally, this formula offers insights into the size and characteristics of the boron-oxygen rings inside the structure (Wear, 1965).

The determination of the structural arrangement of the cyclic boron-oxygen dianion in the crystalline compound borax is achieved by the utilization of X-ray diffraction techniques. The crystal structures of all inorganic borates are composed of both monomeric and polymeric structures that contain boron-oxygen rings of the aforementioned type. The aforementioned structural attributes, which are frequently observed in the crystal structures of inorganic borates, have also been identified in aqueous solutions of commercially significant water-soluble compounds, borate glasses, and alkali borate vapors. The configurations of the soluble borates in an aqueous solution are contingent upon their concentrations and pH levels. Upon dissolution in water, many soluble structures are formed, which are contingent upon the concentration and pH levels (Loomis & Durst, 1992).

Under conditions of elevated pH and reduced concentration, the prevailing species is $B(OH)_4^-$. The formation of the trigonal boric acid molecule, $B(OH)_3$, occurs under conditions of low pH and low concentration. A range of monomeric and polymeric borate anions have been observed at greater concentrations ([B], 0.025 M) within the pH spectrum. This is demonstrated by the examples depicted in Figure 7.2. (In order to engage in a more comprehensive analysis of this particular phenomenon. Boron compounds with reduced valence, such as halides, hydrides, alkyls, and aryls, exhibit a tendency to rapidly undergo oxidation and hydrolysis, unless hindered by steric factors that limit the accessibility of reactive atoms or molecules. The commercial development of reduced boron-containing materials has been impeded by the general reactivity and high cost associated with breaking the strong boron-oxygen bonds present in commonly accessible borates. Hence, the overall level of exposure to these compounds is very low in comparison to inorganic borates (Alfrey, 1983).

7.2.2. Aluminum

Despite being the most prevalent metal found in Earth's rocks, the scientific community did not ascertain the existence of aluminum until the early 19th century. However, it was not until an additional span of 80 years elapsed that scientists successfully determined the methodology for utilizing bauxite ore to extract substantial quantities of refined aluminum. Additionally, it is present in several minerals, such as variscite. In contemporary times, the recycling of aluminum has become a prevalent practice due to the significantly higher energy consumption associated with its primary production. When the metal is flattened through rolling, it acquires a robust and lustrous foil-like appearance, rendering it advantageous for the preservation of food items. The fire protection suit, constructed with a layer of foil, effectively deflects thermal energy (McLean, 1965).

Aluminum is a highly utilized metal, second only to iron, in terms of its widespread application. In comparison to iron's alloy steel, it exhibits a significantly lower weight while maintaining a comparable level of strength. The utilization of aluminum in the construction of domes, shown by the Esplanade Theatre dome in Singapore, allows for the creation of far larger structures compared to those constructed with steel. This is due to the inherent strength-to-weight ratio of aluminum, which enables the dome to support its own weight without collapsing. Aluminum possesses favorable electrical conductivity properties, rendering it suitable for utilization in overhead cables. The production of some airplane components, such as those used in the Boeing 737, utilizes robust aluminum alloys (Becaria et al., 2002).

Did You Know?

JET TURBINE: The jet engine's curved blades grab air accurately and stay stiff at high temperatures. There are various strong metals that meet these requirements, but most are too dense for an airplane engine. Aluminum is the only suitable metal.

Aluminum is the element responsible for making it feasible to fly at great speeds across long distances. It can be shaped easily, weighs one-quarter as much as steel, and never corrodes like steel does. Although steel is more durable, a plane constructed entirely of steel would be too heavy to be able to fly. Instead, aluminum is combined with titanium and steel to create strong alloys that are nevertheless very lightweight. These alloys find employment in the engines and bodies of jet aircraft. Aluminum can be found in almost twice the quantity of the element iron in the rocks of our planet. However, the process of removing impurities from aluminum requires a great deal of energy. However, once it has been cleaned up, it can be recycled in an endless cycle. Therefore, one day these propeller blades might morph into a can of carbonated beverage (Barnhisel & Bertsch, 1983).

7.2.3. Gallium

Gallium exhibits a low melting point of 29°C (84.2°F), resulting in its prompt transition to a liquid state upon contact with the human hand. This particular element is present in trace quantities inside zinc and aluminum ores, such as diaspore. The extraction process of the ore results in the isolation of pure gallium by separating it from the other elements present. Gallium possesses a multitude of applications. The incorporation of indium and tin results in the formation of a liquid alloy known as Galinstan, which finds application in thermometers. Gallium is additionally present in Blu-ray lasers, light-emitting diodes (LEDs), and certain solar panels, including those employed in NASA's Mars rovers (Hoffer, 1980).

7.2.4. Indium

The element indium derives its name from the vibrant hue of indigo, which corresponds to the characteristic emission of light observed when its atoms undergo electrification. The minerals associated with this metal are characterized by their scarcity, with the majority of the metal being extracted from lead and zinc ores, specifically sphalerite. Indium in its pure form has a notable softness, rendering it less commonly employed in its elemental state, with a greater prevalence of utilization in compound configurations. An illustrative instance is the utilization of a chemical known as indium tin oxide on a touchscreen, which facilitates the computer's ability to discern tactile interaction when a finger establishes contact with the screen. Indium is a vital component in the manufacturing of microchips, as well as in the production of welder's goggles and heat- and glare-resistant windows (Alfantazi & Moskalyk, 2003).

Did You Know?

When bent, indium produces a "tin cry" – a sound similar to a scream.

7.2.5. Thallium

The element thallium derived its name from the Greek term "thallos," signifying "green shoot." This nomenclature was assigned due to its initial discovery through the observation of vibrant green luminescence emitted by its flame. The discovery of thallium occurred in 1861 and was credited to William Crookes and Claude-Auguste Lamy. Despite conducting their research independently, both chemists discovered the element through a similar process, namely, the identification of its remnant after the production of potent acids utilizing the mineral pyrite. Subsequently, it was discovered that thallium is present in greater quantities within certain minerals, such as thallium alum. Thallium in its pure form exhibits toxicity and necessitates cautious handling during usage. A thallium-based chlorine chemical is employed in medical imaging techniques to investigate the circulatory system of a patient. Thallium oxide is utilized in the reinforcement of glass, hence enhancing its strength for applications in eyewear and cameras (Peter & Viraraghavan, 2005).

7.2.6. Nihonium

The element nihonium was designated with its name derived from the Japanese term "Nihon," signifying Japan. The metallic element known as nihonium was initially observed in 2003 by research groups investigating the synthetic element moscovium, which possesses an atomic number of 115. It was observed that the atoms of moscovium underwent disintegration within a little time span, transforming into atoms of an element possessing an atomic number of 113. In 2004, Kozuka Morita and colleagues from the RIKEN Nuclear Research Centre in Japan employed an alternative methodology to isolate this particular element by means of fusing bismuth and zinc atoms (Karbowska, 2016).

7.3. THE CARBON GROUP

The constituents of this assemblage possess a total of four valence electrons within the outermost electron shell of every individual atom. These atoms have the capability to form chemical bonds with a maximum of four other atoms. At ambient temperature, all constituent elements within this particular set exist in a solid state. Flerovium (Fl) is an artificially synthesized chemical element, and it is hypothesized to possess solid-state characteristics according to scientific conjecture (Fassler, 2001).

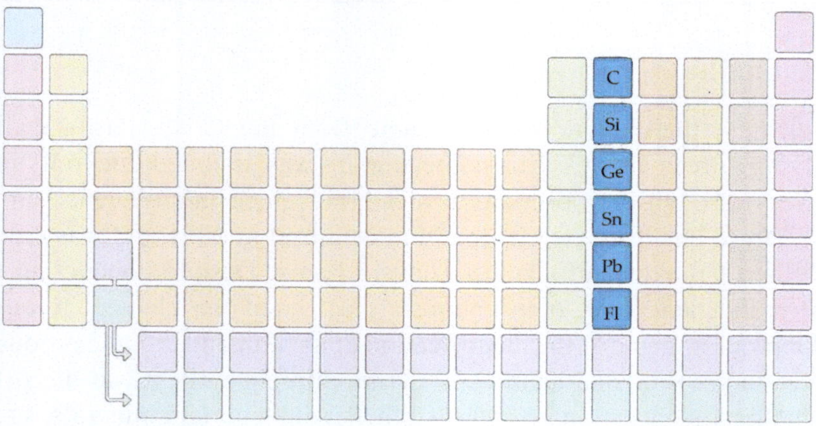

Figure 7.2. The carbon group.

Source: Tom Jackson, creative commons license.

All the elements within this particular group possess the ability to undergo chemical reactions with hydrogen (H). Both carbon (C) and silicon (Si) have the ability to undergo reactions with both metallic and non-metallic elements. These chemical elements undergo a reaction with hydrogen, resulting in the formation of compounds commonly referred to as hydrides. In the process of compound formation, it is observed that each individual element has the potential to relinquish a maximum of four electrons (Yakushev et al., 2021).

7.3.1. Carbon

Carbon possesses the highest quantity of compounds among all elements, surpassing a staggering count of nine million known compounds. Carbon is ranked as the fourth most abundant element in the Universe. Carbon atoms have the capacity to establish four chemical bonds with other atoms, hence enabling the formation of both linear and cyclic structures. On Earth, there are three distinct types of carbon: graphite, diamond, and buckminsterfullerene, which is a carbon-based structure composed of 60 interconnected carbon atoms. Diamond is widely recognized as the most resilient material found in the natural world. This particular item finds frequent application within the realm of jewelry (Sheehan et al., 1994). Certain saws are equipped with diamond-coated blades, enabling them to penetrate various materials with ease. In the realm of gemology, it is widely acknowledged that the sole entity capable of effectively severing a diamond is another diamond. Graphite possesses a significantly lower degree of hardness, hence rendering it suitable for application in the composition of pencil "lead." Furthermore, it finds application in certain battery systems. Coal is presently the primary fuel utilized for energy generation, however it is acknowledged that the emissions it produces have detrimental impacts on both the environment and human health. Crude oil, natural gas, and coal are naturally occurring hydrocarbon molecules composed solely of hydrogen and carbon. These substances has the potential to serve as energy sources and as materials for the production of other items, including polyethylene bags (Lehmann, 2007).

Pink Diamond the Sweet Josephine, a renowned pink diamond, is notable for its substantial size, weighing slightly above 3 grams (0.1 ounces). It stands as one of the most sizeable pink diamonds ever to be successfully traded. Diamonds typically exist as colorless manifestations of pure carbon, with any discernible coloration arising from minute quantities of other substances. As an illustration, the inclusion of boron imparts a blue hue to the gemstone. Curiously, pink diamonds have a notable absence of imperfections, and the underlying cause for their pink hue remains unknown (Roberts &

Grimes, 1997). The Sweet Josephine gemstone was derived from a rough diamond with an estimated age of 1.5 billion years. The aforementioned geological formation originated at a depth of 150 km (93 miles) beneath the Earth's surface and afterwards experienced uplift due to a volcanic eruption. Eventually, it was extracted by mining operations conducted in Australia. The formation of diamonds occurs with the application of pressure and heat above 1,000°C (1,832°F) to carbon (Hofer, 1985). The carbon atoms undergo a rearrangement process, resulting in the formation of a rigid crystal structure, thereby endowing diamond with its exceptional hardness, making it the hardest substance known to mankind. Additionally, this procedure endows diamonds with the capacity to manipulate light, a characteristic that contributes to the captivating brilliance exhibited by these gemstones. Through appropriate cutting and polishing techniques, a diamond can be transformed into an aesthetically pleasing gemstone that is highly valued on a global scale (Gaillou et al., 2010).

7.3.2. Silicon

Approximately 90% of the minerals composing Earth's rocks consist of silicon, which is a prevalent element found in the Earth's crust. The majority of silicon minerals consist of silicates, which are compounds composed of silicon and oxygen. Quartz, which is the mineral manifestation of silicon dioxide or silica, stands as the prevailing silicate compound. Additionally, it is worth noting that sand predominantly consists of this particular material. Amethyst is classified as a kind of quartz. Quartz deposits are commonly encountered inside many rock formations, including granite and sandstone. Opal, a highly prized form of silica, is often utilized in the jewelry industry as a gemstone. The clays utilized in the production of pottery and ceramics are also composed of silicates (Kitawaki et al., 2010).

Both Earth and Mars possess comparable quantities of silicon. This particular element is known to originate within the outermost layer, commonly referred to as the crust, of both celestial bodies. On the other hand, it is noteworthy that the Moon comprises over 50% silicon. According to astronomers, this observation suggests that the Moon originated from the Earth's surface subsequent to a collision event with an asteroid some 4.4 billion years ago (Epstein, 1999).

Silicon finds significant utility in the realm of electronics. Electronic circuits are propelled by thin slices known as silicon wafers. This multifunctional element is additionally employed in the conversion of solar energy into electrical energy through the utilization of solar panels. Aerogel, a durable material with low thermal conductivity, is produced through the utilization of synthetic silica. Fire-fighting outfits incorporate a specialized material that serves the purpose of impeding the propagation of flames towards the firefighter. Another example of a silicon compound is silicone, a versatile material that can be easily molded into many forms. It finds extensive applications across a diverse variety of items, including baking molds and timepieces (Carlisle, 1984).

7.3.3. Germanium

The element in question is designated with a name that is derived from the nation known as Germany. The element in question was first identified in 1886 by scientist Clemens A. Winkler, approximately two decades subsequent to the prediction of its existence and characteristics by Russian chemist Dmitri Mendeleev. Germanite is a mineral that exhibits a notable abundance of germanium, an element primarily obtained through the extraction process of silver, copper, and lead ores. Germanium oxide, a compound of this element, finds application in the manufacturing of wide-angle camera lenses. Furthermore, it finds application in certain microchips and various automotive sensors that contribute to navigation assistance (Scheeper et al., 1994).

7.3.4. Tin

Tin emerged as one of the earliest metals employed by human civilizations. Dating back to approximately 5,000 years ago, the practice of combining tin with copper to produce bronze, an alloy possessing superior strength compared to the individual metals, was prevalent. Cassiterite, an ore, serves as the primary source for obtaining tin in its pure form. Tin possesses a multitude of applications, one of which involves the plating of steel items, such as cans, in order to prevent their corrosion. Tin chloride, a chemical substance, is employed in the process of coloring silk fabrics. This metallic element remains in usage within a diverse range of resilient alloys, such as pewter, soft solder, and bronze (Moskalyk, 2004).

7.3.5. Lead

The chemical symbol for the element lead, Pb, is derived from the Latin term plumbum. The etymology of the term "plumber" can be traced back to ancient Roman civilization, during which plumbing systems utilized malleable metal pipes for the transportation of water. Lead compounds can be observed in several minerals such as crocoite, anglesite, and galena, which serve as the primary reservoirs of lead in its pure form (Winder, 1993). Lead was frequently utilized in previous times as a significant component in paints, hair colors, and pesticides. One prevalent historical application was observed in the field of glassware. Currently, its applications are limited. Lead is capable of absorbing radiation due to its properties. Additionally, it finds application in the manufacturing of diving weights, vehicle batteries, and flexible flashings utilized for roof sealing. Lead lost popularity due to its toxicity (Batzill & Diebold, 2005).

7.3.6. Flerovium

The element known as flerovium derives its name from the renowned Russian scientist, Georgy Flerov. The Joint Institute for Nuclear Research (JINR) in Dubna, Russia, was established by the individual in question, serving as the site where the production of this particular element was initially achieved through the utilization of a particle accelerator, a device designed for the purpose of colliding atoms. Flerovium has a high degree of radioactivity, with its atomic nuclei undergoing disintegration within a brief time span of a few seconds (White et al., 2007).

7.4. THE NITROGEN GROUP

The constituents of this assemblage possess atomic structures characterized by the presence of five electrons in their outermost electron shell. These atoms have the ability to engage in the formation of up to three chemical bonds simultaneously. With the exception of nitrogen (N), all of the components are in a solid state. The density of elements has an increasing trend as one moves down the group (Yakushev et al., 2014). For instance, bismuth (Bi) possesses a density that is approximately 8,000 times greater than that of nitrogen. Phosphorus (P) is characterized by its reactivity, while the other elements within its group exhibit relatively higher stability. When three hydrogen (H) atoms undergo a reaction, all members of the group generate reactive gaseous compounds known as hydrides (Li et al., 2015) (Figure 7.3).

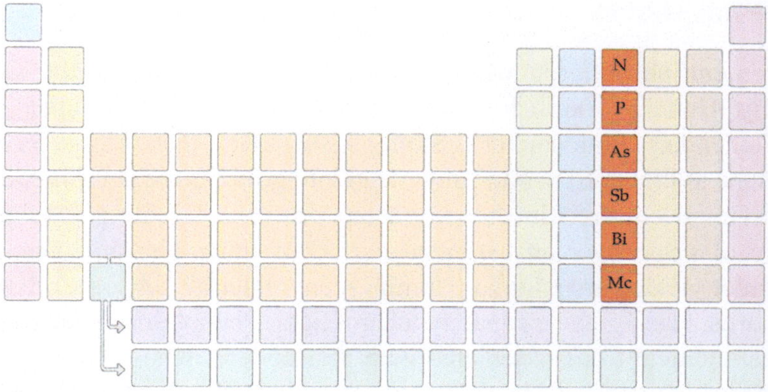

Figure 7.3. The nitrogen group.

Source: Jack Challoner, creative commons license.

7.4.1. Nitrogen

Nitrogen is ubiquitously present in our surroundings due to its status as a transparent gas constituting approximately 75% of Earth's atmosphere. Due to the inherent inertness of pure nitrogen, its liquid state can be effectively employed for cryopreservation purposes, facilitating the freezing and conservation of various substances, including blood and tissue samples. Nitratine is among the limited number of minerals that possess a notable abundance of nitrogen. Industrial techniques can be employed to synthesize nitrogen molecules that possess practical utility. A collection of nitrogen compounds finds application in the field of explosives, encompassing notable examples such as TNT and nitroglycerine (Vitousek et al., 2002). Upon ignition, the substances undergo explosive reactions due to the rapid dissociation of the nitrogen atom bonds. Nitrogen-based fuels, exemplified by nitromethane, are commonly employed in drag bikes due to their superior power output in comparison to carbon-and-hydrogen-based fuels, such as petrol. Hydrazine, a chemical molecule, finds application in spaceship propulsion systems, namely in thrusters employed by missions such as the Phoenix Mars Lander. Certain nitrogen compounds are utilized in the formulation of dyes and adhesives. The Haber process is an industrial methodology that facilitates the conversion of nitrogen and hydrogen gases into ammonia, a liquid substance that is widely employed in the production of nitrogen-based fertilizers. When incorporated into the soil, these fertilizers enhance the development of plants (Saada et al., 2003).

7.4.2. Phosphorus

The element phosphorus was fortuitously encountered by the German alchemist Hennig Brand. In 1669, an individual embarked on a pursuit of the legendary Philosopher's Stone, a substance that was purported to possess the ability to transmute any metal into gold. As part of this endeavor, the individual engaged in the protracted process of boiling a substantial quantity of urine over an extended period of time. The outcome of this experiment yielded an enigmatic luminescent material, subsequently named phosphorus, signifying its capacity to emit light. Phosphorus is the inaugural chemical element to be associated with a discoverer whose name has been historically documented. In the realm of natural occurrences, purity is an elusive quality, as it is seldom seen in its unadulterated form (Macia, 2005). Instead, it manifests itself in many minerals, presenting a diverse range of compositions. Phosphorus exhibits many solid forms that possess incendiary properties, namely red, white, black, and violet. The luminosity observed by Brand was the result of the chemical reaction between white phosphorus and oxygen. Phosphorus mostly exists within phosphate minerals, whereby it forms chemical bonds with oxygen. Apatite, the principal resource of phosphorus, is a notable example of such minerals. Phosphates are found in exquisite china as well as serve as a crucial component in the production of fertilizers. The lateral portions of safety match boxes are comprised only of phosphorous in its pure form. Phosphorus compounds of a more intricate nature, which find application in the realm of insecticides, possess toxic properties (Vessey, 2017).

7.4.3. Arsenic

Arsenic is commonly referred to be the "monarch of toxins." All types of arsenic, whether in its pure form or as part of a compound, exhibit toxicity towards animals. Indeed, the utilization of arsenic as a toxic substance dates back several centuries. This metalloid is present in various minerals exhibiting vivid hues, such as orpiment. The inherent form of arsenic in nature exhibits a lustrous, gray appearance. Certain rat poisons incorporate arsenic compounds in their formulation. The primary application of arsenic in contemporary times is its utilization as a means to enhance the strength and durability of lead materials. The process involves the amalgamation of arsenic and lead to produce a resilient alloy commonly employed in automotive batteries (Guenin & Meziane, 2011).

7.4.4. Antimony

The term "antimony" is derived from the Greek word "anti-monos," which translates as "not alone." This observation may pertain to the inherent absence of the element in its pure form within natural occurrences, since it consistently manifests in conjunction with more substantial metallic elements, such as lead. The symbol of the element, Sb, is derived from stibium, which is the Latin term for kohl, a type of eye cosmetic. Stibnite, a mineral resource, represents the primary and most abundant reservoir of antimony in its pure form. The primary application of its unadulterated state is predominantly observed in the production of rigid alloys, such as those employed in the metallic composition utilized by certain printers. The production of ancient Egyptian kohl involved the utilization of powdered stibnite (Mandal & Suzuki, 2002).

7.4.5. Bismuth

Bismuth is classified as a radioactive element, however its atomic structure has a rather stable nature, enabling its atoms to last for extended periods of time, often spanning millions of years. Bismuth has been recognized by individuals for several centuries. In South America, the Incas included an alloy known as bronze into their weaponry, employing it as a means to enhance their durability. Conversely, the ancient Egyptians employed a mineral containing bismuth to impart a shimmering quality to their cosmetic products. In the presence of atmospheric oxygen, pure bismuth undergoes oxidation, resulting in the formation of visually striking hopper crystals, which exhibit a range of vibrant colors. The aforementioned element exhibits a high degree of brittleness and possesses limited applications in its elemental state. Yellow bismuth pigments find application in the fields of paints and cosmetics, whereas various bismuth compounds are additionally employed in pharmaceutical formulations. Fire sprinklers include an alloy composed of bismuth and tin (Sundar & Chakravarty, 2010).

7.4.6. Moscovium

A limited quantity of approximately one hundred atoms of this synthetic, high-mass element has been produced. Moscovium was initially synthesized at the JINR located in Dubna, Russia. The synthesis of this element was achieved by a group of Russian scientists, under the leadership of Yuri Oganessian, through the collision of americium atoms with fragments of calcium atoms. The name of this entity is derived from the Russian metropolis known as Moscow. This particular element exhibits a high degree of radioactivity, leading to the rapid disintegration of its atomic structure within a brief temporal interval. According to scientific consensus, it is hypothesized that moscovium would exhibit characteristics of a dense, metallic solid. However, due to the limited availability of small samples, researchers are now only able to ascertain the atomic size of moscovium prior to its disintegration (Yang & Sun, 2007).

7.5. THE OXYGEN GROUP

Each atom within the group possesses a total of six electrons in its outermost electron shell. The electron configuration of these elements contributes to their significant reactivity. The constituents of this assemblage primarily exist in solid states, with the exception of oxygen (O), which typically manifests as a gaseous phase under ambient conditions. The density of the elements exhibits an upward trend as one moves down the group. The reactivity of these components exhibits a downward trend throughout the group. The process of combustion consistently involves the presence of oxygen. These elements have the ability to combine and form compounds through chemical reactions. All of them exhibit a reaction with carbon (C) resulting in the formation of compounds, some of which possess potent odors (Oganessian, 2019) (Figure 7.4).

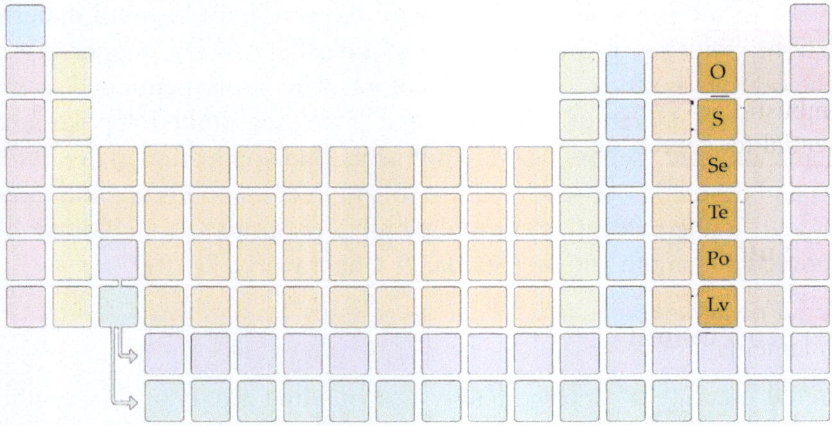

Figure 7.4. The oxygen group.

Source: Jack Challoner, creative commons license.

7.5.1. Oxygen

Oxygen is the predominant constituent in the Earth's crust. Approximately 50% of the composition of rocks and minerals found on Earth consists of oxygen and its many derivatives. Approximately 20% of the Earth's atmosphere consists of pure oxygen. This particular chemical species is characterized by its clear gaseous state. The sustenance of life on Earth is contingent upon the availability of oxygen. Animals respire by inhaling atmospheric air in order to get the oxygen present within it. The oxygen obtained by our bodies' cells is subsequently utilized to metabolize sugars, resulting in the liberation of energy that sustains bodily functions. Combustion, a chemical reaction characterized by the involvement of oxygen, is a process wherein oxygen combines with a fuel, resulting in the production of fire (Lucey & Dangman, 1984). The consumption of oxygen occurs through chemical reactions with various elements, resulting in the formation of compounds known as oxides. Nevertheless, the replenishment of oxygen occurs through the process of photosynthesis, wherein plants release newly generated oxygen. Automobile engines derive their power from the process of combustion, wherein petrol or other fuel sources are consumed. Oxygen has a significant role in the production of steel. Oxygen tanks facilitate respiration for mountaineers in settings characterized by diminished atmospheric oxygen concentrations. Rockets, such as the Atlas V, transport liquid oxygen to facilitate the combustion of fuel in the vacuum of space (Oganessian, 2019).

Did You Know?

Combustion is a chemical reaction that produces heat and light. Oxygen is essential for combustion to take place.

7.5.2. Sulfur

Sulfur, a non-metal, has been recognized since ancient times as one of the rare elements that can be naturally occurring in a pure form. The presence of substantial quantities of this crystalline yellow material is commonly observed in close proximity to volcanic craters. Sulfur, sometimes known as "brimstone," derives its alternative name from the characteristic manner in which its crystals combust, resulting in the transformation of its solid state into a vivid blood-red liquid. Brimstone is believed to serve as the combustible material that sustains the infernal fires in certain religious traditions. The process of extracting elemental sulfur from subterranean reserves involves the utilization of heated water. Subsequently, the heated liquid sulfur is transported to the surface by pumping mechanisms (Bayr, 2005). The aforementioned element is frequently found as a constituent in numerous minerals, including celestine. Numerous sulfur compounds possess unpleasant odors. The malodorous scent emitted by volcanic pools can be attributed to the presence of hydrogen sulfide gas. Additional instances encompass the emission of skunk spray, the release of gaseous compounds from diced onions, and the olfactory perception of the titan arum flower's scent. There exist numerous applications for this non-metallic substance. The compounds derived from it has the ability to enhance the durability of natural rubber, rendering it suitable for application in tire manufacturing. Additionally, these compounds exhibit preservative properties, enabling the preservation of dried fruits. Furthermore, they can be utilized in the production of potent battery acids. The element possesses antibacterial properties and is employed in the production of antibiotic medications, such as penicillin (Semenza, 2007). The Danakil Depression is a geological depression located in the Afar region of northeastern Ethiopia, in the Horn of Africa. It The hot spring located within the Danakil Depression in Africa is encircled by a distinct golden crust composed of sulfur in its pure form. The region located between Ethiopia and Eritrea in East Africa is characterized by its volcanic nature, featuring numerous active craters, dry desert landscapes, geothermal activity such as boiling mud, and pools exhibiting distinctive coloration due to the presence of sulfur and various mineral salts (Tabatabai, 1983).

The Danakil Depression, situated at an elevation of almost 100 meters (330 feet) below sea level, stands as a notable geographical feature characterized by its exceptionally low altitude on the Earth's surface. The region in question experiences minimal to negligible precipitation, accompanied by high temperatures and arid conditions, often exceeding 50°C (120°F). The thermal springs located within the depression are characterized by their high temperature and distinctive green coloration. The water in these springs is composed of pure sulfur, alongside a hazardous sulfur compound known as sulfuric acid. As the process of evaporation occurs, the accumulation of sulfur deposits manifests in aesthetically pleasing formations along the perimeters of the pools, thereby embellishing the expansive terrain. Danakil attracts tourists who are captivated by its extraordinary scenery, despite the hostile conditions that have earned it the reputation as the "most inhospitable place on Earth." (Boyd, 2016).

7.5.3. Selenium

The word "Selenium" is derived from Selene, the Greek deity associated with the Moon. This particular element has characteristics of both metals and non-metals, thereby classifying it as a semi-metal. Selenium exists in two primary elemental states: gray selenium, characterized by its solid and rigid nature, and red selenium, distinguished by its powdery and malleable consistency. Selenium is predominantly utilized as a constituent that imparts pigmentation in glass and ceramics. Selenium has light sensitivity, rendering it suitable for integration into solar cells designed to convert solar radiation into electrical energy. Additionally, it is employed in photocopiers (Bardsley & Lancaster, 1965).

7.5.4. Tellurium

Tellurium is considered to be among the ten least abundant elements found on Earth. The term "Tellus" is derived from the Latin language, specifically from the word "Tellus" which signifies the planet Earth. This particular chemical element is frequently encountered in a combined state with another element, such as the metallic element nickel, as exemplified by the mineral melonite. Tellurium is generated as a byproduct during the refining processes of lead and copper (Parfenov, 1974). There are two distinct physical forms in which pure tellurium can exist: a lustrous, metallic solid or a brown powder. The utilization of this particular element is primarily observed in the composition of glass utilized in the construction of fiber optic cables. These cables are renowned for their ability to transmit significantly larger quantities of information at a notably accelerated rate in comparison to traditional copper cables (Meyer, 1976).

7.5.5. Polonium

Polonium exhibits high levels of radioactivity, as evidenced by the rapid increase in temperature of 1 gram (0.03 ounces) of this metallic element to 500°C (932°F) due to the emission of radiation. The discovery of this element was made by Marie and Pierre Curie in the year 1898. Marie chose to name it after Poland, her country of origin. The occurrence of this phenomenon in the natural world is infrequent, and its typical generation takes place within nuclear reactors (Taylor, 1996). Notwithstanding its inherent radioactivity, this particular element finds application in a limited number of contexts. The detonation of atomic bombs can be initiated by it. The utilization of this technology is observed in the provision of thermal energy and electrical power for spacecraft, exemplified by the Russian Lunokhod rovers that successfully landed on the lunar surface during the 1970s (Reich & Hondal, 2016).

7.5.6. Livermorium

Upon their initial production in the year 2000, livermorium atoms exhibited a rapid disintegration within a very little temporal interval. The initial achievement in generating atoms of this particular element was accomplished at the Joint Institute of Nuclear Research located in Dubna, Russia. The research team collaborated with the Lawrence Livermore National Laboratory, located in California, United States, utilizing the materials provided by the laboratory. The production of this element with high radioactivity involved the use of a particle accelerator, whereby calcium atoms were propelled towards curium atoms, resulting in their collision (Taylor, 1996).

7.6. SUMMARY

This chapter encompasses the Boron Group, Nitrogen Group, Carbon Group, and Oxygen Group. The boron group, alternatively referred to as Group 13 or IIIA, comprises a collection of chemical elements situated within the periodic table. The elements encompassed within this group are boron (B), aluminum (Al), gallium (Ga), indium (In), and thallium (Tl). The elements within the boron group exhibit common characteristics as a result of their comparable electron configurations, namely the presence of three valence electrons in the outermost energy level, notably within the p-orbital. Consequently, these elements exhibit similar chemical characteristics and have a propensity to engage in chemical bonding with other elements through the process of either electron loss or electron sharing, in order to attain a stable electron configuration. The elements belonging to the boron group exhibit a diverse range of uses in the fields of industry, technology, and science. Aluminum, due to its abundance and versatility, is widely utilized in a variety of everyday consumer goods and industrial applications. Gallium and indium possess distinctive features that render them suitable for specialized applications in the fields of electronics and optoelectronics. Boron finds applications in the fields of glass manufacturing and electronics, but thallium, despite its hazardous nature, exhibits restricted utility in specific domains of scientific investigation.

The carbon group, designated as Group 14 or IVA, comprises a collection of chemical elements within the periodic table. The elements encompassed in this group are carbon (C), silicon (Si), germanium (Ge), tin (Sn), and lead (Pb). The carbon group elements exhibit common characteristics as a result of their comparable electron configurations, namely the presence of four valence electrons in the outermost energy level, notably in the p-orbital. As a result, these elements demonstrate shared chemical characteristics, including their capacity to engage in covalent bonding and their tendency to function as tetravalent entities. The elements belonging to the carbon group are of significant importance in a wide range of industries and technological applications. Carbon, as the fundamental element in organic chemistry, plays a vital role in sustaining life and is integral to a wide range of materials and applications. Silicon and germanium are pivotal constituents within the semiconductor sector, serving as fundamental components that propel the advancement of contemporary electronics and computing technologies. Tin and lead, while their potential toxicity, possess a rich historical significance and continue to find restricted current usage in specialized contexts, notably in the form of alloys and coatings.

The nitrogen group, alternatively referred to as Group 15 or VA, comprises a collection of chemical elements within the periodic table. The elements encompassed in this group are nitrogen (N), phosphorus (P), arsenic (As), antimony (Sb), and bismuth (Bi). The elements in the nitrogen group exhibit certain resemblances owing to their analogous electron configurations, namely the presence of five valence electrons in the outermost energy level, notably in the p-orbital. Elements have a proclivity to engage in the formation of covalent and, in certain instances, ionic bonds with other elements in order to attain a state of electron configuration that is stable. These elements exhibit a wide range of applications and play various functions in multiple sectors, encompassing agriculture (as fertilizers), chemistry (in the synthesis of chemical compounds), biology (as vital constituents of life), and specific industrial and technical activities.

The oxygen group, alternatively referred to as Group 16 or VIA, encompasses a collection of chemical elements within the periodic table. The six elements encompassed in this group are oxygen (O), sulfur (S), selenium (Se), tellurium (Te), polonium (Po), and livermorium (Lv). The chemical properties of the elements of the oxygen group are attributed to their comparable electron configurations, characterized by the presence of six valence electrons in the outermost energy level, notably in the p-orbital. There is an observable inclination for these elements to engage in compound formation through the acquisition or sharing of electrons, with the ultimate goal of attaining a stable electron configuration. The oxygen group is of utmost importance in a wide range of domains, including vital functions in sustaining life, facilitating industrial operations, and contributing to the advancement of technology applications.

REFERENCES

1. Alfantazi, A. M., & Moskalyk, R. R., (2003). Processing of indium: A review. *Minerals Engineering, 16*(8), 687–694.
2. Alfrey, A. C., (1983). Aluminum. *Advances in Clinical Chemistry, 23*, 69–91.
3. Atarah, S. A., Egblewogbe, M. N., & Hagoss, G. G., (2020). First principle study of the structural and electronic properties of nihonium. *MRS Advances, 5*(23–24), 1175–1183.
4. Bardsley, C. E., & Lancaster, J. D., (1965). Sulfur. *Methods of Soil Analysis: Part 2 Chemical and Microbiological Properties, 9*, 1102–1116.
5. Barnhisel, R., & Bertsch, P. M., (1983). Aluminum. *Methods of Soil Analysis: Part 2 Chemical and Microbiological Properties, 9*, 275–300.
6. Batzill, M., & Diebold, U., (2005). The surface and materials science of tin oxide. *Progress in Surface Science, 79*(2–4), 47–154.
7. Bayr, H., (2005). Reactive oxygen species. *Critical Care Medicine, 33*(12), S498–S501.
8. Becaria, A., Campbell, A., & Bondy, S. C., (2002). Aluminum as a toxicant. *Toxicology and Industrial Health, 18*(7), 309–320.
9. Bingham, F. T., (1983). Boron. *Methods of Soil Analysis: Part 2 Chemical and Microbiological Properties, 9*, 431–447.
10. Blevins, D. G., & Lukaszewski, K. M., (1998). Boron in plant structure and function. *Annual Review of Plant Biology, 49*(1), 481–500.
11. Bolaños, L., Lukaszewski, K., Bonilla, I., & Blevins, D., (2004). Why boron? *Plant Physiology and Biochemistry, 42*(11), 907–912.
12. Bowen, I. S., (1927). Series spectra of boron, carbon, nitrogen, oxygen, and fluorine. *Physical Review, 29*(2), 231.
13. Boyd, D. A., (2016). Sulfur and its role in modern materials science. *Angewandte Chemie International Edition, 55*(50), 15486–15502.
14. Brocenschi, R. F., Rocha-Filho, R. C., Li, L., & Swain, G. M., (2014). Comparative electrochemical response of estrone at glassy-carbon, nitrogen-containing tetrahedral amorphous carbon, and boron-doped diamond thin-film electrodes. *Journal of Electroanalytical Chemistry, 712*, 207–214.
15. Carlisle, E. M., (1984). Silicon. *Biochemistry of the Essential Ultratrace*

Elements, 1(1), 257–291.

16. Chopra, N. G., Luyken, R. J., Cherrey, K., Crespi, V. H., Cohen, M. L., Louie, S. G., & Zettl, A., (1995). Boron nitride nanotubes. *Science, 269*(5226), 966–967.
17. Clark, H. K., & Hoard, J. L., (1943). The crystal structure of boron carbide. *Journal of the American Chemical Society, 65*(11), 2115–2119.
18. Cubova, K., Semelova, M., Nemec, M., John, J., Milacic, J., Omtvedt, J. P., & Stursa, J., (2018). Extraction of thallium and indium isotopes as the homologues of nihonium into the ionic liquids. *Journal of Radioanalytical and Nuclear Chemistry, 318*, 2455–2461.
19. Dmitriev, S. N., & Popeko, A. G., (2013). Superheavy elements with Z= 114 and 116 have been named flerovium and livermorium. *Mendeleev Communications, 23*(1), 1–2.
20. Epstein, E., (1999). Silicon. *Annual Review of Plant Biology, 50*(1), 641–664.
21. Fässler, T. F., (2001). The renaissance of homoatomic nine-atom polyhedra of the heavier carbon-group elements Si–Pb. *Coordination Chemistry Reviews, 215*(1), 347–377.
22. Fox, G. T., (1973). On the physics of drag racing. *American Journal of Physics, 41*(3), 311–313.
23. Gaillou, E., Post, J. E., Bassim, N. D., Zaitsev, A. M., Rose, T., Fries, M. D., & Butler, J. E., (2010). Spectroscopic and microscopic characterizations of color lamellae in natural pink diamonds. *Diamond and Related Materials, 19*(10), 1207–1220.
24. Golberg, D., Bando, Y., Tang, C. C., & Zhi, C. Y., (2007). Boron nitride nanotubes. *Advanced Materials, 19*(18), 2413–2432.
25. Gordon, J. N., Taylor, A., & Bennett, P. N., (2002). Lead poisoning: Case studies. *British Journal of Clinical Pharmacology, 53*(5), 451.
26. Guenin, E., & Meziane, D., (2011). Microwave assisted phosphorus organic chemistry: A review. *Current Organic Chemistry, 15*(19), 3465–3485.
27. Häussermann, U., Simak, S. I., Ahuja, R., & Johansson, B., (2003). Metal-nonmetal transition in the boron group elements. *Physical Review Letters, 90*(6), 065701.
28. Heidari, A., Schmitt, K., Henderson, M., Besana, E., & Gobato, R., (2020). Pros and cons of livermorium nanoparticles for human cancer

cells, tissues, and tumors treatment under synchrotron radiation using Mathematica 12.0. *Parana Journal of Science and Education (PJSE), 6*, 1–31.

29. Hernberg, S., (2000). Lead poisoning in a historical perspective. *American Journal of Industrial Medicine, 38*(3), 244–254.
30. Hofer, S. C., (1985). Pink diamonds from Australia. *Gems Gemol, 21*(3), 147–155.
31. Hoffer, P., (1980). Gallium: Mechanisms. *J. Nucl. Med.;(United States), 21*(3), 1–20.
32. Karbowska, B., (2016). Presence of thallium in the environment: Sources of contaminations, distribution, and monitoring methods. *Environmental Monitoring and Assessment, 188*, 1–19.
33. Keren, R., (1996). Boron. *Methods of Soil Analysis: Part 3 Chemical Methods, 5*, 603–626.
34. Kitawaki, H., Abduriyim, A., Kawano, J., & Okano, M., (2010). Identification of CVD-grown synthetic melee pink diamond. *Journal of Gemmology, 32*(1), 23.
35. Knight, M. W., King, N. S., Liu, L., Everitt, H. O., Nordlander, P., & Halas, N. J., (2014). Aluminum for plasmonics. *ACS Nano, 8*(1), 834–840.
36. Landrigan, P. J., & Todd, A. C., (1994). Lead poisoning. *Western Journal of Medicine, 161*(2), 153.
37. Lehmann, J., (2007). A handful of carbon. *Nature, 447*(7141), 143–144.
38. Li, B., Sun, X., & Su, D., (2015). Calibration of the basic strength of the nitrogen groups on the nanostructured carbon materials. *Physical Chemistry Chemical Physics, 17*(10), 6691–6694.
39. Loomis, W. D., & Durst, R. W., (1992). Chemistry and biology of boron. *BioFactors (Oxford, England), 3*(4), 229–239.
40. Loss, R. D., & Corish, J., (2012). Names and symbols of the elements with atomic numbers 114 and 116 (IUPAC recommendations 2012). *Pure and Applied Chemistry, 84*(7), 1669–1672.
41. Lucey, J. F., & Dangman, B., (1984). A reexamination of the role of oxygen in retrolental fibroplasia. *Pediatrics, 73*(1), 82–96.
42. Maciá, E., (2005). The role of phosphorus in chemical evolution. *Chemical Society Reviews, 34*(8), 691–701.

43. Mandal, B. K., & Suzuki, K. T., (2002). Arsenic round the world: A review. *Talanta, 58*(1), 201–235.
44. McLean, E. O., (1965). Aluminum. *Methods of Soil Analysis: Part 2 Chemical and Microbiological Properties, 9*, 978–998.
45. Medina-Cruz, D., Tien-Street, W., Vernet-Crua, A., Zhang, B., Huang, X., Murali, A., & Webster, T., (2020). Tellurium, the forgotten element: A review of the properties, processes, and biomedical applications of the bulk and nanoscale metalloid. *Racing for the Surface: Antimicrobial and Interface Tissue Engineering, 1*(1), 723–783.
46. Melloy, G. F., Summon, P. R., & Podgursky, P. P., (1973). Optimizing the boron effect. *Metallurgical Transactions, 4*, 2279–2289.
47. Meyer, B., (1976). Elemental sulfur. *Chemical Reviews, 76*(3), 367–388.
48. Moskalyk, R. R., (2004). Review of germanium processing worldwide. *Minerals Engineering, 17*(3), 393–402.
49. Nehate, S. D., Saikumar, A. K., Prakash, A., & Sundaram, K. B., (2020). A review of boron carbon nitride thin films and progress in nanomaterials. *Materials Today Advances, 8*, 100106.
50. Oganessian, Y., (2019). The making of moscovium. *Nature Chemistry, 11*(1), 98–98.
51. Parfenov, Y. D., (1974). Polonium-210 in the environment and in the human organism. *Atomic Energy Review, 12*(1), 75–143.
52. Parks, J. L., & Edwards, M., (2005). Boron in the environment. *Critical Reviews in Environmental Science and Technology, 35*(2), 81–114.
53. Peter, A. J., & Viraraghavan, T., (2005). Thallium: A review of public health and environmental concerns. *Environment International, 31*(4), 493–501.
54. Recepoglu, Y. K., Goren, A. Y., Vatanpour, V., Yoon, Y., & Khataee, A., (2022). Boron carbon nitride nanosheets in water and wastewater treatment: A critical review. *Desalination, 533*, 115782.
55. Reich, H. J., & Hondal, R. J., (2016). Why nature chose selenium. *ACS Chemical Biology, 11*(4), 821–841.
56. Roberts, J. T., & Grimes, P. E., (1997). Carbon intensity and economic development 1962–1991: A brief exploration of the environmental Kuznets curve. *World Development, 25*(2), 191–198.
57. Saada, A., Breeze, D., Crouzet, C., Cornu, S., & Baranger, P., (2003).

Adsorption of arsenic (V) on kaolinite and on kaolinite–humic acid complexes: Role of humic acid nitrogen groups. *Chemosphere, 51*(8), 757–763.

58. Scheeper, P. R., Van, D. D. A. G. H., Olthuis, W., & Bergveld, P., (1994). A review of silicon microphones. *Sensors and Actuators A: Physical, 44*(1), 1–11.
59. Semenza, G. L., (2007). Life with oxygen. *Science, 318*(5847), 62–64.
60. Sheehan, J. E., Buesking, K. W., & Sullivan, B. J., (1994). Carbon-carbon composites. *Annual Review of Materials Science, 24*(1), 19–44.
61. Sjögren, B., Iregren, A., Montelius, J., & Yokel, R. A., (2015). Aluminum. In: *Handbook on the Toxicology of Metals* (Vol. 1, No. 1, pp. 549–564).
62. Sundar, S., & Chakravarty, J., (2010). Antimony toxicity. *International Journal of Environmental Research and Public Health, 7*(12), 4267–4277.
63. Tabatabai, M. A., (1983). Sulfur. *Methods of Soil Analysis: Part 2 Chemical and Microbiological Properties, 9*, 501–538.
64. Taylor, A., (1996). Biochemistry of tellurium. *Biological Trace Element Research, 55*, 231–239.
65. Vesey, A., (2017). "A way to sell your records": Pop stardom and the politics of drag professionalization on RuPaul's drag race. *Television & New Media, 18*(7), 589–604.
66. Vitousek, P. M., Hättenschwiler, S., Olander, L., & Allison, S., (2002). Nitrogen and nature. *AMBIO: A Journal of the Human Environment, 31*(2), 97–101.
67. Wang, J., Hao, J., Liu, D., Qin, S., Portehault, D., Li, Y., & Lei, W., (2017). Porous boron carbon nitride nanosheets as efficient metal-free catalysts for the oxygen reduction reaction in both alkaline and acidic solutions. *ACS Energy Letters, 2*(2), 306–312.
68. Wang, S., Zhang, W., Jia, F., Fu, H., Liu, T., Zhang, X., & Han, N., (2021). Novel Ag_3PO_4/boron-carbon-nitrogen photocatalyst for highly efficient degradation of organic pollutants under visible-light irradiation. *Journal of Environmental Management, 292*, 112763.
69. Wear, J. I., (1965). Boron. *Methods of Soil Analysis: Part 2 Chemical and Microbiological Properties, 9*, 1059–1063.
70. White, L. D., Cory-Slechta, D. A., Gilbert, M. E., Tiffany-Castiglioni, E., Zawia, N. H., Virgolini, M., & Basha, M. R., (2007). New and

evolving concepts in the neurotoxicology of lead. *Toxicology and Applied Pharmacology, 225*(1), 1–27.

71. Winder, C., (1989). Reproductive and chromosomal effects of occupational exposure to lead in the male. *Reproductive Toxicology, 3*(4), 221–233.
72. Winder, C., (1993). Lead, reproduction, and development. *Neurotoxicology, 14*(2, 3), 303–317.
73. Yakushev, A., Gates, J. M., Türler, A., Schädel, M., Düllmann, C. E., Ackermann, D., & Wiehl, N., (2014). Superheavy element flerovium (element 114) is a volatile metal. *Inorganic Chemistry, 53*(3), 1624–1629.
74. Yakushev, A., Lens, L., Düllmann, C. E., Block, M., Brand, H., Calverley, T., & Yakusheva, V., (2021). First study on nihonium (Nh, element 113) chemistry at TASCA. *Frontiers in Chemistry, 9*, 753738.
75. Yang, N., & Sun, H., (2007). Biocoordination chemistry of bismuth: Recent advances. *Coordination Chemistry Reviews, 251*(17–20), 2354–2366.
76. Yuan, C., Suzuki, T., Otsuka, T., Xu, F., & Tsunoda, N., (2012). Shell-model study of boron, carbon, nitrogen, and oxygen isotopes with a monopole-based universal interaction. *Physical Review C, 85*(6), 064324.
77. Zhang, X., Davidson, E. A., Mauzerall, D. L., Searchinger, T. D., Dumas, P., & Shen, Y., (2015). Managing nitrogen for sustainable development. *Nature, 528*(7580), 51–59.

CHAPTER 8

Halogens and Noble Gases

Contents

8.1. INTRODUCTION

The halogens are recognized as highly reactive elements within the periodic table and display a wide range of chemical properties, resulting in the formation of several compounds. Conversely, noble gases exhibit the lowest degree of chemical reactivity among all elements. Prior to the 1960s, chemists commonly denoted these elements as inert gases due to the prevailing belief that they displayed negligible chemical reactivity. However, it has been shown that krypton and xenon possess the ability to form chemical bonds with other elements, primarily fluorine. To this day, the formation of compounds comprising helium, neon, or argon remains elusive (Kendrick et al., 2013).

The subsequent enumeration delineates the overarching chemical characteristics of halogens and noble gases, elucidating both their shared attributes and distinctive features:

- Halogens are rarely encountered in their elemental form in natural settings, whereas noble gases exclusively exist in nature as elemental entities (Kobayashi, et al., 2017).
- As elemental substances, halogens exclusively occur in the form of diatomic molecules, such as F_2, Cl_2, and Br_2. The noble gases only consist of monatomic species, such as helium (He), neon (Ne), and argon (Ar) (Kendrick et al., 2015).
- In an aqueous solution, halogens have the tendency to generate negatively charged ions known as halide ions. These ions readily engage in the formation of ionic compounds with nearly all metallic elements. In aqueous solutions, noble gases exhibit a complete absence of ion formation (Kendrick & Burnard, 2013).
- Except for fluorine, which only produces the F– ion, halogens have the ability to form polyatomic or complex negative ions.

  Polyatomic ions such as ClO^-, ClO_4^-, and BrO_3^- serve as illustrative instances. Once more, it should be noted that noble gases do not possess the ability to form ions (Kendrick et al., 2018).
- Halogens exhibit the tendency to engage in covalent chemical bonding with other nonmetallic elements. As a result, compounds composed of nonmetals frequently manifest as diminutive molecules. Halogens have the capability to engage in covalent bonding with certain metals or metalloids. In contrast, only the

elements krypton and xenon have the capability to create such molecules. Radon is also likely to exhibit similar behavior, however, its study is limited among chemists due to its radioactive nature (Broadley et al., 2016).

- With the exclusion of fluorine, halogens have the ability to exist in both positive and negative oxidation states. This implies, for instance, that halogens exhibit a propensity to rapidly establish chemical bonds with both hydrogen and oxygen (Rohrbacher et al., 2000).

8.2. THE HALOGENS

Nonmetals are allocated among five distinct groups of elements within the periodic table, namely groups IVB, VB, VIB, VIIB, and VIII. The nonmetals located in groups IVB, VB, and VIB are discussed in detail in the Nonmetals section, which can be found in a separate volume within this comprehensive set of many volumes. The elements exhibit a diverse array of chemical and physical properties, resulting in fewer discernible tendencies among groups. The elements contained within this volume, specifically group VIIB, which consists of the halogens, and group VIII, which comprises the noble gases, demonstrate a notable resemblance to other elements within their respective groups (Cicerone, 1981). This resemblance is characterized by the prominent manifestation of shared group features. Halogens are not typically encountered in their elemental form, although upon isolation, they manifest as diatomic gases, characterized by molecular structures comprising two atoms. Additionally, these substances have potent oxidizing properties, indicating their high chemical reactivity and propensity to engage in reactions with metals and other elements in their neutral state (Trueman, 1971). The term "halogen" originates from Greek terms that pertain to the capacity of certain elements to generate salts. The subsequent five elements are classified as halogens:

- Fluorine (F);
- Chlorine (Cl);
- Bromine (Br);
- Iodine (I); and
- Astatine (At).

Astatine is found in extremely small levels in its natural state on Earth. All isotopes of astatine exhibit significantly brief half-lives. Consequently, even in cases when astatine samples are intentionally generated, the decay process occurs prior to the substantial accumulation of astatine (Aiuppa et al., 2009).

The chemical properties of fluorine, chlorine, bromine, and iodine exhibit significant similarities. The observed variations can be mostly attributed to disparities in the diameters of the halogen atoms. Fluorine atoms exhibit a relatively diminutive size, but the atomic dimensions progressively increase as one down the column, culminating in iodine atoms which possess the biggest size. The somewhat smaller size of fluorine results in distinct chemical properties that differentiate it from the other three elements, which exhibit a higher degree of similarity amongst themselves (Correns, 1956; Politzer et al., 2007). As an illustration, hydrofluoric acid (HF) can be classified as a weak acid, in contrast to hydrochloric acid (HCl), hydrobromic acid (HBr), and hydroiodic acid (HI), which are considered strong acids. The fluorine atom exhibits a valency of one, limiting its capacity to create a single chemical connection with other atoms. In contrast, chlorine, bromine, and iodine atoms possess the ability to engage in multiple interactions with other atoms (Rohrbacher et al., 2000). Fluorine exclusively forms the ion fluoride, but the remaining halogens have the ability to generate oxyanions, including hypochlorite (ClO^-), perchlorate (ClO_4^-), bromate ($\pm \quad _4^-$), and iodate (IO_4^-).

8.2.1. Fluorine

Fluorine is classified as the ninth element in the periodic table. This element is of moderate prevalence, ranking 13th in terms of its abundance. Fluoride ion (F–) is predominantly present in minerals within the Earth's crust. The reduction of the fluoride ion to neutral fluorine, which exists as gaseous diatomic molecules (F_2), poses significant challenges in terms of safety (Cicerone, 1981). The handling of fluorine gas necessitates specialized training and meticulous control measures due to its very hazardous nature, hence limiting its manipulation to individuals with appropriate expertise (Liang et al., 2013) (Figure 8.1).

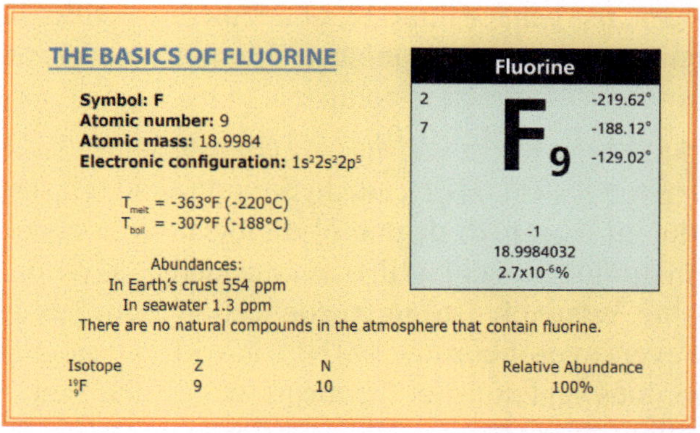

Figure 8.1. Basics of fluorine.

Source: Joe Palca, creative commons license.

The elements chlorine, bromine, and iodine exhibit a significant range of compound diversity due to their ability to make multiple bonds with various additional atoms (Correns, 1956; Trueman, 1971). Fluorine exhibits distinct characteristics, as its atoms possess the capacity to form a singular bond with another atom exclusively. Several significant compounds that incorporate fluorine atoms include hydrofluoric acid (HF), sulfur hexafluoride (SF_6), uranium hexafluoride (UF_6), polytetrafluoroethylene (a carbon and fluorine polymer known to consumers as Gore-Tex and Teflon), sodium fluoride (NaF), stannous fluoride (SnF_2), and chlorofluorocarbons (CFCs) like $CCl_2 F_2$.

The characteristics of substances are significantly impacted by the types of intermolecular forces of attraction present among their constituent components (Böhm et al., 2004). HF exhibits notable distinctions in both its physical and chemical characteristics when compared to the remaining hydrogen halides (HCl, HBr, HI). Hydrogen fluoride is a volatile liquid with a boiling point of approximately 20°C, a temperature commonly associated with ambient conditions. Conversely, the comparatively weaker intermolecular interactions among the remaining molecules result in their relatively facile vaporization and significantly lower boiling temperatures, which fall below the standard ambient temperature (Brewer, 1965). The liquid state of HF, as opposed to a gaseous state, can be linked to the presence of hydrogen bonding between HF molecules, a phenomenon that is not observed in other compounds (Politzer et al., 2007). The phenomenon of hydrogen bonding pertains to a notably robust attractive interaction that occurs between the hydrogen atom of one molecule and an atom, typically nitrogen, oxygen, or fluorine, of another molecule. The absence of hydrogen bonding in chlorine, bromine, and iodine atoms results in reduced intermolecular forces of attraction compared to hydrogen fluoride molecules (Aiuppa et al., 2009; Li et al., 2020).

HCl, hydrobromic acid (HBr), and hydroiodic acid (HI) are classified as strong acids due to their high degree of dissociation in aqueous solutions. Conversely, hydrofluoric acid (HF) is considered a weak acid, mostly attributed to the relatively stronger connection between hydrogen and fluorine. However, hydrofluoric acid (HF) is widely recognized as one of the most hazardous and caustic chemicals within the field of chemistry. Chemical reactions occur between this substance and glass, metals, as well as organic matter (Purser et al., 2008; Gouverneur & Seppelt, 2015). The primary application of hydrofluoric acid is in glass etching. However, it is imperative to exercise caution while handling this acid due to its high dermal absorption potential, which can result in severe burns to the skin, deeper tissue damage, and harm to bone structures. If left untreated in a timely manner, mortality may ensue. Furthermore, it should be noted that hydrofluoric acid (HF) is not suitable for storage in glass containers. Prior to the development of plastics, hydrofluoric acid (HF) was typically held in vessels made of platinum. Currently, the prevailing practice involves the storage of the substance in polyethylene containers, generally referred to as plastic bottles (Fuge, 1988).

Sulfur hexafluoride (SF_6) is a gas that lacks color, odor, and toxicity. The substance is employed as a contrast agent in the context of ultrasonic imaging. Within the electrical industry, the utilization of this particular gas is mostly attributed to its exceptional insulating characteristics, surpassing those of air. Additionally, it has the potential to substitute the environmentally detrimental polychlorinated biphenyls (PCBs) that have been extensively employed by the industry over an extended period of time. Regrettably, SF_6 exhibits a high potency as a greenhouse gas, therefore prompting the likelihood of its gradual elimination due to mounting apprehensions regarding the phenomenon of global warming (Deshmukh et al., 1995; Purser et al., 2008). A single unit of SF_6, weighing one pound, possesses a warming potential that is similar to the warming potential generated by 11 tons of CO_2. While the Environmental Protection Agency (EPA) presently lacks a regulatory obligation concerning restrictions on emissions of SF_6, several corporations have proactively undertaken measures to decrease their SF_6 emissions voluntarily. The EPA acknowledged the commendable efforts of Consolidated Edison Company of New York and Arizona Public Service in February 2009. These efforts primarily focused on enhancing their handling and maintenance practices, leading to the successful prevention of about 250,000 pounds of SF_6 emissions into the atmosphere (Lommerse et al., 1996; Dolbier, 2005).

UF_6 was initially employed in the creation of the atomic bomb during the Manhattan Project, which took place from 1941 to 1946, in the context of World War II. Uranium is naturally found as a combination of isotopes. The isotope ${}^{238}_{92}U$, which is the most prevalent in natural uranium, accounts for approximately 99.3% of its composition. However, due to its non-fissionable nature, it is unsuitable for utilization in atomic weaponry or conventional nuclear power facilities. The fissionable isotope, specifically ${}^{235}_{92}U$, which possesses use in power plants or weapons, constitutes a mere 0.7 percent of the overall uranium composition (O'Hagan, 2010; Gillis et al., 2015). The chemical separation of isotopes within the same element is deemed impractical; nevertheless, physical methods such as gaseous diffusion and gas centrifugation can effectively achieve isotopic separation. Enrichment occurs as a sample of uranium undergoes either procedure, resulting in an increase in the concentration of ${}^{235}_{92}U$. Once the concentration of uranium-235, denoted as ${}^{235}_{92}U$, reaches approximately 3–5 percent, it becomes suitable for the fabrication of fuel intended for use in a light-water nuclear reactor. The uranium used in weapons, specifically uranium-235, can have a concentration of up to 85%. The current method employed for isotope separation involves the utilization of UF_6 (Rzepa, 1997; Hagmann, 2008).

Teflon and Gore-Tex are widely utilized consumer goods derived from polytetrafluoroethylene, a compound characterized by elongated and linear carbon chains, each of which is bonded to two fluorine atoms. Teflon is widely recognized for its characteristic of being "nonstick" and is frequently employed as a nonstick coating on culinary utensils due to its ability to prevent the formation of chemical bonds with food substances. Teflon exhibits exceptional lubricating properties due to its inability to create chemical bonds, which is analogous to its non-reactivity in the context of chemical bonding. Gore-Tex materials are renowned for their hydrophobic properties. Gore-Tex is frequently included in outdoor apparel and footwear in order to provide protection against moisture and wet environments for the wearer (Brammer et al., 2001; Strunecká et al., 2004). Fluorine is frequently found in chemical compounds in the form of the fluoride ion

(F^-), which is present in salts such as NaF and stannous fluoride (SnF_2). Both of these chemicals have the potential to be incorporated into toothpaste formulations for the purpose of combating dental caries. While fluoride has demonstrated efficacy in combating tooth decay, it is imperative to exercise several precautions. Excessive exposure to fluoride can lead to dental fluorosis, characterized by the appearance of tooth discoloration or mottling (Fuge & Andrews, 1988). Certain individuals exhibit sensitivity to the consumption of table salt, scientifically referred to as NaCl, in their dietary intake. It is worth noting that sodium fluoride may potentially induce similar adverse effects. The utilization of stannous fluoride has been associated with the potential occurrence of tooth discoloration, leading to a yellowish appearance. Due to the aforementioned concerns, the majority of prominent toothpaste manufacturers provide consumers with the option to choose between fluoride-containing and fluoride-free toothpaste variants (Wang et al., 2014).

CFCs were initially formulated for use as refrigerants and marketed under the trade name Freon. CFCs were employed for an extended period of time in refrigeration systems and automotive air-conditioning apparatus. Nevertheless, during the 1980s, it was determined that chlorine included in CFCs posed a significant threat to the ozone layer in the stratosphere, which serves as a crucial shield safeguarding life on Earth against the detrimental effects of ultraviolet radiation emitted by the Sun. Subsequently, CFCs have been progressively substituted with hydrofluorocarbons (HFCs), which are devoid of chlorine (Shah & Westwell, 2007).

8.2.2. Chlorine

Chlorine is classified as the seventeenth element in the periodic table. Although chlorine is found in large quantities in saltwater and the Earth's crust, it ranks as the 20th most prevalent element. Chlorine is predominantly encountered in its natural state as the chloride ion (Cl^-) within various salt compounds. The substance commonly referred to as table salt is scientifically known as NaCl, and it has been utilized for culinary purposes since ancient times to enhance the taste of food (Evans, 2005). Sodium chloride holds significant importance as an electrolyte for various living species, to the extent that animals are observed to regularly visit salt licks, which are outcroppings containing sodium chloride, in order to fulfill their salt demands (Figure 8.2).

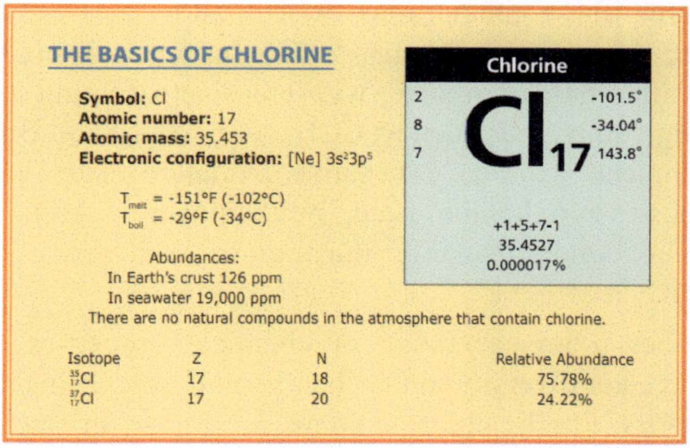

Figure 8.2. Basics of chlorine.

Source: Howard Bryant, creative commons license.

Chlorine, in its unadulterated state, manifests as a fragrant, faintly verdant gaseous substance composed of diatomic molecules with the chemical formula Cl_2. Chlorine gas and other chlorine derivatives has significant oxidizing properties, rendering them widely employed in several countries globally for the purpose of disinfecting municipal drinking water sources and treating swimming pool water (Sedla, & von Gunten, 2010). The utilization of chlorine as a water disinfectant during the 19th century has yielded significant advantages for the human population. By effectively eradicating detrimental bacteria such as E. coli, chlorine has played a crucial role in mitigating the spread of waterborne infectious diseases, including cholera and giardia. Moreover, chlorine can be identified in a wide array of domestic commodities, predominantly as disinfectants and bleach (Johnson et al., 1957; Öberg, 2002).

With the exclusion of certain noble gases, it is likely that nearly all elements present in the periodic table have the capability to create chemical compounds with chlorine. Metal cations possessing charges of +1 or +2, namely Na^+, Ag^+, $Ca^{2+,}$ and $Fe^{2+,}$ have the ability to create ionic compounds when combined with the chloride anion. Molecular compounds consisting of metal atoms in elevated oxidation states and chlorine atoms can be observed, with notable instances including $FeCl_3$, $TiCl_4$, and $AlCl_3$. Metalloids in positive oxidation states exhibit a pronounced tendency to form molecular compounds with chlorine. Notable instances of such compounds are BCl_3 (alternatively represented as B_2Cl_6), $SiCl_4$, $AsCl_3$, and $SbCl_3$. Nonmetals have the ability to engage in the formation of molecular compounds with chlorine, wherein the nonmetal may exist in either a positive or negative oxidation state (Hom, 1972; Brungs, 1973). Instances where the nonmetal exhibits a positive oxidation state, with chlorine assuming a -1-oxidation state, including compounds such as CCl_4, NCl_3, SCl_2, and BrCl. Instances when the nonmetal exhibits a negative oxidation state, while chlorine assumes a positive oxidation state, predominantly pertain to oxygen-containing compounds. Notable illustrations encompass Cl_2O, ClO_2, $HClO_3$, and HOCl.

Chlorine has a greater range of chemical properties compared to fluorine. The greater diversity in the chemistry of chlorine can be attributed to several factors. Firstly, chlorine atoms are bigger in size compared to fluorine atoms. This disparity in size allows chlorine atoms to accommodate a larger number of electrons, nearly twice as many as fluorine atoms. Additionally, the electronic configuration of chlorine atoms enables the promotion of electrons to d orbitals, which in turn facilitates the formation of more bonding sites. Therefore, a chlorine atom has the ability to form many chemical bonds with other atoms, whereas a fluorine atom is limited to forming only a single link with another atom. As previously mentioned, chlorine is widely recognized for its exceptional efficacy as an oxidizing agent. In contrast to fluorine, which primarily acts as an oxidizing agent in its elemental form (F_2), chlorine exhibits various common forms in which it functions as an oxidizing agent. These forms include chlorine gas (Cl_2), as well as several salts such as NaClO, $NaClO_2$, $NaClO_3$, $NaClO_4$, and $Ca(ClO)_2$. Additionally, chlorine can act as an oxidizing agent in the presence of acids such as HClO, $HClO_2$, $HClO_3$, and $HClO_4$, as well as in the compound ClO_2 (Vasconcelos et al., 1997; Spicer et al., 1998).

Chlorine exhibits a wide range of oxidation states, encompassing -1, 0, +1, +3, +5, and +7. Hence, chlorine serves as an illustrative instance of an element capable of undergoing disproportionation, also referred to as autooxidation, reactions. Disproportionation is a chemical reaction in which a single component fulfills the dual role of both oxidizing agent and reducing agent. One example of a disproportionation reaction involving chlorine is as follows:

$$3Cl_2 + 6KOH \rightarrow 5KCl + KClO_3 + 3H_2O,$$

Chlorine originates in the 0 oxidation state, often known as the neutral state. In the compound $KClO_3$, a portion of chlorine atoms undergo oxidation to the +5 oxidation state, while the remainder chlorine atoms experience reduction to the -1 oxidation state in KCl (Hua & Reckhow, 2007).

Chemical reactions can also take place between compounds when chlorine exists in distinct oxidation states, leading to changes in these states. For instance, under the context of the chemical reaction

$2HCl + NaClO \rightarrow Cl_2 + NaCl + H_2O$,

Chlorine atoms initially exist in the -1 and +1 oxidation states, and subsequently undergo a transition to the 0 oxidation state upon becoming Cl_2. This reaction exemplifies the perils associated with the combination of everyday household substances in the absence of comprehensive knowledge regarding potential outcomes (Molina, & Molina, 1987). Vinegar comprises acetic acid, which, while not as potent as hydrochloric acid, nonetheless exhibits acidic properties. Sodium hypochlorite (NaClO) is a chemical compound found in bleach. Consequently, the combination of these two substances inside a domestic setting has the potential to generate harmful chlorine gas. Another frequently observed error, which can have severe consequences, involves the combination of bleach and household ammonia. It is imperative to avoid the simultaneous use of the two aforementioned cleaners, since their combination results in the formation of an exceedingly hazardous gas that can be fatal when confined within an enclosed space, such as a bathroom. Numerous fatalities have occurred as a consequence of individuals committing this error (Furukawa et al., 1978; Korich et al., 1990)

Due to its toxic nature, the transportation and on-site storage of chlorine gas pose significant risks. Consequently, chlorine in compound form is commonly employed for various applications such as bleaching agents and the purification or disinfection of water sources. The solution commonly known as home bleach, such as the brand Clorox, consists of a 5 percent concentration ofNaClO. The implementation of chlorine as a means to sterilize communal water sources has emerged as a notable achievement in the realm of public health throughout history. The implementation of chlorination has significantly decreased the incidence rates of typhoid, cholera, hepatitis, and giardiasis. The chemicals most frequently employed include sodium hypochlorite, calcium hypochlorite ($Ca(OCl)_2$), and chlorine dioxide (ClO_2). Alternative disinfectants, such as ozone (O_3), can be employed; however, one drawback of ozone is its lack of residual efficacy. To clarify, while the majority of infections are effectively eradicated upon exposure to ozone, it is important to note that no residual ozone persists (Keswick et al., 1985; Thornton & Campbell, 2001). Consequently, a minority of pathogens manage to survive this process, leading to their potential resurgence in numbers. In the presence of chlorine, residual amounts of chlorine persist in the water, so impeding the resurgence of infections.

The predominant form in which chlorine is found in naturally occurring inorganic compounds is as chloride ions (Cl^-). Sodium chloride, commonly referred to as table salt (NaCl), is a prevalent chemical component that can be obtained through two primary methods: evaporation of seawater or extraction from the mineral halite, also known as rock salt. With the exception of a few compounds such as AgCl, $PbCl_2$, and Hg_2Cl_2, the majority of compounds containing chloride ions exhibit a high degree of solubility in water. As a result, the availability of chloride compounds from terrestrial sources is often limited, with rock salt being the predominant source (Gohil, & Suresh, 2017). Furthermore, it is worth noting that the chloride ion exhibits less inherent chemical reactivity. In the context of hydrochloric acid (HCl), a chemically significant reagent incorporating the chloride ion, the predominant focus lies on the hydrogen ion (H^+), rather than the chloride ion.

Chlorine is widely recognized due to its notorious reputation. The development of chlorinated hydrocarbons took place during the initial decades of the 20th century, primarily for various applications including solvents, dry-cleaning agents, pesticides, plastics, refrigerants, and electrical insulators. Moreover, a significant proportion of naturally occurring chlorinated hydrocarbons is generated through metabolic activities in plants and natural phenomena such as volcanic eruptions. Regrettably, the majority of these compounds exhibit a high degree of persistence, indicating their resistance to degradation in the environment. Consequently, they survive in the environment for extended periods, frequently leading to bioaccumulation within the food chain (Rossman et al., 1994; Deborde & Von Gunten, 2008).

A considerable number of these chemicals have carcinogenic properties, pose risk to human health, or inflict damage onto wildlife. One prominent illustration of a pesticide that has gained infamy is dichlorodiphenyltrichloroethane (DDT), which was initially developed due to its convenient mass administration and its comparatively low toxicity to humans, particularly when administered in little quantities (De Beer, 1994; Falconer & Bidleman, 1994). The aforementioned characteristics render DDT an economically viable tool in the fight against insect-borne illnesses, such as malaria. Regrettably, DDT exhibits limited biodegradability in the environment, leading to its accumulation within the tissues of organisms rather than transforming into benign chemicals. As DDT undergoes biomagnification, it accumulates in the tissues of animals, leading to a disruption of reproductive processes in vertebrate species (Deng et al., 2014).

Chlorine is utilized in the production of numerous types of polymers. One notable illustration is polyvinyl chloride (PVC). PVC exhibits a wide range of applications, encompassing several fields such as plumbing, vinyl siding, magnetic stripe cards, upholstery, and flooring.

The qualitative identification of chloride ions and the quantitative determination of chloride concentration in a solution are both straightforward procedures. The solubility of chloride salts in water is a characteristic feature of most metals, with the exception of AgCl, which exhibits limited solubility. Silver chloride is a whitish substance that exhibits high solubility in weak solutions of aqueous ammonia, as evidenced by the next chemical reaction:

$$AgCl\ (s) + 2NH_3(aq) \rightarrow Ag(NH_3)^{+2}(aq) + Cl^-.$$

Silver is capable of forming precipitates when combined with bromide and iodide ions. In contrast, it is seen that silver bromide exhibits solubility just in concentrated aqueous ammonia, while silver iodide demonstrates complete insolubility. The process of titration involves the utilization of potassium chromate (K_2CrO_4) as an end-point indicator to determine the concentration of chloride ions in a solution. This is achieved by titrating the chloride solution with silver nitrate (Macauley et al., 2006). The solubility of reddish brown silver chromate (Ag_2CrO_4) is comparatively lower than that of silver chloride, resulting in the precipitation of silver chloride prior to silver chromate. Upon complete precipitation of chloride, the subsequent precipitation of silver chromate occurs, as demonstrated in the ensuing chemical reaction:

$$2AgNO_3(aq) + K_2CrO_4(aq) \rightarrow Ag_2CrO_4(s) + 2KNO_3(aq).$$

The mixture undergoes a color change from white to reddish brown as Ag_2CrO_4 precipitates. The quantity of $AgNO_3$ introduced is utilized for the determination of the concentration of chloride ions present in the initial solution.

8.2.3. Bromine

Bromine, denoted by its atomic number 35, ranks as the 47th most prevalent element within the Earth's crust. Similar to chlorine, bromine exhibits a higher prevalence in seawater compared to its occurrence in terrestrial rocks and minerals. In contrast to fluorine and chlorine, which exist as gaseous states under typical atmospheric circumstances, bromine is found in a liquid state. Indeed, bromine is one of the two elements in the periodic table that exists in a liquid state under standard atmospheric pressure and at normal temperature. The other element that shares this characteristic is mercury (Wofsy et al., 1975; Adriano & Doner, 1983). Liquid bromine is a volatile, odorous, and very corrosive liquid with a deep red color. It possesses a viscous consistency and is known to be poisonous, making it extremely hazardous to handle. When bromine comes into touch with the skin, it immediately induces damage. It is of paramount importance to refrain from inhalation of the emitted gases. Indeed, the fumes emitted by bromine are comparatively more hazardous than those of chlorine due to bromine's higher density, enabling it to penetrate the respiratory system more profoundly (Figure 8.3).

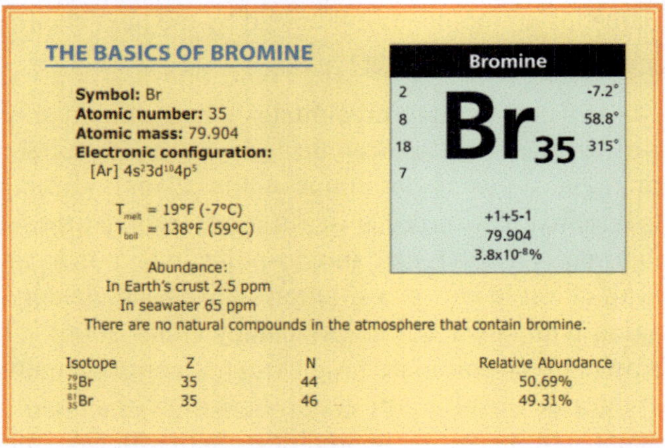

Figure 8.3. Basics of bromine.

Source: Greg Nobles, creative commons license.

The chemical properties of bromine closely resemble those of chlorine. Bromine functions as an oxidizing agent, albeit with a lesser degree of oxidizing capability compared to chlorine or fluorine. Moreover, bromine exhibits the ability to generate a multitude of chemical compounds that serve as analogs to their respective chlorine counterparts (Saikia et al., 2016).

The names of chlorine and iodine are derived from their distinctive colors. Bromine, in its elemental state, manifests as a dark-red, volatile liquid characterized by a pungent odor, from which it derives its nomenclature. The handling of pure bromine presents significant hazards, as it has the potential to cause skin burns and inhalation of its red, dense vapors can result in toxicity (Adriano & Doner, 1983; Wennberg, 1999).

In the United States, the primary source of bromine is its extraction from seawater or from subterranean salt deposits formed by ancient oceans. In each 1,800 gallons (6,813.7 liters) of seawater, an estimated quantity of 1 pound (0.454 kilogram) of bromine can be found. One of the primary applications of bromine in contemporary times is its utilization in the process of synthesizing organic molecules (Winid, 2015).

The chemical properties of bromine exhibit notable similarities to those of chlorine, mostly due to its shared ability to assume a diverse range of oxidation states, including -1, 0, +1, +3, +5, and +7. In contrast to chlorine, the occurrence of compounds in various oxidation states is more limited for bromine. While chlorine compounds can be found in multiple oxidation states, bromine compounds are predominantly observed in the -1 and +5 oxidation states. Bromine compounds in alternative oxidation states have been synthesized; nevertheless, their practical applications are often limited (Westerhoff et al., 2004; Haller & Riedel, 2014).

In industry, both inorganic and organic bromide compounds are employed as brominating agents. In essence, their function entails the introduction of bromine atoms into other chemical compounds. As an illustration, within the category of inorganic substances, there exist several compounds such as phosphorus tribromide (PBr_3), bromine chloride (BrCl), aluminum bromide ($AlBr_3$), ammonium bromide (NH_4Br), and thionyl bromide ($SOBr_2$). The process of organic synthesis often involves a sequential progression of reactions, wherein intermediate compounds are generated in one stage and subsequently transformed into other entities in the subsequent step. The addition of bromine is an often employed practice due to its facile addition and removal properties (Akamatu et al., 1954; Lim et al., 1977). Chemists frequently introduce bromine atoms into carbon chains that possess double or triple bonds, so generating intermediates that can afterwards undergo conversion into the intended final products. An illustrative instance is presented in the subsequent equation:

$$CH_2 = CH_2(aq) + Br_2(aq) \rightarrow CH_2Br—CH_2Br\ (aq).$$

A prevalent category of brominating chemicals comprises alkyl bromides, wherein an alkyl group denotes a hydrocarbon moiety such as $CH_3\cdot$, $CH_3CH_2\cdot$, or $CH_3CH_2CH_2\cdot$ (the sign [·] represents an unpaired electron; this notation is frequently employed for molecular fragments). For instance, in the case where the alkyl group is a methyl group ($CH_3\cdot$), the appropriate brominating agent to be employed would be CH_3Br (Sander et al., 2003; Von Glasow, 2004).

Hydrobromic acid (HBr) is a strong acid formed by the dissociation of bromide salts. Nevertheless, the utilization of the weak acid is significantly less prevalent compared to its counterpart, the potent acid known as hydrochloric acid (HCl). While there are numerous compounds that contain both chlorine and bromine, it is worth noting that chlorine is more cost-effective to manufacture and, despite its inherent risks, it is comparatively safer to handle than bromine. Consequently, there is a notable scarcity in the demand for compounds containing bromine (Clayton & Mayeda, 1963; Eustace, 1980).

The determination of bromide ions in solution is based on the observation that the solubility of silver bromide is between that of silver chloride and silver iodide. Silver ions have the ability to form precipitates with various negative ions, including halides as well as ions such as phosphate (PO_4^{3-}), carbonate (CO_3^{2-}), and sulfide (S^{2-}). The process of acidification involves the addition of nitric acid (HNO_3) to an unidentified solution, which effectively eliminates the presence of other ions, thereby isolating the halides (Beckwith et al., 1996; Bobrowski et al., 2003). As previously discussed in the preceding section titled "The Chemistry of Chlorine," it is seen that AgCl exhibits a high solubility in dilute aqueous ammonia, whereas AgBr only dissolves in concentrated aqueous ammonia (Wang et al., 1994). Notably, AgI does not exhibit any solubility in ammonia. Hence, a conclusive method to ascertain the existence of bromide ions in a solution involves acidifying the solution using nitric acid and subsequently introducing silver nitrate (Pratt et al., 2013). In the event that a precipitate is observed, it is quite probable that the precipitate will consist of one of the halides. Confirmation of the presence of bromide ions in the solution can be established if the precipitate exhibits insolubility in dilute ammonia, but solubility in concentrated ammonia.

These reactions are shown in the following equations:

$AgNO_3(aq) + KBr\ (aq) \rightarrow AgBr\ (s) + KNO_3(aq)$;

$AgBr\ (s) + 2NH_3\ (aq) \rightarrow Ag(NH_3)_2^+\ (aq) + Br^-\ (aq)$.

Alternatively, in the event that the solid exhibits a yellow hue and exhibits insolubility in concentrated ammonia, it is quite probable that the substance in question is AgI. Consequently, one may deduce that the initial unidentified solution contains ions of iodide (Yang et al., 2005).

8.2.4. Iodine and Astatine

Iodine, which is identified as element 53 on the periodic table, does not occur naturally in its elemental form. However, there exist a limited number of rocks that do contain this particular chemical species. Similar to other halogens, iodine exhibits oxidizing properties. Iodine plays a crucial role in maintaining human health by effectively preventing the occurrence of goiters. Indeed, iodine is the most massive element that is important for maintaining good health (Downs & Adams, 2017). Artificial isotopes of iodine are routinely employed in the domain of nuclear medicine, a specialized branch of medical practice that leverages radioactive substances for the purpose of disease diagnosis and therapy (Figures 8.4 and 8.5).

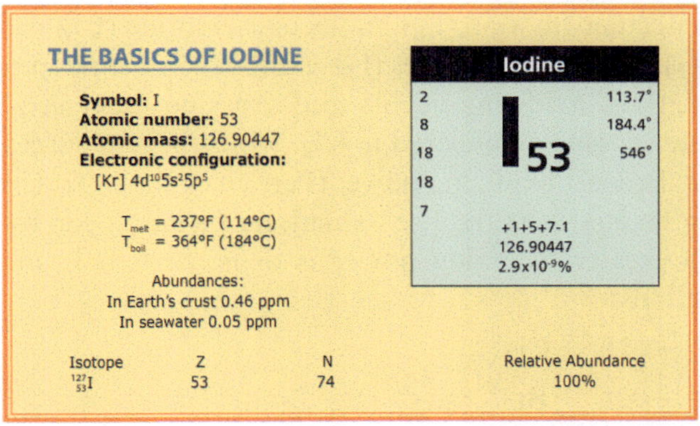

Figure 8.4. Basics of iodine.

Source: Monica Halka, creative commons license.

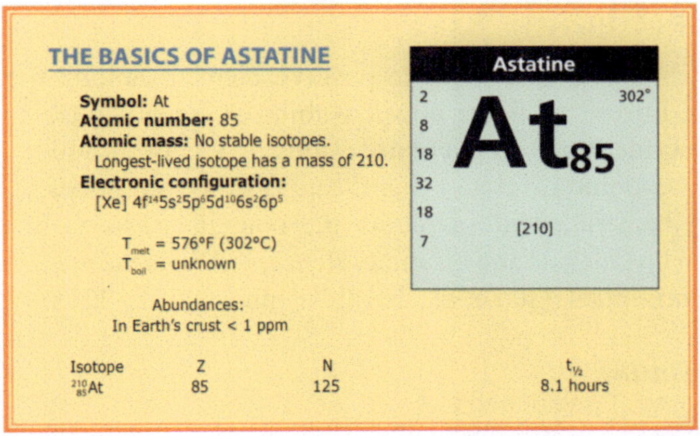

Figure 8.5. Basics of astatine.

Source: Brian Nordstorm, creative commons license.

On the contrary, astatine (element 85), which is recognized as the most massive halogen, lacks any stable isotopes. The presence of astatine in Earth's crust is limited, as it is mostly generated by the radioactive decay of other elements, resulting in only trace levels (Meyer, 2018; Tanudji et al., 2023).

8.2.4.1. Iodine

Due to the same characteristics of chlorine, bromine, and iodine, iodine exhibits a chemistry that closely resembles that of its triad counterparts. Similar to molecular fluorine (F_2), molecular chlorine (Cl_2), and molecular bromine (Br_2), molecular iodine (I_2) exhibits oxidizing properties. The relative strength of oxidizing agents in decreasing order is as follows: F_2 exhibits greater potency compared to Cl_2, which in turn surpasses Br_2, and finally, I_2 has the least oxidizing power (Petrich et al., 2002; Hermann et al., 2013; Lindegren et al., 2020). The oxidation of a halide ion by a molecular halogen with greater oxidizing power is demonstrated in the subsequent series of reactions:

$$F_2 + 2Cl^- \rightarrow 2F^- + Cl_2$$

$$F_2 + 2Br^- \rightarrow 2F^- + Br_2$$

$$F_2 + 2I^- \rightarrow 2F^- + I_2$$

$$Cl_2 + 2Br^- \rightarrow 2Cl^- + Br_2$$

$$Cl_2 + 2I^- \rightarrow 2Cl^- + I_2$$

$$Br_2 + 2I^- \rightarrow 2Br^- + I_2.$$

Iodine, being the most electropositive element among the halogens, exhibits the lowest oxidizing power within the series. I_2 has no reactivity towards fluoride ions (F^-), chloride ions (Cl^-), or bromide ions (Br). Nevertheless, iodine exhibits a significant disinfectant potency, effectively eliminating bacteria, while avoiding the potential risks to human health associated with chlorine and bromine. Hence, iodine is the preferred halogen for use in antiseptics (Guo et al., 2018; Nishinaka et al., 2018).

8.2.4.2. Astatine

The half-lives of the two most stable isotopes of astatine, namely ^{211}At and ^{210}At, are quite short at 7.2 hours and 8.3 hours, respectively. However, it is important to note that these isotopes are not naturally present on Earth (Vasaros et al., 1981). Artificial preparation of these substances is feasible; however, the yield is limited to minute quantities. Astatine is found in its elemental state as diatomic molecules (At_2), and its most stable ionic form is astatide (At^-).

8.3. THE NOBLE GASES

In the year 1869, Dmitri Mendeleev produced his periodic table, which did not include the noble gases as they had not yet been identified at that time. Indeed, their presence had not been conjectured. Due to the absence of the atomic number idea during Mendeleev's time, the periodic table he constructed did not exhibit any discernible "voids" in the regions presently occupied by the noble gases. The sequential arrangement of elements, where sodium succeeded fluorine, potassium succeeded chlorine, rubidium succeeded bromine, and so on, exhibited a highly logical pattern. The progression from the halogen family to the alkali metal family demonstrated a systematic and organized relationship (Kipfer et al., 2002; Wieler, 2002).

In the 19th century, a group of elements known as the noble gases, including helium (He), neon (Ne), argon (Ar), krypton (Kr), xenon (Xe), and radon (Rn), were identified. The presence of all noble gases on Earth has been discovered by scientists, who have determined that these gases are formed as byproducts of the radioactive decay of certain elements. All noble gases are constituents of the Earth's atmosphere, but with the exception of radon isotopes due to their significantly limited lifespan, resulting in little presence of radon inside the atmosphere. The noble gases exhibit melting and boiling points that are significantly lower than the average temperature of a room, and they exist in a liquid state within a narrow temperature range (Kestin & Wakeham, 1972; Farley & Neroda, 1998). Due to their absence in the atmosphere in any chemical amalgamation with other elements, it was a logical assumption to consider them as chemically inert or non-reactive. Indeed, it is worth noting that the noble gases only occur in the Earth's atmosphere as monatomic gases. This implies that a noble gas molecule is composed of a solitary atom, in contrast to diatomic molecules such as N_2 and O_2. Furthermore, for an extended period, the absence of any instances when noble gases engage in chemical bonding with other elements, whether observed in natural occurrences or synthesized in controlled laboratory settings, has been widely documented. The electrical configurations of noble gases are responsible for their chemical inertness (Lewis et al., 1975; Burnard et al., 2013). Helium, being the least dense noble gas, possesses an electron configuration of 1s2, indicating that its sole electrons occupy a completely filled energy level. Consequently, helium does not engage in the formation of chemical bonds. The noble gases that remain exhibit electron configurations terminating in the ns2 np6 pattern, wherein the variable n represents the principal energy level. The eight electrons collectively establish an energetically stable configuration known as a "octet," resulting in minimal or negligible inclination of the atoms to engage in chemical bonding. These elements exhibit a tendency to neither donate electrons to other atoms nor acquire electrons from other atoms. Additionally, they do not readily engage in the sharing of electrons through covalent bonding with other atoms (Sano et al., 2013).

In 1962, Neil Bartlett, a scientist from Canada affiliated with the University of British Columbia, successfully synthesized XeF_2 by a meticulous process involving the combination of xenon gas with fluorine gas under the influence of sunlight. Subsequently, Bartlett relocated to the University of California at Berkeley. He, along with scientists around, later achieved the synthesis of several xenon and krypton chemicals, primarily fluorides, oxides, and oxyfluorides. As of present, no chemical compounds have been synthesized involving helium, neon, or argon (Najafi et al., 1983; Stute & Schlosser, 2000).

8.3.1. Helium

Helium, which is identified as element number 2 in the periodic table, is recognized as the second most prevalent element in the universe, following hydrogen. The stability of hydrogen atoms does not necessitate the existence of a neutron. Nevertheless, due to the presence of two protons in helium nuclei, which experience a significant electrostatic repulsion, it is imperative for a helium nucleus to include a minimum of one neutron. The phenomenon of the strong nuclear force, which is responsible for the binding of helium nuclei, counteracts the repulsive force between protons and establishes an attractive interaction between neutrons and protons. All the remaining elements in the periodic table share the same criteria, which is the presence of at least one neutron in their nuclei. An intriguing characteristic of helium, along with neon and argon, is the absence of any known chemical compounds using these elements. In the gaseous phase, helium has the capability to undergo ionization, resulting in the formation of He^{+} and He^{2+} ions. However, it is important to note that helium atoms do not engage in chemical bonding with either other helium atoms or atoms of any other element (Van Sciver et al., 2012). As a result, the element helium does not exhibit any chemical reactivity. The examination of helium encompasses an investigation into its physical characteristics, whereas its chemical attributes are notably absent (Figure 8.6).

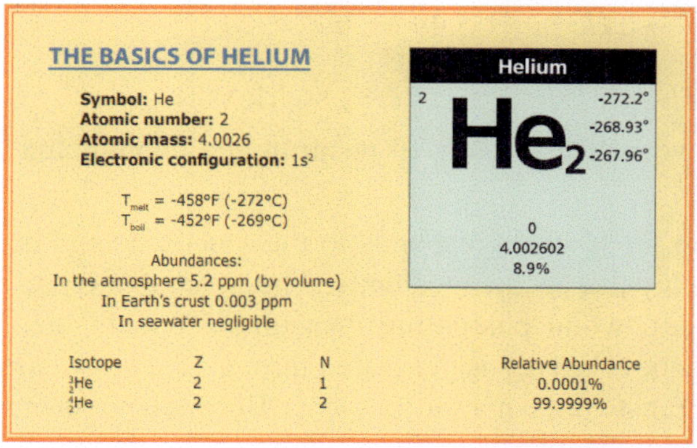

Figure 8.6. Basics of helium.

Source: Brian Nordstorm, creative commons license.

Hydrogen and helium, being the two gases with the lowest molecular weights, have historically been favored for employment in the context of balloons and airships. Hydrogen, on the other hand, possesses flammability properties, and its utilization has led to numerous incidents involving explosions, with the most renowned occurrence being the catastrophic loss of the Hindenburg airship in 1937. Despite the identification of helium in the Earth's atmosphere during the 1890s, the quantity of helium accessible for industrial purposes remained inadequate (Slater, 1928; Tisza, 1947). The economic recovery of helium from subsurface reserves did not become feasible until the 1930s. Subsequently, helium was substituted for hydrogen in dirigibles.

In stars such as the Sun and others with comparable mass and age, the sustained fusion of hydrogen into helium occurs within the stellar core due to the presence of high temperatures and pressures. The aforementioned mechanism is commonly referred to as the proton-proton chain, often abbreviated as the PP chain. Hydrogen nuclei, characterized by their rapid motion, exhibit a high degree of proximity and engage in frequent collisions, occurring at a rate of millions per second. Certain entities combine to generate deuterium, which subsequently undergoes collision with a proton, resulting in the production of 3He or light helium (Tisza, 1938). The latter can then undergo further collision with another helium atom of the same kind, ultimately leading to the formation of 4He. The chain reaction is represented as follows:

$$ {}_{1}^{1}p + {}_{1}^{1}p \rightarrow {}_{2}^{2}H + e^{+} + \nu_{e} + 0.42\ \text{MeV} $$

$$ {}_{1}^{2}H + {}_{1}^{1}p \rightarrow {}_{2}^{3}He + \gamma + 5.49\ \text{MeV} $$

$$ {}_{2}^{3}He + {}_{2}^{3}He \rightarrow {}_{2}^{4}He + 2\ {}_{1}^{1}p + 12.86\ \text{MeV} $$

where p symbolizes proton, e^{+} positron, γ a photon, and ν_e an electron neutrino.

The energy in MeV corresponds to the kinetic energy acquired by the product particles due to their velocity. The symbol γ represents a photon with little mass, while possessing measurable energy that is contingent upon its frequency. Energy is generated through the conversion of mass, as described by Einstein's equation $E = mc^2$. Both kinetic energy and photon energy play a significant role in the regulation of temperature within the star core. All fusion reactions involving atoms with atomic numbers lower than that of iron release energy in various forms (Anderson, 1960; O'nions & Oxburgh, 1983).

The proton-proton (PP) chain is well recognized as the primary mechanism for helium production within the Sun. However, it is worth noting that there are a few supplementary processes that also contribute to the overall generation of helium (Nuttall et al., 2012). The PPII chain, which is a sequence of nuclear reactions, is responsible for approximately one-third of the helium content found within the Sun:

$$ {}_{2}^{3}\text{He} + {}_{2}^{4}\text{He} \rightarrow {}_{2}^{7}\text{Be} + \gamma $$

$$ {}_{4}^{7}\text{Be} + e^{-} \rightarrow {}_{3}^{7}\text{Li} + \nu_{e} $$

$$ {}_{3}^{7}\text{Li} + {}_{1}^{1}\text{H} \rightarrow {}_{2}^{4}\text{He} + {}_{2}^{4}\text{He} $$

The subsequent PPIII chain, which accounts for a mere 0.1 percent of helium production in the Sun, is infrequent. However, it becomes prevalent in stars of greater mass, where the temperature in the stellar core surpasses 23 million Kelvin. This elevated temperature facilitates the fusion of beryllium with hydrogen:

$$ {}_{4}^{7}\text{Be} + {}_{1}^{1}\text{H} \rightarrow {}_{5}^{8}\text{B} + \gamma $$

$$ {}_{5}^{8}\text{B} \rightarrow {}_{4}^{8}\text{Be} + e^{-} + \nu_{e} $$

$$ {}_{4}^{8}\text{Be} \rightarrow 2\ {}_{2}^{4}\text{He}. $$

Helium existed prior to the formation of stars, though. In the immediate aftermath of the big bang, the prevailing conditions within the cosmos were sufficiently hot and dense to facilitate the formation of helium through various mechanisms (Penrose, 1951; Feynman, 1955; Atkins, 1959). The fusion events involving the collision of free-flying neutrons (n), protons (p), and deuterons (d) in significant quantities can result in the formation of helium-3 (3 He) and helium-4 (4 He) as the products:

$^{2}_{1}d + ^{2}_{1}d \rightarrow ^{3}_{2}He + ^{1}_{1}n$ followed by $^{2}_{1}d + ^{3}_{2}He \rightarrow ^{4}_{2}He + ^{1}_{1}p$

Or

$^{2}_{1}d + ^{2}_{1}d \rightarrow ^{3}_{1}H + ^{1}_{1}p$ followed by $^{2}_{1}d + ^{3}_{1}H \rightarrow ^{4}_{2}He + ^{1}_{0}n$.

There were also slower processes involving photon emission:

$^{2}_{1}d + ^{1}_{0}n \rightarrow ^{3}_{1}H + \gamma$ followed by $^{3}_{1}H + ^{1}_{1}p \rightarrow ^{4}_{2}He + \gamma$

Or

$^{2}_{1}d + ^{1}_{1}p \rightarrow ^{3}_{2}He + \gamma$ followed by $^{3}_{2}He + ^{1}_{0}n \rightarrow ^{4}_{2}He + \gamma$.

The abundance of helium, the second most prevalent element in the universe, can be attributed to its production through a diverse range of mechanisms. However, the unintentional release of this natural gas into the atmosphere occurs throughout the refining process. Similar to oil and natural gas, helium is considered a nonrenewable resource that is projected by certain experts to be depleted within a span of 25 years. The potential occurrence of such an event would provide a significant challenge for numerous sectors and scientific researchers that rely on the crucial cooling qualities of helium (Allen & Misener, 1976; Li et al., 2022).

8.3.2. Neon

Neon, with an atomic number of 10, is the second most prevalent noble gas found in the Earth's atmosphere. Similar to helium, neon has limited chemical reactivity. Nevertheless, the emission spectrum of neon is widely recognized for its highly luminous lines throughout the red and orange portions of the visible spectrum (Craig & Lupton, 1976; Grossberg & Mingolla, 1985). This characteristic has led to its prevalent application in the illumination of outdoor signage (Figure 8.7).

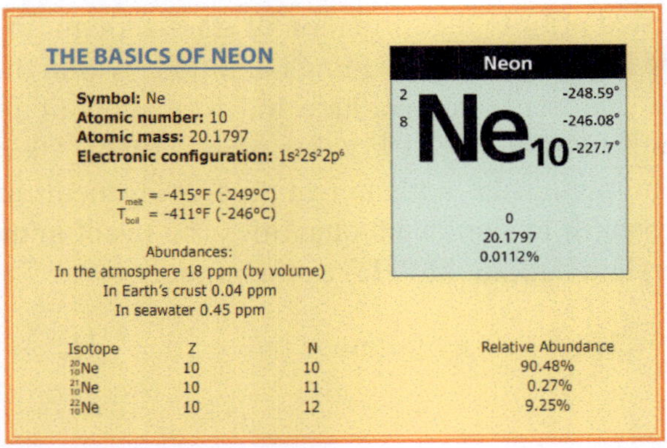

Figure 8.7. Basics of neon.

Source: Monica Halka, creative commons license.

Neon is produced through the process of synthesis within stars that possess a mass beyond approximately four solar masses. Once the fusion process within the core of a star has converted all the helium into carbon, the core will undergo another collapse (Suárez-Figueroa et al., 2011). This collapse will continue until the temperature, which is elevated by the gravitational in fall of matter, reaches a value of 6×10^8 K. At this temperature, carbon fusion can occur through the subsequent reaction:

$$^{12}_{6}C + ^{12}_{6}C \rightarrow ^{20}_{10}Ne + ^{4}_{2}He.$$

The majority of neon that is generated undergoes fusion reactions within the star, resulting in the formation of heavier atomic nuclei. Consequently, any neon currently detected must have originated from an alternative source (Sarda et al., 1988).

The origins of contemporary neon within the cosmos are characterized by intermittent occurrences, however they can be deemed dependable. Eventually, the majority of stars will undergo fusion and gravitational collapse, culminating in the supernova phase where they will have a powerful explosion. Numerous stars have experienced this shift in previous instances, and an indeterminate number will continue to endure such a transformation in forthcoming periods. During the intense and turbulent process of a supernova explosion, a diverse array of particles and nuclei undergo collisions, resulting in the formation of numerous heavy atomic species (Wiese et al., 1966; Black & Pepin, 1969). The creation of neon in supernovae is attributed to an alpha process that occurs when oxygen particles collide with alpha particles, as described below:

$$^{16}_{8}O + ^{4}_{2}He \rightarrow ^{20}_{10}Ne + \gamma$$

While the majority of astrophysicists agree that this particular reaction is responsible for the presence of neon in the interstellar medium (ISM), as well as in recently created stars, there exists a contentious debate on its prevalence in the Sun (Aston, 1920). It is well acknowledged that neon plays a significant role in solar convection, and having knowledge of its abundance within the Sun facilitates the prediction of the convection zone's magnitude. The calculations pertaining to the behavior of convective zones do not align with the previously acknowledged neon abundance value, which poses challenges in terms of experimental measurement (Rothman, 2015).

8.3.3. Argon

Argon, which is identified as element number 18 in the periodic table, is generated through the process of radioactive decay of potassium isotopes that exhibit instability. Due to the large presence of potassium in the Earth's composition, substantial quantities of argon can be generated. Argon constitutes around 1% of the Earth's atmosphere, rendering it the most prevalent noble gas in the atmosphere and the third most abundant gas overall. Similar to helium and neon, argon exhibits limited chemical reactivity. Hence, the existence of this compound in the Earth's atmosphere remained an enigma for a considerable period of time. In contrast to nitrogen and oxygen, which constitute approximately 80% and 20% of the Earth's atmosphere, respectively, and are susceptible to removal through chemical reactions, argon does not participate in such reactions (Khriachtchev et al., 2000; Zenker, 2008). During the late 18th century, chemists achieved a notable accomplishment by successfully identifying a significant portion, specifically 99%, of the gases present in the Earth's atmosphere. However, they encountered considerable perplexity when confronted with the remaining 1 percent of gases that eluded their comprehension. The inability to chemically react to this particular constituent of air with any other substances hindered its discovery for a period of one hundred years. Ultimately, the issue was resolved as scientists developed an understanding that a group of inert gases exists, which demonstrates resistance to engaging in chemical interactions (Turner, 1973). Following the development of air liquefaction techniques in the 1890s, spectroscopic methods were employed to extract and identify each of the inert gases present in air, such as argon. Chemists successfully assigned these gases to their respective column adjacent to the halogens by accurately estimating their atomic weights (Figure 8.8).

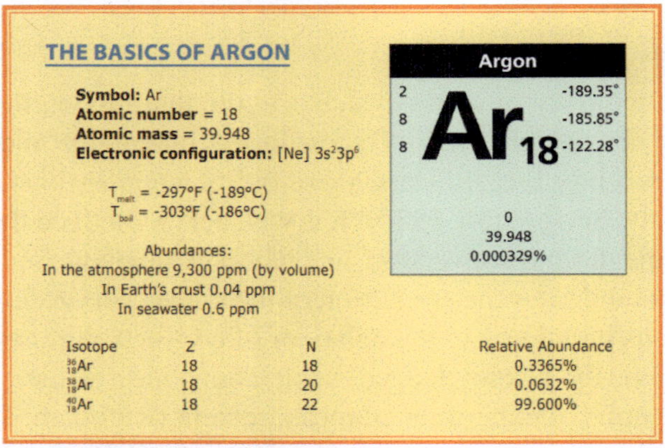

Figure 8.8. Basics of argon.

Source: Nicole Leonard, creative commons license.

Despite its minimal presence in Earth's atmosphere, argon is widely distributed across the universe. Argon-40 is the prevailing isotope of argon found in the Earth's atmosphere, but argon-36 is the isotope that is frequently detected in stellar spectra, notably that of the Sun (Ginsberg et al., 2002; Vargo, 2004). This is due to the fact that argon is classified as an alpha-process element, which implies that its synthesis mostly occurs by the capture of a helium nucleus onto sulfur 32, as previously mentioned:

$$^{32}_{16}\mathrm{S} + ^{4}_{2}\alpha \rightarrow ^{36}_{18}\mathrm{Ar}\,.$$

The aforementioned phenomenon takes place during the main sequence phase of a star's existence, resulting from interactions within its core, as well as during supernova events that disperse argon into the ISM. The presence of argon has been observed in the spectra of various celestial entities, including the Sun, the hot center stars of planetary nebulae, and white dwarfs (Aldrich & Nier, 1948; Deslattes et al., 1983). However, the most noteworthy findings pertain to the process of new star formation within blue compact galaxies. The prevalence of a blue hue in these galaxies suggests a predominance of youthful and high-temperature stars (Peck & Fisher, 1964). Due to their relatively compact nature, these galaxies exhibit a high mass density, resulting in frequent and numerous collisions. This heightened level of activity facilitates the accumulation of gas collections, which in turn leads to the formation of new stars. The significance of these celestial locations lies in their capacity to yield the most dependable data pertaining to the abundance of argon in stars characterized by low metallicities (Murphy & Arundell, 1994; Wijbrans et al., 1995).

8.3.4. Krypton and Xenon

Krypton and xenon are chemical elements with atomic numbers 36 and 54, respectively. Trace amounts of these gases can be detected in the environment. The process involves liquefying air and subsequently distilling it to isolate each constituent gas. Minerals and meteorites also contain minute quantities of krypton and xenon (Barker et al., 1974; Basford et al., 1973). Similar to all noble gases, the presence of noble gases on Earth can be attributed to the process of radioactive decay occurring in other elements (Figures 8.9 and 8.10).

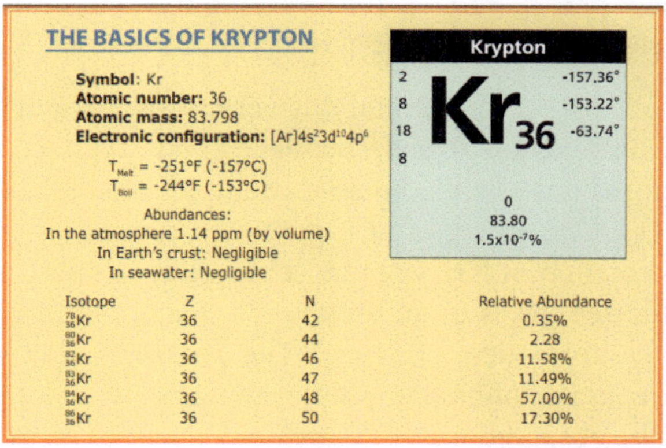

Figure 8.9. Basics of krypton.

Source: Greg Nobles, creative commons license.

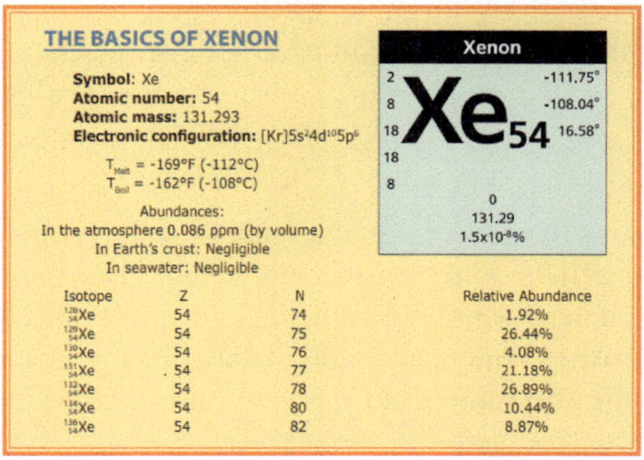

Figure 8.10. Basics of xenon.

Source: John Phillips, creative commons license.

Chemical compounds can indeed be formed by the elements krypton and xenon. The dissimilarity in chemical behavior can be attributed to the comparatively greater atomic sizes of krypton and xenon. Both krypton and xenon possess valence electrons in the "d" subshells. These "d" electrons are primarily responsible for forming chemical bonds with other elements, particularly fluorine (Varney, 1952; Baldini, 1962).

The initial creation of a noble gas compound by Neil Bartlett (1932–2008) was driven by the observation of a reaction involving the gases PtF_6 and oxygen, resulting in the formation of a solid substance with a crystalline structure, namely $[O_2^+][PtF_6^{\blacklozenge}]$. By replacing oxygen with xenon gas, the researcher successfully synthesized a red crystalline solid that is commonly denoted as $[Xe^+]$ $[\pm\ _6^{\blacklozenge}]$. Currently, a variety of compounds have been identified in which xenon forms chemical bonds with either fluorine or oxygen. Within the solid state, notable examples encompass XeF_2, XeF_4, XeF_6, and XeO_3. One example of a gaseous chemical is xenon tetroxide (XeO_4). The formation of KrF_2 can be achieved by subjecting a mixture of krypton and fluorine gases to an electric discharge, similar to the process observed for the formation of XeF_2. The resulting compound, krypton difluoride, is a volatile solid composed of transparent, colorless crystals. In contrast, it can be observed that KrF_2 exhibits a higher degree of instability in comparison to XeF_2. Numerous endeavors have been undertaken to generate compounds comprising helium, neon, and argon. However, the resulting entities have exhibited a fleeting nature at low temperatures and have proven to be entirely unstable when exposed to ambient conditions. Chemists argue that radon has a comparable level of chemical behavior to xenon due to the similarity in features observed among elements within the same column of the periodic table. The study of radon is hindered by its scarcity and radioactivity (Whalley & Schneider, 1955; Velazco, 1978).

Krypton and xenon, two elements with greater atomic mass than iron, are produced through the r-process in supernovae. The r-process involves the rapid absorption of neutrons by iron nuclei. A portion of the heavy noble gases gradually accumulates over extended periods of time, spanning thousands of years, within the atmospheres of massive stars. This accumulation occurs through the process of neutron capture and subsequent electron release, necessitating the presence of iron-56 nuclei as initial particles. These iron-56 nuclei originate from leftovers of previous supernova explosions. The nomenclature "s-process" is attributed to the more sluggish progression of this synthesis, mostly as a result of the diminished neutron density (Saunders et al., 1994; Aprile, 2017). Although the quantities of krypton and xenon are quite low, they can still be quantified, and recent investigations have shown intriguing irregularities in their distribution. Due to their limited chemical reactivity and low propensity for mechanical bonding, the incorporation of these elements into interstellar dust grains, a phenomenon observed in the case of numerous other elements, is highly improbable. Therefore, it may be inferred that the abundance of these elements has likely stayed stable during the billions of years following the birth of the solar system. The latest analysis of spectra obtained from the extrasolar but near-interstellar space has revealed a notable disparity in the accumulation of krypton between the solar system and its surrounding region (Suzuki & Kubota, 1979).

Furthermore, investigations into the atmospheres of other planets have yielded diverse findings regarding their composition. The Martian atmosphere exhibits a noteworthy composition of xenon isotopes that deviates from that found in Earth's atmosphere. The analysis of the atmospheric composition of Titan, a moon of Jupiter, reveals an unexpected scarcity of xenon and krypton compared to Jupiter's atmosphere. This discrepancy may be attributed to the entrapment of these gases within clathrate hydrates, which are crystalline structures created by water molecules and are currently in a frozen state on Titan's surface. Isotopically anomalous xenon has been detected in meteorites and lunar samples. The discipline of noble gas astronomy persistently offers captivating insights into the chronicles of the solar system's past and the progression of the universe (Errandonea et al., 2002).

8.3.5. Radon

Radon is classified as the 86th element in the periodic table. Due to the relatively brief half-lives exhibited by all isotopes of radon, their decay process occurs at a rapid rate. Due to the limited availability of radon for research purposes, there exists a scarcity of knowledge regarding its chemical properties. The radioactive decay properties of radon are of significant interest (Weigel, 1978; Beylkin, 1987). The three isotopes of radon with the longest half-lives are Rn-219 (with a half-life of 3.96 seconds), Rn-220 (with a half-life of 55.6 seconds), and Rn-222 (with a half-life of 3.8 days) (Figure 8.11).

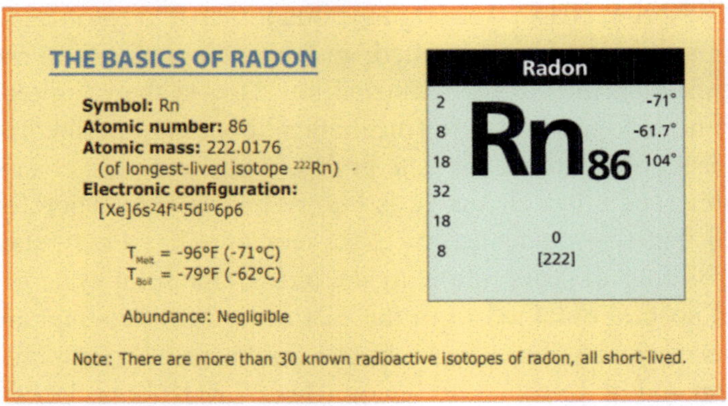

Figure 8.11. Basics of radon.

Source: John Phillips, creative commons license.

Each of these isotopes is an emitter of alpha particles, as illustrated, for example, by the decay of Rn-222:

$$^{222}_{86}\mathrm{Rn} \rightarrow {}^{4}_{2}\mathrm{He} + {}^{218}_{84}\mathrm{Po},$$

where $^{4}_{2}He$ is an alpha particle.

Most of the radon on Earth comes not from the stars, but from the spontaneous radioactive decay of terrestrial uranium 238 and its daughter products, protactinium (Pa), thorium (Th), and radium (Ra), in the following sequence.

$$^{238}\mathrm{U} \rightarrow {}^{234}\mathrm{Th} + \alpha$$

$$^{234}\mathrm{Th} \rightarrow {}^{234}\mathrm{Pa} + \beta$$

$$^{234}\mathrm{Pa} \rightarrow {}^{234}\mathrm{U} + \beta$$

$$^{234}\mathrm{U} \rightarrow {}^{230}\mathrm{Th} + \alpha$$

$$^{230}\mathrm{Th} \rightarrow {}^{226}\mathrm{Ra} + \alpha$$

$$^{226}\mathrm{Ra} \rightarrow {}^{222}\mathrm{Rn} + \alpha$$

The concentration of uranium in ground rock and soil typically averages around 2 parts per million (ppm). However, in regions with favorable geological conditions, such as granite, volcanic rock, and ancient shale, the concentration of uranium can be significantly higher. This is observed in western Ohio, where the uranium content often exceeds twice the average value (Cecil et al., 2000; Al-Zoughool & Krewski, 2009). Uranium and radium exist in solid form and are typically found inside the geological formations they occupy, whereas radon naturally occurs as a gaseous element. During the process of nuclear fission in a stationary ^{226}Ra nucleus, the emission of an alpha particle and radon nucleus occurs in opposite directions. Consequently, there exists a possibility for the radon nucleus to escape from the solid portion of the rock into a porous region, with an approximate probability of 30%. Subsequently, the radon is capable of facilely traversing the pores till it ultimately dissipates into the atmosphere (Turekian et al., 1977; Samet, 1989; Nazaroff, 1992).

Radon, a radioactive element, exhibits a relatively brief half-life of 3.8 days. Consequently, the fraction of radon that remains within the rock undergoes decay through a series of alpha and beta emissions, ultimately resulting in the formation of stable lead 206. The radioactive transitions elucidated in this context exhibit minimal or no impact on human beings, unless the radon gas that is released manages to infiltrate human dwellings via their foundations (Barrett,1984; Markkanen & Arvela, 1992).

8.4. SUMMARY

Halogen, there are six nonmetallic elements, namely fluorine, chlorine, bromine, iodine, astatine, and element 117, which exhibit comparable chemical properties. They are typically found in the penultimate column of the periodic table, as conventionally organized. All of these substances exhibit strong reactivity as oxidizing agents, as described in the context of oxidation-reduction reactions. They possess a valence of 1, with the exception of fluorine, which has a single valence. Transition metals quickly form compounds with both metals and nonmetals, exhibiting a diverse range of chemical reactions. It is important to note that these elements are not found in their pure, uncombined state in nature. Astatine, an element with radioactive properties, is found in trace quantities in nature as an intermediate result of decay. It lacks any stable isotopes that are nonradioactive. Element 117 has been synthetically generated in limited quantities, with no natural occurrences observed. Metal halides, which are halogen salts produced with metal atoms, exhibit high stability. The most well-known example of a metal halide is sodium chloride. The nomenclature of the halogen lamp is derived from the presence of halogens in the gas composition contained within its tungsten-filament bulb. These halogens are intentionally used to enhance the longevity of the filament and augment luminosity.

A noble gas, often known as an inert gas, the rightmost group of the periodic table, as typically organized, consists of seven chemical elements:

helium, neon, argon, krypton, xenon, radon, and oganesson. All of these elements lack color, odor, and flammability. With the exception of oganesson, they are present in minuscule quantities in the atmosphere. It is worth noting that helium is the second most abundant element in the universe, following hydrogen. The elements in question possess stable electronic configurations, characterized by the absence of unpaired electrons available for bonding. Consequently, they exhibit a high degree of reactivity, or lack thereof, leading to their classification as "noble" or inert. However, it is worth noting that krypton, xenon, and radon, due to their relatively weaker hold on outer electrons, are capable of forming compounds, primarily in conjunction with fluorine. These gases exhibit a relatively simpler mechanism of absorbing and emitting electromagnetic radiation compared to other substances. This characteristic is effectively utilized in their application within fluorescent lighting devices and discharge lamps. When placed in a transparent container at low pressure and subjected to an electric current, they exhibit a distinct coloration. The utilization of substances with significantly low boiling and melting points is advantageous in the field of low-temperature research, as they serve as effective refrigerants.

REFERENCES

1. Adriano, D. C., & Doner, H. E., (1983). Bromine, chlorine, and fluorine. *Methods of Soil Analysis: Part 2 Chemical and Microbiological Properties, 9*, 449–483.
2. Aiuppa, A., Baker, D. R., & Webster, J. D., (2009). Halogens in volcanic systems. *Chemical Geology, 263*(1–4), 1–18.
3. Akamatu, H., Inokuchi, H., & Matsunaga, Y., (1954). Electrical conductivity of the perylene–bromine complex. *Nature, 173*(4395), 168–169.
4. Aldrich, L. T., & Nier, A. O., (1948). Argon 40 in potassium minerals. *Physical Review, 74*(8), 876.
5. Allen, J. F., & Misener, A. D., (1938). Flow of liquid helium II. *Nature, 141*(3558), 75.
6. Al-Zoughool, M., & Krewski, D., (2009). Health effects of radon: A review of the literature. *International Journal of Radiation Biology, 85*(1), 57–69.
7. Anderson, P. W., (1966). Considerations on the flow of superfluid helium. *Reviews of Modern Physics, 38*(2), 298.
8. Aprile, E., Aalbers, J., Agostini, F., Alfonsi, M., Amaro, F. D., & Anthony, M., (2017). Removing krypton from xenon by cryogenic distillation to the ppq level. *The European Physical Journal C, 77*, 1–12.
9. Aston, F. W., (1920). XLIV. The constitution of atmospheric neon. *The London, Edinburgh, and Dublin Philosophical Magazine and Journal of Science, 39*(232), 449–455.
10. Atkins, K. R., (1959). Ions in liquid helium. *Physical Review, 116*(6), 1339.
11. Baldini, G., (1962). Ultraviolet absorption of solid argon, krypton, and xenon. *Physical Review, 128*(4), 1562.
12. Balibar, S., Alles, H., & Parshin, A. Y., (2005). The surface of helium crystals. *Reviews of Modern Physics, 77*(1), 317.
13. Ball, T. K., Cameron, D. G., Colman, T. B., & Roberts, P. D., (1991). Behavior of radon in the geological environment: A review. *Quarterly Journal of Engineering Geology and Hydrogeology, 24*(2), 169–182.
14. Barker, J. A., Watts, R. O., Lee, J. K., Schafer, T. P., & Lee, Y. T., (1974). Interatomic potentials for krypton and xenon. *The Journal of*

Chemical Physics, 61(8), 3081–3089.

15. Barrett, H. H., (1984). Iii the radon transform and its applications. In: *Progress in Optics* (Vol. 21, pp. 217–286). Elsevier.
16. Basford, J. R., Dragon, J. C., Pepin, R. O., Coscio, M. R., & Murthy, V. R., (1973). Krypton and xenon in lunar fines. In: *Lunar Science: Revised Abstracts of Papers Presented at the. Lunar Science Conference* (Vol. 4, p. 53). Lunar Science Institute.
17. Beckwith, R. C., Wang, T. X., & Margerum, D. W., (1996). Equilibrium and kinetics of bromine hydrolysis. *Inorganic Chemistry, 35*(4), 995–1000.
18. Beylkin, G., (1987). Discrete radon transform. *IEEE Transactions on Acoustics, Speech, and Signal Processing, 35*(2), 162–172.
19. Black, D. C., & Pepin, R. O., (1969). Trapped neon in meteorites—II. *Earth and Planetary Science Letters, 6*(5), 395–405.
20. Bobrowski, N., Hönninger, G., Galle, B., & Platt, U., (2003). Detection of bromine monoxide in a volcanic plume. *Nature, 423*(6937), 273–276.
21. Bohart, G. S., & Adams, E. Q., (1920). Some aspects of the behavior of charcoal with respect to chlorine. *Journal of the American Chemical Society, 42*(3), 523–544.
22. Böhm, H. J., Banner, D., Bendels, S., Kansy, M., Kuhn, B., Müller, K., & Stahl, M., (2004). Fluorine in medicinal chemistry. *ChemBioChem, 5*(5), 637–643.
23. Brammer, L., Bruton, E. A., & Sherwood, P., (2001). Understanding the behavior of halogens as hydrogen bond acceptors. *Crystal Growth & Design, 1*(4), 277–290.
24. Brewer, R. F., (1965). Fluorine. *Methods of Soil Analysis: Part 2 Chemical and Microbiological Properties, 9*, 1135–1148.
25. Broadley, M. W., Ballentine, C. J., Chavrit, D., Dallai, L., & Burgess, R., (2016). Sedimentary halogens and noble gases within Western Antarctic xenoliths: Implications of extensive volatile recycling to the sub continental lithospheric mantle. *Geochimica et Cosmochimica Acta, 176*, 139–156.
26. Brungs, W. A., (1973). Effects of residual chlorine on aquatic life. *Journal (Water Pollution Control Federation)*, 2180–2193.
27. Burnard, P., Zimmermann, L., & Sano, Y., (2013). *The Noble Gases as*

Geochemical Tracers: History and Background (pp. 1–15). Springer Berlin Heidelberg.

28. Cecil, L. D., & Green, J. R., (2000). Radon-222. In: *Environmental Tracers in Subsurface Hydrology* (pp. 175–194). Boston, MA: Springer US.
29. Cicerone, R. J., (1981). Halogens in the atmosphere. *Reviews of Geophysics, 19*(1), 123–139.
30. Clayton, R. N., & Mayeda, T. K., (1963). The use of bromine pentafluoride in the extraction of oxygen from oxides and silicates for isotopic analysis. *Geochimica et Cosmochimica Acta, 27*(1), 43–52.
31. Correns, C. W., (1956). The geochemistry of the halogens. *Physics and Chemistry of the Earth, 1*, 181–233.
32. Craig, H., & Lupton, J. E., (1976). Primordial neon, helium, and hydrogen in oceanic basalts. *Earth and Planetary Science Letters, 31*(3), 369–385.
33. De Beer, D., Srinivasan, R., & Stewart, P. S., (1994). Direct measurement of chlorine penetration into biofilms during disinfection. *Applied and Environmental Microbiology, 60*(12), 4339–4344.
34. Deborde, M., & Von, G. U. R. S., (2008). Reactions of chlorine with inorganic and organic compounds during water treatment—Kinetics and mechanisms: A critical review. *Water Research, 42*(1, 2), 13–51.
35. Deng, L., Huang, C. H., & Wang, Y. L., (2014). Effects of combined UV and chlorine treatment on the formation of trichloronitromethane from amine precursors. *Environmental Science & Technology, 48*(5), 2697–2705.
36. Deshmukh, A. N., Wadaskar, P. M., & Malpe, D. B., (1995). Fluorine in environment: A review. *Gondwana Geol. Mag, 9*, 1–20.
37. Deslattes, R. D., LaVilla, R. E., Cowan, P. L., & Henins, A., (1983). Threshold studies of a multivacancy process in the K β region of argon. *Physical Review A, 27*(2), 923.
38. Dolbier, Jr. W. R., (2005). Fluorine chemistry at the millennium. *Journal of Fluorine Chemistry, 126*(2), 157–163.
39. Downs, A. J., & Adams, C. J., (2017). *The Chemistry of Chlorine, Bromine, Iodine and Astatine: Pergamon Texts in Inorganic Chemistry* (Vol. 7). Elsevier.
40. Errandonea, D., Schwager, B., Boehler, R., & Ross, M., (2002). Phase

behavior of krypton and xenon to 50 GPa. *Physical Review B, 65*(21), 214110.

41. Eustace, D. J., (1980). Bromine complexation in zinc-bromine circulating batteries. *Journal of the Electrochemical Society, 127*(3), 528.
42. Evans, R. B., (2005). Chlorine: State of the art. *Lung, 183*(3), 151–167.
43. Falconer, R. L., & Bidleman, T. F., (1994). Vapor pressures and predicted particle/gas distributions of polychlorinated biphenyl congeners as functions of temperature and ortho-chlorine substitution. *Atmospheric Environment, 28*(3), 547–554.
44. Farley, K. A., & Neroda, E., (1998). Noble gases in the earth's mantle. *Annual Review of Earth and Planetary Sciences, 26*(1), 189–218.
45. Feynman, R. P., (1955). Chapter II application of quantum mechanics to liquid helium. In: *Progress in Low Temperature Physics* (Vol. 1, pp. 17–53). Elsevier.
46. Fuge, R., & Andrews, M. J., (1988). Fluorine in the UK environment. *Environmental Geochemistry and Health, 10*, 96–104.
47. Fuge, R., (1988). Sources of halogens in the environment, influences on human and animal health. *Environmental Geochemistry and Health, 10*, 51–61.
48. Furukawa, K., Tomizuka, N., & Kamibayashi, A., (1979). Effect of chlorine substitution on the bacterial metabolism of various polychlorinated biphenyls. *Applied and Environmental Microbiology, 38*(2), 301–310.
49. Furukawa, K., Tonomura, K., & Kamibayashi, A., (1978). Effect of chlorine substitution on the biodegradability of polychlorinated biphenyls. *Applied and Environmental Microbiology, 35*(2), 223–227.
50. Gillis, E. P., Eastman, K. J., Hill, M. D., Donnelly, D. J., & Meanwell, N. A., (2015). Applications of fluorine in medicinal chemistry. *Journal of Medicinal Chemistry, 58*(21), 8315–8359.
51. Ginsberg, G. G., Barkun, A. N., Bosco, J. J., Burdick, J. S., Isenberg, G. A., Nakao, N. L., & Kelsey, P. B., (2002). The argon plasma coagulator: February 2002. *Gastrointestinal Endoscopy, 55*(7), 807–810.
52. Gohil, J. M., & Suresh, A. K., (2017). Chlorine attack on reverse osmosis membranes: Mechanisms and mitigation strategies. *Journal of Membrane Science, 541*, 108–126.

53. Gouverneur, V., & Seppelt, K., (2015). Introduction: Fluorine chemistry. *Chemical Reviews, 115*(2), 563–565.

54. Grossberg, S., & Mingolla, E., (1985). Neural dynamics of form perception: Boundary completion, illusory figures, and neon color spreading. *Psychological Review, 92*(2), 173.

55. Guo, N., Maurice, R., Teze, D., Graton, J., Champion, J., Montavon, G., & Galland, N., (2018). Experimental and computational evidence of halogen bonds involving astatine. *Nature Chemistry, 10*(4), 428–434.

56. Hagmann, W. K., (2008). The many roles for fluorine in medicinal chemistry. *Journal of Medicinal Chemistry, 51*(15), 4359–4369.

57. Haller, H., & Riedel, S., (2014). Recent discoveries of polyhalogen anions–from bromine to fluorine. *Zeitschrift für Anorganische und Allgemeine Chemie, 640*(7), 1281–1291.

58. Hermann, A., Hoffmann, R., & Ashcroft, N. W., (2013). Condensed astatine: Monatomic and metallic. *Physical Review Letters, 111*(11), 116404.

59. Hom, L. W., (1972). Kinetics of chlorine disinfection in an ecosystem. *Journal of the Sanitary Engineering Division, 98*(1), 183–194.

60. Hua, G., & Reckhow, D. A., (2007). Comparison of disinfection byproduct formation from chlorine and alternative disinfectants. *Water Research, 41*(8), 1667–1678.

61. Hua, G., & Reckhow, D. A., (2012). Evaluation of bromine substitution factors of DBPs during chlorination and chloramination. *Water Research, 46*(13), 4208–4216.

62. Johnson, C. M., Stout, P. R., Broyer, T. C., & Carlton, A. B., (1957). Comparative chlorine requirements of different plant species. *Plant and Soil, 8*, 337–353.

63. Kendrick, M. A., & Burnard, P., (2013). Noble gases and halogens in fluid inclusions: A journey through the Earth's crust. *The Noble Gases as Geochemical Tracers,* 319–369.

64. Kendrick, M. A., Honda, M., & Vanko, D. A., (2015). Halogens and noble gases in mathematician ridge meta-gabbros, NE Pacific: Implications for oceanic hydrothermal root zones and global volatile cycles. *Contributions to Mineralogy and Petrology, 170*, 1–20.

65. Kendrick, M. A., Honda, M., Pettke, T., Scambelluri, M., Phillips, D., & Giuliani, A., (2013). Subduction zone fluxes of halogens and

noble gases in seafloor and forearc serpentinites. *Earth and Planetary Science Letters, 365*, 86–96.

66. Kendrick, M. A., Scambelluri, M., Hermann, J., & Padrón-Navarta, J. A., (2018). Halogens and noble gases in serpentinites and secondary peridotites: Implications for seawater subduction and the origin of mantle neon. *Geochimica et Cosmochimica Acta, 235*, 285–304.
67. Kestin, J., Knierim, K., Mason, E. A., Najafi, B., Ro, S. T., & Waldman, M., (1984). Equilibrium and transport properties of the noble gases and their mixtures at low density. *Journal of Physical and Chemical Reference Data, 13*(1), 229–303.
68. Kestin, J., Ro, S. T., & Wakeham, W. A., (1972). Viscosity of the noble gases in the temperature range 25–700° C. *The Journal of Chemical Physics, 56*(8), 4119–4124.
69. Keswick, B. H., Satterwhite, T. K., Johnson, P. C., DuPont, H. L., Secor, S. L., Bitsura, J. A., & Hoff, J. C., (1985). Inactivation of Norwalk virus in drinking water by chlorine. *Applied and Environmental Microbiology, 50*(2), 261–264.
70. Khriachtchev, L., Pettersson, M., Runeberg, N., Lundell, J., & Räsänen, M., (2000). A stable argon compound. *Nature, 406*(6798), 874–876.
71. Kipfer, R., Aeschbach-Hertig, W., Peeters, F., & Stute, M., (2002). Noble gases in lakes and ground waters. *Reviews in Mineralogy and Geochemistry, 47*(1), 615–700.
72. Kobayashi, M., Sumino, H., Nagao, K., Ishimaru, S., Arai, S., Yoshikawa, M., & Ballentine, C. J., (2017). Slab-derived halogens and noble gases illuminate closed system processes controlling volatile element transport into the mantle wedge. *Earth and Planetary Science Letters, 457*, 106–116.
73. Korich, D. G., Mead, J. R., Madore, M. S., Sinclair, N. A., & Sterling, C., (1990). Effects of ozone, chlorine dioxide, chlorine, and monochloramine on *Cryptosporidium parvum* oocyst viability. *Applied and Environmental Microbiology, 56*(5), 1423–1428.
74. Larsen, S., & Widdowson, A. E., (1971). Soil fluorine. *Journal of Soil Science, 22*(2), 210–221.
75. Lewis, R. S., Srinivasan, B., & Anders, E., (1975). Host phase of a strange xenon component in Allende: Most of the noble gases reside in three rare minerals comprising less than 0.5 percent of the meteorite. *Science, 190*(4221), 1251–1262.

76. Li, X., Wu, X., Liu, S., Li, Y., Fan, J., & Lv, K., (2020). Effects of fluorine on photocatalysis. *Chinese Journal of Catalysis, 41*(10), 1451–1467.

77. Li, Y., Li, J., Zhou, J., Zhao, F. H., & XU, D., (2022). Research progress and new views on evaluation of helium resources. *Journal of Earth Sciences and Environment, 44*(3), 363–373.

78. Liang, T., Neumann, C. N., & Ritter, T., (2013). Introduction of fluorine and fluorine-containing functional groups. *Angewandte Chemie International Edition, 52*(32), 8214–8264.

79. Lim, H. S., Lackner, A. M., & Knechtli, R. C., (1977). Zinc-bromine secondary battery. *Journal of the Electrochemical Society, 124*(8), 1154.

80. Lindegren, S., Albertsson, P., Bäck, T., Jensen, H., Palm, S., & Aneheim, E., (2020). Realizing clinical trials with astatine-211: The chemistry infrastructure. *Cancer Biotherapy & Radiopharmaceuticals, 35*(6), 425–436.

81. Lommerse, J. P., Stone, A. J., Taylor, R., & Allen, F. H., (1996). The nature and geometry of intermolecular interactions between halogens and oxygen or nitrogen. *Journal of the American Chemical Society, 118*(13), 3108–3116.

82. Macauley, J. J., Qiang, Z., Adams, C. D., Surampalli, R., & Mormile, M. R., (2006). Disinfection of swine wastewater using chlorine, ultraviolet light, and ozone. *Water Research, 40*(10), 2017–2026.

83. Markkanen, M., & Arvela, H., (1992). Radon emanation from soils. *Radiation Protection Dosimetry, 45*(1–4), 269–272.

84. Meyer, G. J., (2018). Astatine. *Journal of Labelled Compounds and Radiopharmaceuticals, 61*(3), 154–164.

85. Molina, L. T., & Molina, M. J., (1987). Production of chlorine oxide (Cl_2O_2) from the self-reaction of the chlorine oxide (ClO) radical. *Journal of Physical Chemistry, 91*(2), 433–436.

86. Murphy, A. B., & Arundell, C. J., (1994). Transport coefficients of argon, nitrogen, oxygen, argon-nitrogen, and argon-oxygen plasmas. *Plasma Chemistry and Plasma Processing, 14*, 451–490.

87. Najafi, B., Mason, E. A., & Kestin, J., (1983). Improved corresponding states principle for the noble gases. *Physica A: Statistical Mechanics and its Applications, 119*(3), 387–440.

88. Nazaroff, W. W., (1992). Radon transport from soil to air. *Reviews of*

Geophysics, 30(2), 137–160.

89. Nishinaka, I., Hashimoto, K., & Suzuki, H., (2018). Thin layer chromatography for astatine and iodine in solutions prepared by dry distillation. *Journal of Radioanalytical and Nuclear Chemistry, 318*, 897–905.
90. Nuttall, W. J., Clarke, R. H., & Glowacki, B. A., (2012). Stop squandering helium. *Nature, 485*(7400), 573–575.
91. O'Hagan, D., (2010). Fluorine in health care: Organofluorine containing blockbuster drugs. *Journal of Fluorine Chemistry, 131*(11), 1071–1081.
92. O'nions, R. K., & Oxburgh, E. R., (1983). Heat and helium in the earth. *Nature, 306*(5942), 429–431.
93. Öberg, G., (2002). The natural chlorine cycle–fitting the scattered pieces. *Applied Microbiology and Biotechnology, 58*, 565–581.
94. Pauling, L., & Marsh, R. E., (1952). The structure of chlorine hydrate. *Proceedings of the National Academy of Sciences, 38*(2), 112–118.
95. Peck, E. R., & Fisher, D. J., (1964). Dispersion of argon. *JOSA*, *54*(11), 1362–1364.
96. Penrose, O., (1951). CXXXVI. On the quantum mechanics of helium II. *The London, Edinburgh, and Dublin Philosophical Magazine and Journal of Science, 42*(335), 1373–1377.
97. Petrich, T., Helmeke, H. J., Meyer, G., Knapp, W., & Pötter, E., (2002). Establishment of radioactive astatine and iodine uptake in cancer cell lines expressing the human sodium/iodide symporter. *European Journal of Nuclear Medicine and Molecular Imaging, 29*, 842–854.
98. Politzer, P., Lane, P., Concha, M. C., Ma, Y., & Murray, J. S., (2007). An overview of halogen bonding. *Journal of Molecular Modeling, 13*, 305–311.
99. Pratt, K. A., Custard, K. D., Shepson, P. B., Douglas, T. A., Pöhler, D., General, S., & Stirm, B. H., (2013). Photochemical production of molecular bromine in arctic surface snowpacks. *Nature Geoscience, 6*(5), 351–356.
100. Purser, S., Moore, P. R., Swallow, S., & Gouverneur, V., (2008). Fluorine in medicinal chemistry. *Chemical Society Reviews, 37*(2), 320–330.
101. Rittenhouse, G., (1967). Bromine in oil-field waters and its use in

determining possibilities of origin of these waters. *AAPG Bulletin, 51*(12), 2430–2440.

102. Rohrbacher, A., Halberstadt, N., & Janda, K. C., (2000). The dynamics of noble gas—Halogen molecules and clusters. *Annual Review of Physical Chemistry, 51*(1), 405–433.
103. Rossman, L. A., Clark, R. M., & Grayman, W. M., (1994). Modeling chlorine residuals in drinking-water distribution systems. *Journal of Environmental Engineering, 120*(4), 803–820.
104. Rothman, H., (2015). *Neon Metropolis: How Las Vegas Started the Twenty-First Century*. Routledge.
105. Rzepa, H., (1997). Some influences of fluorine in bioorganic chemistry. *Chemical Communications*, (7), 645–652.
106. Saikia, I., Borah, A. J., & Phukan, P., (2016). Use of bromine and bromo-organic compounds in organic synthesis. *Chemical Reviews, 116*(12), 6837–7042.
107. Samet, J. M., (1989). Radon and lung cancer. *JNCI: Journal of the National Cancer Institute, 81*(10), 745–758.
108. Sander, R., Keene, W. C., Pszenny, A. A. P., Arimoto, R., Ayers, G. P., Baboukas, E., & Van, D. R., (2003). Inorganic bromine in the marine boundary layer: A critical review. *Atmospheric Chemistry and Physics, 3*(5), 1301–1336.
109. Sano, Y., Marty, B., & Burnard, P., (2013). Noble gases in the atmosphere. *The Noble Gases as Geochemical Tracers,* 17–31.
110. Sarda, P., Staudacher, T., & Allègre, C. J., (1988). Neon isotopes in submarine basalts. *Earth and Planetary Science Letters, 91*(1, 2), 73–88.
111. Saunders, M., Jimenez-Vazquez, H. A., Cross, R. J., Mroczkowski, S., Gross, M. L., Giblin, D. E., & Poreda, R. J., (1994). Incorporation of helium, neon, argon, krypton, and xenon into fullerenes using high pressure. *Journal of the American Chemical Society, 116*(5), 2193–2194.
112. Sedlak, D. L., & Von, G. U., (2011). The chlorine dilemma. *Science, 331*(6013), 42–43.
113. Shah, P., & Westwell, A. D., (2007). The role of fluorine in medicinal chemistry. *Journal of Enzyme Inhibition and Medicinal Chemistry, 22*(5), 527–540.

114. Slater, J. C., (1928). The normal state of helium. *Physical Review, 32*(3), 349.
115. Smart, B. E., (1996). Introduction: Fluorine chemistry. *Chemical Reviews, 96*(5), 1555–1556.
116. Spicer, C. W., Chapman, E. G., Finlayson-Pitts, B. J., Plastridge, R. A., Hubbe, J. M., Fast, J. D., & Berkowitz, C. M., (1998). Unexpectedly high concentrations of molecular chlorine in coastal air. *Nature, 394*(6691), 353–356.
117. Strunecká, A., Patočka, J., & Connett, P., (2004). Fluorine in medicine. *J. Appl. Biomed., 2*, 141–150.
118. Stute, M., & Schlosser, P., (2000). Atmospheric noble gases. In: *Environmental Tracers in Subsurface Hydrology* (pp. 349–377). Boston, MA: Springer US.
119. Suárez-Figueroa, M. C., Gómez-Pérez, A., & Fernández-López, M., (2011). The NeOn methodology for ontology engineering. In: *Ontology Engineering in a Networked World* (pp. 9–34). Berlin, Heidelberg: Springer Berlin Heidelberg.
120. Suzuki, M., & Kubota, S., (1979). Mechanism of proportional scintillation in argon, krypton, and xenon. *Nuclear Instruments and Methods, 164*(1), 197–199.
121. Tanudji, J., Aspera, S. M., & Kasai, H., (2023). Theoretical comparison study of iodine and astatine adsorption on Au (111) surface. *e-Journal of Surface Science and Nanotechnology*.
122. Taylor, R. H., Falkinham, III. J. O., Norton, C. D., & LeChevallier, M. W., (2000). Chlorine, chloramine, chlorine dioxide, and ozone susceptibility of *Mycobacterium avium. Applied and Environmental Microbiology, 66*(4), 1702–1705.
123. Thornton, J., & Campbell, D., (2001). *Pandora's Poison: Chlorine, Health, and a New Environmental Strategy* (Vol. 322, No. 7284, p. 497). British Medical Journal Publishing Group.
124. Tisza, L., (1938). Transport phenomena in helium II. *Nature, 141*(3577), 913–913.
125. Tisza, L., (1947). The theory of liquid helium. *Physical Review, 72*(9), 838.
126. Trueman, J. R., (1971). The halogens. *Inhibition and Destruction of the Microbial Cell,* 137–183.

127. Turekian, K. K., Nozaki, Y., & Benninger, L. K., (1977). Geochemistry of atmospheric radon and radon products. *Annual Review of Earth and Planetary Sciences, 5*(1), 227–255.

128. Turner, G., Cadogan, P. H., & Yonge, C. J., (1973). Argon selenochronology. In: *Proceedings of the Lunar Science Conference* (Vol. 4, p. 1889).

129. Van, S. S. W., Timmerhaus, K. D., & Clark, A. F., (2012). *Helium Cryogenics* (pp. 197–217). New York, NY: Springer New York.

130. Vargo, J. J., (2004). Clinical applications of the argon plasma coagulator. *Gastrointestinal Endoscopy, 59*(1), 81–88.

131. Varney, R. N., (1952). Drift velocities of ions in krypton and xenon. *Physical Review, 88*(2), 362.

132. Vasaros, L., Norseyev, Y. V., Nhan, D. D., & Khalkin, V. A., (1981). Nucleophilic replacement of bromine in bromobenzene by iodine and astatine in homogeneous media. *Radiochemical and Radioanalytical Letters, 47*(5), 313–322.

133. Vasconcelos, J. J., Rossman, L. A., Grayman, W. M., Boulos, P. F., & Clark, R. M., (1997). Kinetics of chlorine decay. *Journal-American Water Works Association, 89*(7), 54–65.

134. Velazco, J. E., Kolts, J. H., & Setser, D. W., (1978). Rate constants and quenching mechanisms for the metastable states of argon, krypton, and xenon. *The Journal of Chemical Physics, 69*(10), 4357–4373.

135. Von, G. R., Von, K. R., Lawrence, M. G., Platt, U., & Crutzen, P. J., (2004). Impact of reactive bromine chemistry in the troposphere. *Atmospheric Chemistry and Physics, 4*(11, 12), 2481–2497.

136. Wang, J., Sánchez-Roselló, M., Aceña, J. L., Del Pozo, C., Sorochinsky, A. E., Fustero, S., & Liu, H., (2014). Fluorine in pharmaceutical industry: Fluorine-containing drugs introduced to the market in the last decade (2001–2011). *Chemical Reviews, 114*(4), 2432–2506.

137. Wang, T. X., Kelley, M. D., Cooper, J. N., Beckwith, R. C., & Margerum, D. W., (1994). Equilibrium, kinetic, and UV-spectral characteristics of aqueous bromine chloride, bromine, and chlorine species. *Inorganic Chemistry, 33*(25), 5872–5878.

138. Wang, W. L., Wu, Q. Y., Huang, N., Wang, T., & Hu, H. Y., (2016). Synergistic effect between UV and chlorine (UV/chlorine) on the degradation of carbamazepine: Influence factors and radical species. *Water Research, 98*, 190–198.

139. Weigel, F., (1978). Radon. *Chem. Ztg, 102*(9), 287–299.
140. Wennberg, P., (1999). Bromine explosion. *Nature, 397*(6717), 299–301.
141. Westerhoff, P., Chao, P., & Mash, H., (2004). Reactivity of natural organic matter with aqueous chlorine and bromine. *Water Research, 38*(6), 1502–1513.
142. Whalley, E., & Schneider, W. G., (1955). Intermolecular potentials of argon, krypton, and xenon. *The Journal of Chemical Physics, 23*(9), 1644–1650.
143. Wieler, R., (2002). Noble gases in the solar system. *Reviews in Mineralogy and Geochemistry, 47*(1), 21–70.
144. Wiese, W. L., Smith, M. W., & Glennon, B. M., (1966). *Atomic Transition Probabilities, Hydrogen Through Neon* (Vol. I). Institute for Basic Standarts National Bureau of Standarts, Washington DC.
145. Wijbrans, J. R., Pringle, M. S., Koppers, A. A. P., & Scheveers, R., (1995). Argon geochronology of small samples using the Vulkaan argon laserprobe. In: *Proceedings of the Royal Netherlands Academy of Arts and Sciences* (Vol. 2, No. 98, pp. 185–218).
146. Winid, B., (2015). Bromine and water quality–selected aspects and future perspectives. *Applied Geochemistry, 63*, 413–435.
147. Wofsy, S. C., McElroy, M. B., & Yung, Y. L., (1975). The chemistry of atmospheric bromine. *Geophysical Research Letters, 2*(6), 215–218.
148. Yang, X., Cox, R. A., Warwick, N. J., Pyle, J. A., Carver, G. D., O'Connor, F. M., & Savage, N. H., (2005). Tropospheric bromine chemistry and its impacts on ozone: A model study. *Journal of Geophysical Research: Atmospheres, 110*(D23).
149. Zenker, M., (2008). Argon plasma coagulation. *GMS Krankenhaushygiene Interdisziplinar, 3*(1).
150. Zhang, X., Liu, Y., & Huang, Q., (2020). Stable halogen 2D materials: The case of iodine and astatine. *Journal of Physics: Condensed Matter, 32*(33), 335301.

INDEX

G

H

I

K

L

M